(1) VIRGIN AND CHRIST CHILD, ENTHRONED — 'The Blue Virgin Widow', Chartres Cathedral.
  12th century stained glass set in 13th century south ambulatory window.

(2) FATHER, SON AND HOLY SPIRIT: THE CRUCIFIXION — oil painting by the Austrian School (15th C). The National Gallery, London.

THE FRANCIS BACON RESEARCH TRUST JOURNAL
Series I, Volume 4:
Festival of Promise

THE GREAT VISION

by
Peter Dawkins

The Judaic-Christian Mysteries

The Vision and Birth of the New Rosicrucianism

(The Life and Times of Francis Bacon, 1572-1579)

Based on edited transcripts of lectures first given at the Festival of Promise
Conferences, March 1980, 1982 and 1984, plus other material.

Copyright © The Francis Bacon Research Trust, 1985

First published by the Francis Bacon Research Trust in 1985
Printed by Coventry Printers, Coventry

(Series 1) ISSN: 0262 8228
(Vol. 4) ISBN: 0 86293 007 3

ACKNOWLEDGEMENTS

The author wishes to acknowledge with gratitude the help given him by Mr. Thomas Bokenham, Cmdr. Martin Pares, R. N., Mrs. D. Brameld, Mr. Noel Fermor, Mr. Paul Lucas, Mr. Ewen MacDuff Strachan and other members of the Francis Bacon Society; Donna James for the typing, help in proof reading and collection of material; the Earl and Countess of Verulam, and the staff and trustees of the various Museums, Art Galleries and Art Collections who have allowed us to reproduce their paintings, drawings, and other works of art; the Marquess of Northampton, the Tiger Trust, and all those many others who have helped, contributed to or supported in one way or another the work of the Francis Bacon Research Trust.

CONTENTS

Contents Page 5
List of Illustrations 7
The Francis Bacon Research Trust: General Information 10

PREFACE 13

THE JUDAIC-CHRISTIAN MYSTERIES
    The Spring Festival of Promise and the Water Initiation 19
    AN and ANNA: Divine Mind and Matter 24
    The Virgin Mary and Divine Thought 26
    Man 28
    Father, Son and Holy Ghost 29
    The Three, Seven and Twelve 29
    BAL, the Word of God 30
    Divine Transcendance and Immanence 30
    John, the Man sent from God 31
    John and Jesus: Baptism by Water and Baptism by Holy Spirit 34
    Jesus and his Christ Mission 38
    Jesus and the Great Initiations 43
    The Birth of the Christ Child, Jesus 51
    The Crucified Serpent: Michael and Lucifer: St. George and the Dragon 53
    John the Beloved and the Three Maries 58
    The Last Supper 58
    The Old and New Covenants 60

THE VISION AND BIRTH OF THE NEW ROSICRUCIANISM
(*THE LIFE AND TIMES OF FRANCIS BACON, 1572-1579*)
    Supernova in Cassiopiea 67
    At University 73
    Vision of Universal Reformation and Enlightenment 83
    Early Elizabethan Chivalry and Pageantry 103
    The Kenilworth and Woodstock Entertainments 120
    Hemetes and the New Rosicrucian Birth 128
    Royal Revelation and Banishment 139
    Discovery of High Culture and Romance in France 153
    Return to France 180
    The French Academy 195
    Love's Labour Lost 205
    Anthony Bacon, Secret Agent 223

APPENDIX
    The Kenilworth Festivities: Laneham's Letter 254
    Pierre Amboise's *Life* of Francis Bacon 256
    Humanism 262
    The Pléiade 264
    Sir Nicholas Bacon's Will 266

NOTES AND BIBLIOGRAPHY
    Notes: Part I 270
    Notes: Part II 276
    Select Bibliography 297

ILLUSTRATIONS

Page	Ref.	**Diagrams:**
22	A.	The Cycle of Seasons
23	B.	The Cycle of Initiation
36	C.	The Celestial Chart
46	D.	The Ladder of Initiation (Jesus)
62	E.	Israel's National History
66	F.	Francis Bacon: Historical Chart (1568-80)
105	G.	The Royal Line of British Kings — House of Judah
234	H.	Map of Europe
246	I.	Tudor-Stewart Family Tree
247	J.	Knolly-Carey Family Tree
248	K.	Boleyn-Carey Family Tree
248	L.	Wyatt-Lee Family Tree
249	M.	Spanish and Austrian Habsburg Family Tree
250	N.	Valois-Navarre Family Tree
251	O.	Bourbon-Condé Family Tree
252	P.	Guise-Lorraine Family Tree

Page	Ref.	**Portraits:**
72	10.	Queen Elizabeth I, aged 39 (1572) — miniature by Nicholas Hilliard. *Reproduced by courtesy of the National Portrait Gallery (NPG 108).*
115	12.	Queen Elizabeth I, woodcut, final page of *Genealogie of the Kinges of England*, printed by Gylles Godet (1560). *Reproduced by courtesy of The John Rylands University Library of Manchester.*
116	13.	Sir Henry Lee, aged 35 (1568) — by Antonio Mor. *Reproduced by courtesy of the National Portrait Gallery (NPG 2095).*
124	14.	Robert Dudley, Earl of Leicester, aged 44 (1576) — miniature by Nicholas Hilliard. *Reproduced by courtesy of the National Portrait Gallery (NPG 4197).*
138	16.	William Cecil, Lord Burghley, and Robert Cecil, Earl of Salisbury (c. 1606) — artist unknown. *Reproduced by courtesy of the Marquess of Salisbury (NPG 1833).*
146	17.	Lady Anne Bacon, aged 54 (1580) — attributed to George Gower.
152	18.	Sir Amyas Paulett, with Garter George (c. 1575) — drawing by G.P. Harding. *Reproduced by courtesy of the National Portrait Gallery (NPG 2399).*
157	19.	Catherine de Medici, Queen of France — drawing by François Clouet. *Reproduced by courtesy of the Trustees of the British Museum.*
158	20.	Charles IX, King of France — from studio of François Clouet. *Reproduced by kind permission of Sotheby's. Private Collection.*
159	21.	Henri III, King of France — after an engraving by l'Armessin. *Reproduced by kind permission of the Bibliothèque Nationale, Paris.*
160	22.	Marguerite de Valois, Queen of Navarre — engraving from *l'Histoire de la maison de Bourbon* (1779). *Reproduced by courtesy of the Trustees of the British Museum.*
161	23.	François de Valois, Duke of Alençon and Anjou, aged 31 — by François Clouet. *Reproduced by kind permission of Sotheby's. Private Collection.*
162	24.	Henri IV, King of France and Navarre — from engraving attrib. to Pierre du Moustier. *Reproduced by kind permission of the Bibliothèque Nationale, Paris.*
163	25.	Henri de Lorraine, Duke of Guise — from engraving by Le Blond. *Reproduced by courtesy of the Trustees of the British Museum.*
183	28.	Sir Francis Walsingham (c. 1585-9) — attrib. John de Critz, senior. *Reproduced by courtesy of the National Portrait Gallery (NPG 1807).*
209	32.	Nicholas Hilliard, aged 30 (1577) — miniature, self portrait. *Reproduced by courtesy of the Board of Trustees, Victoria & Albert Museum, London (P155-1910).*

Page	Ref.	
209	33.	Alice Hilliard (*née* Brandon), wife of Nicholas Hilliard, aged 22 (1578) — miniature by N. Hilliard.
210	34.	Francis Bacon, aged 18 (1578/9) — miniature by Nicholas Hilliard. *Reproduced from* Baconiana Vol. L. No. 167, *Journal of the Francis Bacon Society.*
224	35.	Sir Nicholas Bacon, Lord Keeper of the Great Seal, aged 68 (1579 — the year of his death) — artist unknown.
230	36.	Anthony Bacon, aged 36 (1594) — artist unknown.
237	37.	Don Juan, Archduke of Austria — engraving from a contemporary Spanish print. *Reproduced by courtesy of the Trustees of the British Museum.*
241	38.	Queen Elizabeth I, aged 42, with Garter George (*c.* 1575) — artist unknown. *Reproduced by gracious permission of Her Majesty the Queen, The Royal Collection.*
243	39.	The Right Honourable, Francis Lord Verulam, Viscount Saint Alban — frontispiece to first edition, *Sylva Sylvarum* (1627). *Reproduced by kind permission of the British Library.*
245	40.	Lord Francis Bacon, Baron of Verulam, Viscount St. Alban, Counsillor of Estate and Lord Chancelor of England — frontispiece to his *Of the Advancement and Proficience of Learning* (1640).

Page	Ref.	**Illustrations Generally:**
65 & Cover	Front Spenser.	St. George and the Dragon — woodcut from *The Faerie Queene* (1590) by Edmund Spenser.
1	1.	The child Jesus seated on the Virgin Mary's lap — 12th century stained glass window, Chartres Cathedral.
2	2.	The Holy Trinity with Christ crucified — painting, Austrian School (15th C). *Reproduced by courtesy of the Trustees, The National Gallery, London (ref. 3662).*
12	3.	Titlepage to the 'Bishop's Bible' (1602) — woodcut by Rowland Lockey and Christopher Switzer. *Reproduced by kind permission of the British Library.*
17	4.	Crucified Serpent — woodcut from titlepage to *The Faerie Queen,* Pt. II (1617).
45	5.	The child Jesus seated on the Virgin Mary's lap — 12th century sculpture, southern doorway, 'Royal' West Porch, Chartres Cathedral.
49	6.	Jesus enthroned in glory — 12th century sculpture, central doorway, 'Royal' West Porch, Chartres Cathedral.
52	7.	The child Jesus seated on the Virgin Mary's lap — painting, Margarito of Arezzo. *Reproduced by courtesy of the Trustees, The National Gallery, London (ref. 564).*
54	8.	Crucifixion — painting ascribed to the Master of San Francesco. *Reproduced by courtesy of the Trustees, The National Gallery, London (ref. 6361).*
57	9.	Crucifixion — from painted altarpiece by Raphael. *Reproduced by courtesy of the Trustees, The National Gallery, London (ref. 3943).*
112	11.	Capital letter 'C' from first page of *Acts and Monuments* (1563) by John Foxe. *Reproduced by kind permission of the British Library.*
131	15.	Frontispiece to *Hemetes the Heremyte* — pen and ink drawing, presented to Her Majesty Queen Elizabeth I in manuscript form after the Kenilworth Entertainment of 1575, first published 1579.
175	26.	The Speaker of the Prologue — woodcut from *L'Amfiparnaso, Comedia harmonica* (Venice, 1597) by Horatio Vecchi.
176	27.	The Quintain — drawing by Antoine Caron, part of design for the Valois Tapestries (*c.* 1582). *Reproduced by courtesy of the Courtauld Institute Galleries (Witt Collection drawing 4647).*
194	29.	Titlepage of *The Whole Booke of Psalmes* (1583).
196	30.	Capital letter 'S', first letter in the Dedication to *The French Academie* (first English edition, 1586).
196	31.	'Fish & Honeysuckle' emblem from titlepage to 1st (1603) and 2nd (1604) Quartos of *The Tragicall Historie of Hamlet, Prince of Denmark*, by William Shake-speare.

Page	**Headpieces:**
13	"Archer" headpiece, "Prologo Lelio", *L'Amfiparnaso, Comedia harmonica* (Venice, 1597), by Horatio Vecchi.
19	"IHS" headpiece, from *The Shepheards Calendar*, 1st Folio London, (1617), by Edm. Spenser.
67	"AA" headpiece, "Dedication", *De Furtivis Literarum Notis Vulgo. De Ziteris* (Naples, 1563), by Ioan. Baptisa Porta. (1st appearance of the "AA" emblem?)
73	"AA" headpiece, *The Art of English Poesie* (London, 1589). (Earliest English use of "AA" emblem).
83	"Grail" headpiece, preface to "The Great Instauration", *Advancement and Proficience of Learning* (London, 1640), by Francis Bacon.
103	"AA" headpiece, "The Argument", *Orlando Furioso* (London, 1591), by Sir John Harington.
120	"AA" headpiece, *A Discourse upon the Meanes of wel governing and maintaining in good peace, a Kingdome or Other Principalitie* (London, 1602), translated by Simon Patericke.
128	"AA" headpiece, "Preface to the Reader", *Daemonology* (1603).
139	"AA" headpiece, *Tragicall Historie of Hamlet*, 1st Quarto (London, 1603), by William Shake-speare.
153	"AA" headpiece, *Seneca* (London, 1614), translated by Lodge.
180	"AA" headpiece, "Upon the Lines and Life of the Famous Scenicke Poet, Master William Shakespeare", *Mr. William Shakespeares Comedies, Histories and Tragedies*, 1st Folio (London, 1623).
195	"Archer" headpiece, "The First Booke", *The Faerie Queen*, 1st Folio (London, 1611), by Edm. Spenser.
205	"Bride" headpiece, titlepage, *The Genealogies Recorded In The Sacred Scriptures*, 1st Quarto (London, 1612), by I.S.
223	"AA" headpiece, "Dedication", *The Countess of Pembroke's L'Arcadie* (London, 1625), by Sir Philip Sydney.
254	"YHVH" headpiece, from *The Faerie Queen*, 1st Folio (London, 1611), by Edm. Spenser.
268	"IHS" headpiece, from *The Faerie Queen*, 1st Folio (London, 1611), by Edm. Spenser.
297	"Mosaic Tablets" headpiece, from *The Faerie Queen*, 1st Folio (London, 1611) by Edm. Spenser.

Page	**Tailpieces:**
63, 261	"Tau, Anchor and Serpent" emblem, titlepage, *New Atlantis* (London, 1631), by Sir Francis Bacon.
71, 102, 127, 151, 193, 222	"Pan" tailpiece, *Philostrate* (Paris, 1578), by Blaise de Vigenère. (1st appearance of the "Pan" tailpiece.)
82, 119, 204, 240, 299	"Pan" tailpiece, *Histoire du Monde* (Lyon, 1583), by Plinius.

THE FRANCIS BACON RESEARCH TRUST

GENERAL INFORMATION

AIMS OF THE TRUST	To study the life and works of Sir Francis Bacon and further the Great Instauration — the universal and general reformation of the whole wide world through the renewal of all arts and sciences — a program and method for the steady and gradual enlightenment of mankind during the Aquarian Age inaugurated, inspired and guided by those poet-philosophers and true artists of the English Renaissance who are referred to as 'the Brethren of the Rose Cross' because of their principal symbol of the rose of England and the cross of St. George.
TRUSTEES	The Most Hon. the Marquess of Northampton Lt. Col. T. S. M. Welch T. D. Bokenham J. D. Maconochie J. M. Shaw
DIRECTOR	Peter Dawkins, MA (Cantab), Dip. Arch., Architect.
OFFICE ADDRESS & TELEPHONE	The Old Rectory, Alderminster, Nr. Stratford-upon-Avon, Warwickshire, England CV37 8NY. *Alderminster (078-987) 633*
CONFERENCES STUDY GROUPS & OTHER EVENTS	Regular weekend Conferences are held at the major Festival times (*i.e.* the Solstices, Equinoxes and Quarters). In addition various study groups, lectures, seminars and other events are organised by the F.B.R.T.
FRIENDSHIP	Friendship is open to all those who are able to support, contribute to or take part in the work of the Trust. Friends receive the F.B.R.T. Newsletters and one Journal free each year, plus advance notice of and invitations to all conferences, lectures, courses, seminars, and exhibitions that the Trust is able to organise. Some events are arranged specially for Friends only. In addition substantial discounts on all F.B.R.T. publications and events are offered to Friends.
INFORMATION	All information regarding the work and Friendship of the F.B.R.T. is available from the Secretary at the Office address. Enquiries and bookings should be made via the Secretary. Friends of the Trust will receive priority booking. General Programmes and Publications Lists will be sent to those who ask to be put on the Mailing List.

Registered Charitable Trust No. 280616

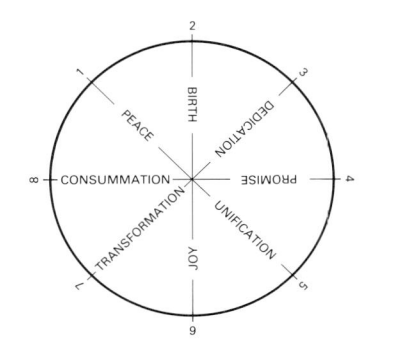

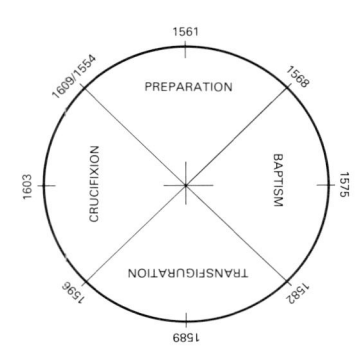

The F.B.R.T. Journals

The purpose of the F.B.R.T. Journals is to give a perspective of the complete initiatory path that all the great Religions or Wisdom Traditions have taught from time immemorial, as being the Way, the Truth and the Life. The scheme of the Journals is to cover the whole cycle of initiation in a series of sixteen volumes or Journals. Each Journal will mark a particular point in the initiatory cycle, which the Solar Festivals of each year celebrate — eight of them marking the end of an initiation and the beginning of the next, and eight marking the mid- or high-points of the initiations. (See F.B.R.T. Journal I/1: *The Pattern of Initiation*).

The series of Journals includes studies of:

(A) the festivals or initiatory focal points that each Journal commemorates.

(B) the various principal Wisdom Traditions or Religious paths, all of which teach and practice the same fundamental knowledge of Truth, set apart from each other only because each tradition has naturally been designed and has evolved to suit the particular time, society and circumstances in which it is born and bred, including its specific role in the overall drama of human evolution in this world.

(C) the life and times of the great initiate who is known to the world as Francis Bacon, whose work specifically and purposefully laid the foundations for our present modern societies and future more enlightened societies to be built upon, and whose life demonstrated the initiatory path. The work has an esoteric basis deeply rooted in a profound knowledge and understanding of the Ancient Wisdom Traditions and Mysteries, and a clear revelation of future times and the purpose of humanity. The descriptive symbol that the workers adopted is that of the Grail knight, the 'Seeker-after-Truth' — the Rose Cross.

(D) Baconian teachings and interpretations of the Ancient Wisdom teachings.

(E) The Great Instauration or Rosicrucian scheme for the complete renovation of all science (knowledge) and art (practice) the whole world over.

The first two Journals of this series prepare the ground for these studies, (a) with an overall view of the pattern of initiation in the evolution of human consciousness (see F.B.R.T. Journal I/1: *The Pattern of Initiation*), and (b) with a study of the meaning and role of virginity or purity in human life (see F.B.R.T. Journal I/2: *The Virgin Ideal*). The third Journal begins the study of a particular Wisdom Tradition — the Bardic Mysteries — and the story of Francis Bacon's life (see F.B.R.T. Journal I/3: *Dedication to the Light*).

11

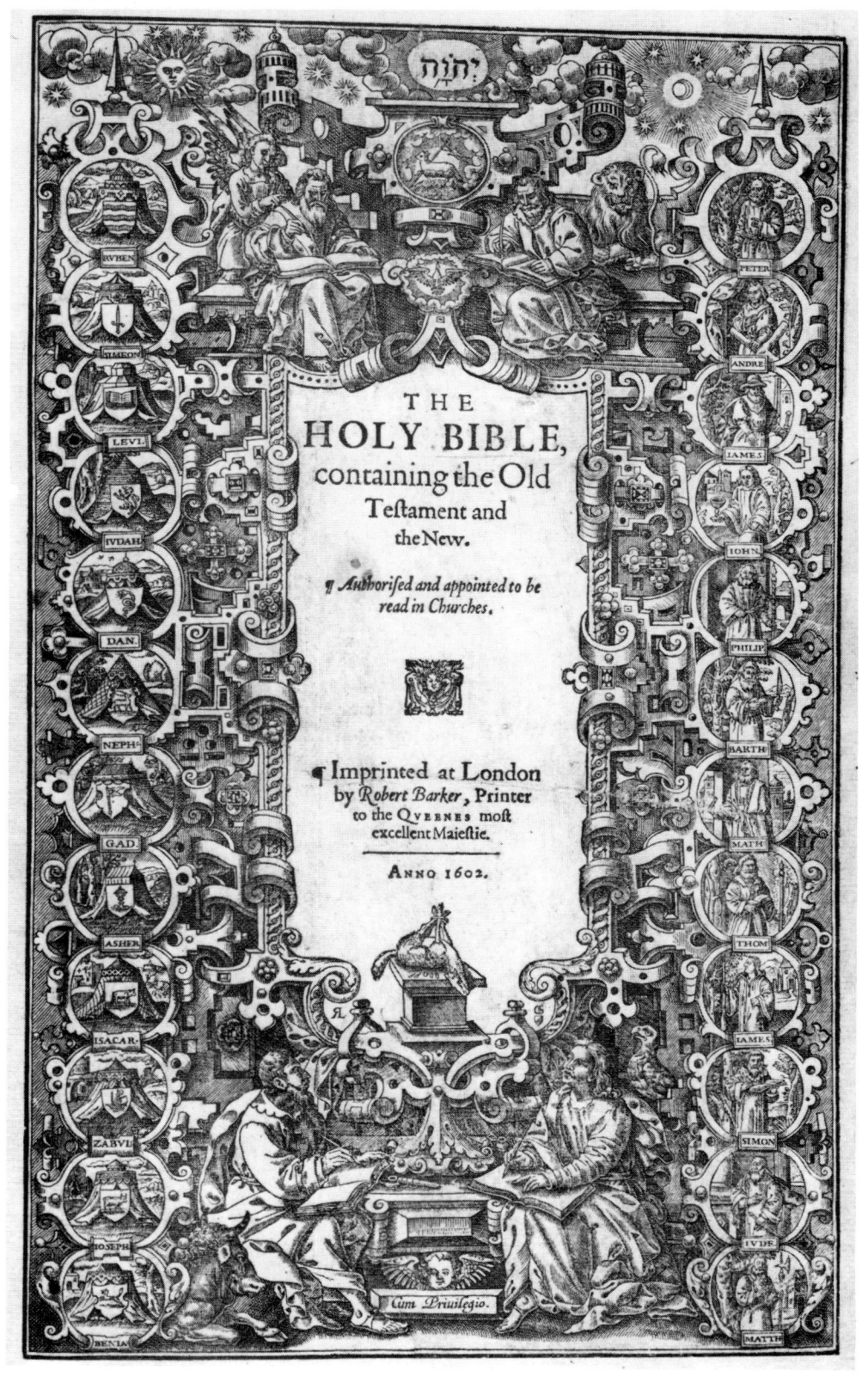

(3) TITLEPAGE TO THE "BISHOP'S BIBLE" (1602) — woodcut by Rowland Lockey and Christopher Switzer.

PREFACE

"A man doth vainly boast of loving God whom he never saw, if he love not his Brother whom he hath seen."

These words quoted above, written by Francis Bacon, perhaps summarise the real knowledge and driving force of that herald of our dawning Age, as they also summarise the essence of the Christian revelation and teachings. Christianity is the name given to what is also described as the New Testament or Covenant, which is wholly to do with loving one's neighbour as oneself. The New Covenant is founded upon the Old Covenant, not superceding the Old but fulfilling it by completing it. The Old Covenant is to do with loving God with all one's heart, soul, mind and strength. Both Covenants or Laws existed from the beginning of time, but the so-called 'Old' Covenant is the first to be realised by man. But this is not enough: man must then realise and learn to keep the 'New' Covenant also, for if he loves not his fellow man (and all of life) then his love for God is not only incomplete, but misunderstood and "vain". Jesus came to demonstrate dramatically just what this New Covenant really means; others have followed his example ever since, Francis Bacon being an outstanding one.

The Great Vision that Christianity gives is that of a supremely great man loving the world so much that he gave, and still continues to give, his life totally in service for the good of the world. This is also the Great Vision that Francis Bacon saw, understood, and was so motivated by that he founded a world-wide philanthropic work that would by degrees help man to see, understand and achieve what Jesus had so palpably demonstrated.

This particular Journal, the fourth in the series, is about this Great Vision; thus the book deals with some of the Christian Mysteries and with the Vision and Birth of the new Rosicrucianism, which is the name given esoterically to Francis Bacon's work and fellowship of true Christians. The first part of the book is devoted entirely to the Judaic-Christian Mysteries, whilst the second part deals with the life of Francis Bacon between the years 1572-1579, when he was aged 11-18 years old, and during which he experienced his great vision or initial illumination and founded the new Rosicrucian work. The Christian Mysteries continue to be dealt with throughout the second part of the book, as they are inseparable from Francis Bacon, his teachings and his life.

Born a prince, but concealed and fostered, Francis Bacon was compelled to live a secret or 'masked' life: but in doing this he acted out in his own life much of the Ancient Mysteries themselves, and used all this as a means to educate and initiate future generations. Man has always loved a mystery and a treasure hunt. He is by nature inquisitive and, once his interest is aroused, usually wants to discover the secret. In this he plays hide-and-seek with his Creator, Who concealed the living truth in matter, in nature, so that it might be discovered by man. Great initiates do likewise.

The third Journal, *Dedication to the Light*, dealt with the Bardic Mysteries, and with the birth and adoption of Francis Bacon. This fourth Journal is a natural continuation of the third Journal, covering Francis Bacon's youth, his university experience and great vision, his involvement with 'good entertainment', his sudden and distressing discovery of his royal parentage, his ensuing banishment to the Court of France, his love affair with Marguerite de Valois, his occupation in learning all he could in France, associating with the poets and sages of France, and busying himself with cipher work, and his eventual return to England to put into general practice all that he had learnt, tried and tested.

A large part of Francis' fascinating and exciting life is only discoverable through a lot of hard work, and being able to piece together and interpret all the various clues, historical details and cipher records. Ciphers of all kinds, from pictorial to mathematical, play a substantial part in recording many of the key experiences as well as some of the profoundest teachings in his life. Francis Bacon was an expert on cipher, and used many ciphers throughout his life-time in order to conceal from contemporaries and yet record for posterity the most intimate details of his life and of those around him. He intended that posterity should discover the story and teachings, so he always left the necessary signposts and keys. However, it is a slippery path, testing one's capacities to be both open-minded, highly perceptive and also critical every step of the way. Both intuition and reason are required, which is what Francis Bacon purposely set out to teach and develop to a high degree in mankind. The 'treasure hunt' to discover his real life and genius, and his work, comprises a universal education and an initiatory experience, developing the powers and the talents of each researcher stage by stage. What is shared in these Journals is only a fraction of what may be discovered and made good use of.

The unravelling of the mystery requires, as Francis Bacon almost certainly intended, a group effort and a reasonably lengthy period of time in which to discover the truth step by step, a little at a time, to test it, to assimilate it, and to comprehend what it in fact means and what use it might possibly be to know that truth. In this book I offer a few new discoveries which I have stumbled across, or been led to; but, like the pygmy who sits upon the giant's shoulders, and who is thereby enabled to see a little further than the giant, the few new discoveries (if discoveries they be) were made possible because of the enormous amount of dedicated research and discovery that has already been done, recorded and generously shared.

To name all those researchers to whom I am directly indebted would be too cumbersome in this preface, but many of their names may be found in the bibliography and the text. Two discoveries which I consider to be outstanding and which I have adopted, gratefully, in this story of Francis Bacon's life, are the decipherments of the Biliteral and Word Ciphers used by Francis Bacon. I have checked them as far as I have been able, together with a mass of other evidence, cryptic and open, with history as at present known and recorded; and I have come to accept the decipherments as being genuine and reasonably accurate, giving a remarkable and highly illuminating account of Francis' life, as written by himself. The decipherments are not complete, nor are they always arranged in the correct historical order by Dr. Owen and Mrs. Gallup, the principal decipherers of the Word and Biliteral Ciphers respectively. By cross-checking the historical data from many sources, I have been able to place the deciphered stories in their right order, establish their validity and fill in gaps.

When quotations from the decipherments of the Word and Biliteral Ciphers are given in the text, a different fount type from the main text is used so that the cipher passages can easily be identified. It should also be noted that the Word Cipher — Francis Bacon's "greatest" cipher — is used by him to convey the truth in a **dramatic** way: he has written up his life experiences, and those of others he was closely associated with, to be read as a play, a great drama, with all the passion, emotion and poetry of the stage. Thus he has undoubtedly, as it were, 'gilded the lily' where appropriate. In addition, his youthful euphuistic language flows throughout his early cipher accounts, embellishing and poeticising the actual history; yet at the same time never changing or corrupting the essential truth. In contrast to this dramatic Word Cipher history, the Biliteral Cipher stands as a sober, almost dispassionate key to and summary of the history contained in the Word Cipher, cross-checking every part, helping to explain obscurities, and offering additional details. The Word Cipher is more the emotional expression of Francis as a poet, lover and participator in the joys, sorrows and excitement of life. The Biliteral is more an expression of his keenly tempered mind, methodical, careful and precise. The mechanics of each cipher helps to determine these two forms of expression. There is no doubt that Francis preferred the Word Cipher, not only for its freedom in expression, but also because in it he could record his and others "appetites and passions", which he considered to be the "principles, fountains, causes, and forms of motion", and thus the "proper objects of philosophy".

It might be useful for me to indicate here some of the 'new' discoveries. The notable ones are:

(i)   that Francis Bacon appears to have gone to France in September 1575, banished by the Queen his mother when he was first made aware of his royal birth. He was then aged 14 years. The generally accepted date of his first journey into France is September 1576. Several researchers have noticed that there is evidence for his having travelled earlier, but have never been able to fix the date.

(ii) that the 14-year-old Francis Bacon may have played a major role in the story and drama of the grand Entertainments put on for the Queen at Kenilworth Castle and Woodstock Manor during the late Summer months of 1575, immediately prior to his banishment.

(iii) that the French Academy, made famous in the *Academie Francoise* first published in France in 1578, was probably a private Academy run by Michel Eyquen Sieur de Montagne, the well-known and popular French essaist, and that Francis Bacon had the privilege of joining the Academy during the Christmas of 1577. He subsequently recorded his experiences and the discourses in Montaigne's Academy, publishing the record anonymously (but with good cipher clues) the following February, in French, as *Academie Francoise*. Other Baconian scholars have discovered Francis Bacon's authorship of the book, but not, to my knowledge, the identity of the Academy, nor why both Francis and Anthony became such good friends of Michel Montagne.

(iv) that the secret agent of Queen Elizabeth's who worked under the *alias* of "Edward Burnham" and who made at least four dangerous missions to gather highly important intelligence for the Queen, would appear to have been Anthony Bacon, Francis' foster-brother and close friend.

PART I

THE JUDAIC-CHRISTIAN MYSTERIES

(4) CRUCIFIED SERPENT — woodcut emblem from titlepage to *The Faerie Queen*, Part II (published 1617 but dated 1613).

The Judaic-Christian Mysteries

THE SPRING FESTIVAL OF PROMISE AND THE 'WATER' INITIATION

The Spring Festival, or Festival of Promise, is a solar festival occurring at the Vernal Equinox, March 21st-22nd. It is closely associated with Lady Day on March 25th. Lady Day is the calendrical time marking what is known as **the Anunciation of the Virgin**. This is the occasion when the Virgin Mary receives a fore-knowledge or vision of the child that she will one day give birth to, whose name will be (or become) Jesus, meaning 'the embodied Light', the Christ child.

In terms of nature and the seasons of the year, Spring-time is the general period of plants beginning to push out of the ground and grow in profusion, strong and vital, unfurling their leaves to gather in the sunlight and transform it into useable plant energy, and drinking in all the watery sustenance that can be given them. In this vibrant and profuse leafing process is revealed a vision of the future — a promise of something glorious that all this activity will result in. It is a vision or promise that all Nature, come the Autumn, will give birth to her golden child, the fruit of all her growth and endeavours.

In terms of man's individual soul, the Spring Equinox for each of us occurs when, having passed the point of dedication,[1] and entered the path of initiation, we come to a vision, a fore-knowledge or promise of the future that lies before us. We glimpse the divine potential hidden within us, that is even then stirring within our breasts, vibrant with life and activity, which one day we will surely give birth to, revealing it and living it in all its fullness. Once we have that vision, we never forget it nor lose sight of it, even though it may be temporarily eclipsed from time to time by cloudy emotions and shadowy thoughts. From that moment on we endeavour to bring that vision into full manifestation. It is a promise — a covenant made betwixt oneself and God. As all divine covenants are always kept, as surely as anything can be that promise is inexorably fulfilled, even though we do not grasp or perceive all the ramifications and processes of life that we will be subjected to in order to reach its fulfillment. No pregnant mother ever knows beforehand the full implications of the eventual birth, nor what experiences she will actually be subjected to during the course of the pregnancy and labour; but she does have some idea given to her, both by the joy of life quickening inside her womb, and by knowledge passed on to her as the result of other people's (or her own) previous motherhood experiences. She has a vision, a promise, given to her. So it is with each person in the metaphysical, initiatory sense.

Sometimes we may seem to lose sight of the vision when difficulties are hurled upon us; but after the difficulties are over, the problems resolved, the vision will come back to us in renewed form, and eventually we will attain that vision. When we attain it we give birth to the Golden Child, which is also known as the Holy Grail. The Holy Grail is that part of our soul which reaches a state of perfect illumination and knowledge of God — radiant and reflective. It is the only part of our personal soul which remains truly immortal. The rest of our psyche, our natural soul (sometimes referred to as "the body"), is in a state of constant development and change, continually generating, destroying and regenerating its character and form. This, our lower, natural self, is mortal, corruptible, subject to endless cycles of birth, growth, decay, rebirth, *etc.;* whether it is to do with our thoughts and opinions, our desires and feelings, or the cells of our physical bodies or the whole bodies themselves. But, **because** of this mortality, which is no more that the alchemy of life, the 'gold' is produced that is immortal, deathless, so perfect that it is beyond change. Just as the candle under the influence of heat is made to change from a solid to a liquid, then from a liquid to a gaseous state, in order that it can then transmute into the flame that reveals the light, so our lower selves are continually transforming so that they can eventually transmute into the flame of the spiritual soul, *radiant* with Christ light.

Man's psyche (physical, emotional and mental) is in a continuous state of development and transformation solely to produce its fruit, in which is the golden seed of life and light — the Holy Grail. The immortal, changeless Grail — our spiritual soul — would not exist without the ever-changing, ever-mortal natural soul or psyche to give it birth and to continue nurturing or feeding it (*i.e.* adding to it.) The symbol of the burning candle is a beautifully simple and vivid revelation of this truth, as also is the perpetual cycle of the seasons and nature — the end of each year producing the fruit of the cycle in which something wholly good and life-giving is to be found. Another useful symbol used to explain the truth is that of the building of Solomon's Temple. 'Solomon' means 'peaceful' — a state of peace, which is a celestial state of perfect illumination, beyond change because it has no more need of change, even though 'change' produced it. Solomon's Temple represents the celestial soul or Holy Grail. It is built up, stage by stage, from stones that are quarried from the ground, cut, shaped, ground and polished, gradually being transformed and trued to the architect's design, and which are then raised up to be cemented into the growing Temple. But as each "living stone",[2] each a 'jewel', can be said to be a little temple in its own right, so Solomon's Temple is always in a state of perfection (in terms of quality), with only its 'size' increasing. In short, the mortal, corruptible and reincarnationary state of the **natural soul** is as essential to the manifestation of Divinity as is the immortal, incorruptible and eternal state of the **celestial soul**, and the **spirit** or archetype that generates the soul. These three are one and together form the manifestation of Divinity.

The spiritual archetype or spirit of Man has all the divine attributes focussed within it, such that it is divine love, wisdom and power. These three work within matter to generate the natural soul and evolve it through an ever-lasting series of transformations in which it learns to love, understand and serve truth. For every truth that is loved, understood and served, a 'stone' of the natural world is spiritualised and becomes added to the perfectly beautiful state of the celestial soul. The loving, understanding, serving soul is itself an embodiment of Divinity.

The Spring Festival is a time of courting, of love-making, a time when young lovers seek and find each other and make their first rosy-eyed vows or love-troths. The plighting of love, one to another, is a form of the divine promise or covenant, particularly if the love comes from the heart. It embodies a wish, or desire, and a vision of what the future might bring — a vision of glorious, light-filled promise to young, eager lovers. And so it may be, if the vision is true (*i.e.* from the heart) and the lovers are constant to their vows. The real meaning of Christian betrothal and marriage is to be found in this — the key being in the heart-centredness of the couple. Other degrees or types of marriage are possible, but they are not Christian and not lasting; only the heart-based love is true and eternal. Older societies used to know and respect this fact of life, and lovers were not pushed into a type of marriage which did not suit them. Society had various grades of marriage to suit every condition possible, and usually no-one was required to make vows which they could not possibly keep. When priests were true initiates, they were able to perceive the real desires or motivations of each couple, and only if these desires were seen and known to be strongly heart-based would the initiate-priest advise and consent to performing the 'church' wedding with its glorious blessing and severe vows.

The love-troth of a courting couple is an outwardly expressed form of the esoteric love troth made inwardly between each human soul and God — our psyche being the young 'maiden' and the Spirit of God being the 'male' lover. The love-troth, vow or promise made by each individual with God is a natural progression from and result of the dedication that the person concerned made previously — a dedication to trying to follow a life of good, to seek for truth, to undertake initiation. The dedication, marked in the seasonal calendar by **Candlemas**, allows the candidate to enter the first initiation, the first step on the seven-runged ladder to perfection. This first initiation is symbolised by the element of water, as it is to do with the development and control of the emotions, gradually raising the baser desires and feelings until they become heart-based and peaceful. True loving comes when the emotions become heart-based. As our emotion is our very life-force or motivating power (*E-motion* being the only motivating power in the universe), it is important to begin with the right motives — which means the right desires, emanating from a pure heart as perfect love. Until we have governed something of our emotional nature we cannot progress much further. But once the heart nature is touched and begins to unfold, then the promise is given, the covenant made. This occurs at the height or mid-point of the Water Initiation.

This Water Initiation is summed up in the first part of the Druidic exhortation that, in order to achieve *Awen* (the holy state of Christhood), man must learn to "**love truth, understand truth, and serve truth**." In Christian gospel terms these three stages are called "**faith, hope, and charity**". In all sacred traditions that I know of, the emotions are symbolised by water. Like water, they can be very turbulent, or they can be quiet and peaceful. They can be dark and fearful, or brightly reflect and refract the light of the heavens. They can be muddy and obscure, or they can become clear as a crystal that one can see right through. Perfect faith is perfect emotion or love, unshakeable, the true foundation and starting point of all else.

Out of water emerged all living creatures — a manifestation of the primaeval Creation. For "in the beginning" God made Matter, referred to as the "Waters" or "Deep", and inpregnated these universal 'Waters' with his living Word or Idea, which is Life and Light. Vivified and organised by the Word, forms of life were (and are) generated out of this 'Sea' of Matter, the 'Ocean of life', and gradually (*i.e.* stage by stage) brought to a perfection of life expression.

But even this generation of life forms out of the Universal Sea of Matter was (and is) but an expression of and natural progression from the very first act of God, the Unmanifest and Absolute Cause, which was to DESIRE to become manifest; which desire was an emanation of love from the 'Heart' of the Absolute. The Divine Emotion was (and is) the life force of all Creation — the motivating power that brings all things into being; which generates, sustains, evolves and perfects all living form according to a divinely conceived and perceived Idea or Plan, and which dissolves each form when it has fulfilled its purpose.

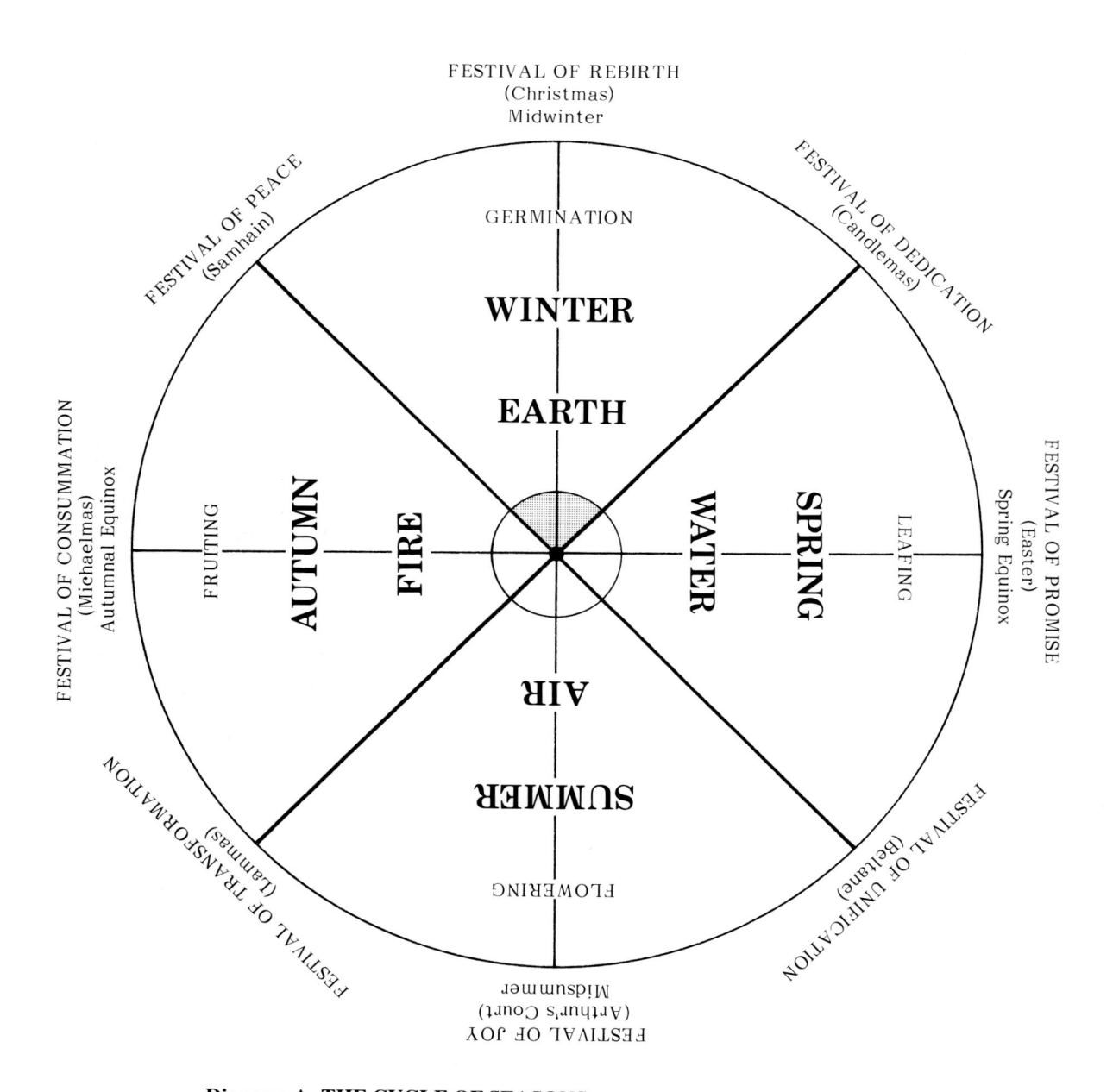

Diagram A: THE CYCLE OF SEASONS

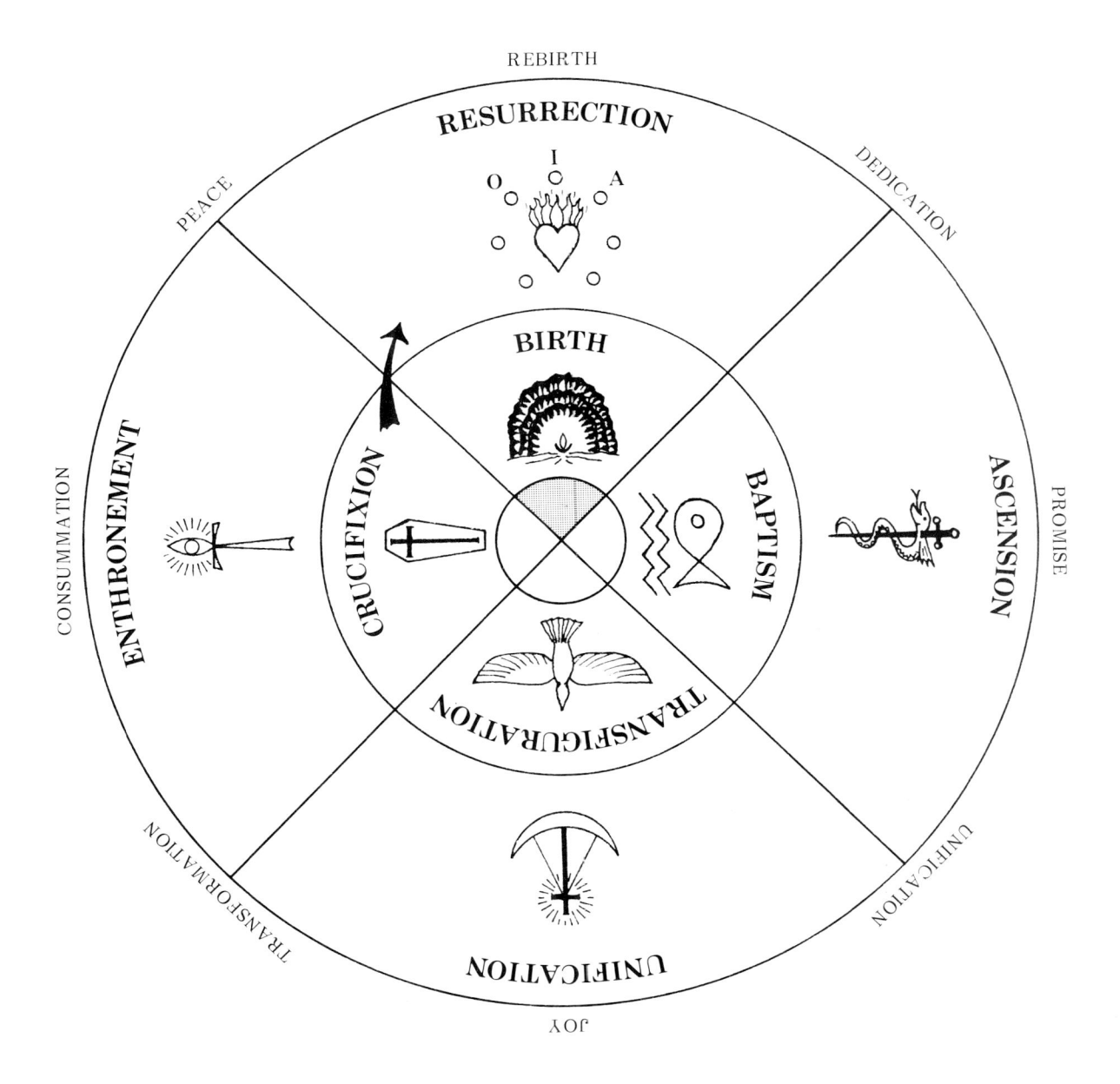

Diagram B: THE CYCLE OF INITIATION

AN AND ANNA: DIVINE MIND AND MATTER

The initial Desire of God — the emanation of Divine Love from its Absolute Source — immediately caused a polarity to occur or to manifest. Firstly a polarity between Non-Being and Being, between the Supra-Absolute and the Absolute, between Non-Existence and Existence; and, secondly, an inherent polarity within the Absolute BEING-ness, which is the essential condition for any manifestation or existence to be possible. That is to say, for a BEING to exist as such, there must be a condition of polarity (or duality), *i.e.* a positive and a negative, a crown and a root, *etc.*, together with the relationship between the two poles. It is in this relationship that love, as well as power, can be expressed — as a force, sensitivity and a harmony between the two 'loves'. Thus, in the opening sentence of the Hebraic and Christian Bible, it states:

"In the beginning[3] God[4] made [became] the Heaven and the Earth."[5]

As Francis Bacon pointed out, this statement does not refer to the creation of heavens and earths which occurred later, but to the creation by the Divine Will or DESIRE-TO-BE of the first fundamental polarity; namely, the heavenly Mind of God and the Universal Matter. In sacred tradition this first and fundamental condition of BEING has been described as the Holy Trinity: the Heavenly Father (the Divine Mind), the Earthly Mother (the Divine Matter), and the Son (the relationship between the Divine Parents)[6]. All Three are One, and came into Being together, simultaneously, from Non-Being. No one Person of the Trinity has precedence over either of the other two Persons, but all Three are "co-existent and co-eternal", each dependant upon each other and together constituting the whole BEING that we call 'GOD', the All-Good. Thus the true statement that: "the One became the Three" — not Two, or Four, but **Three**. The problems that various people and sects have had with so-called 'Dualism' stems from a misunderstanding of this basic principle of life.

This first existing, Absolute Being, that we call God, and which in the East (in the Vedic teachings) is referred to as "the One True Existence, *Paramathika*", or *Brahm*[7], contains the basic Three Principles of Creativity known as Power, Wisdom and Love, all Three being inherent in the Son, the Relationship of the Divine Parents; but of the Three, Love is the first to be expressed, as an emanation from the ubiquitous Heart.

It should be understood that the primal, pure Matter is not matter or substance as we commonly know it. What we normally see and experience as matter is in fact a form of life that is built out of the fundamental and divine Matter, but not pure Matter itself. Divine Matter cannot be seen, neither is it separate from the Divine Mind, but is simply the polarity to Divine Mind and the root principle of all manifest form. It cannot even be held within our experience or our consciousness, because it lies at the extreme limit of any infinite experience or consciousness we might ever achieve. Divine Mind and Divine Matter contain all living experience and consciousness: they themselves are not contained by anything. One of the best philosophical attempts to describe this condition of Absolute Being is as 'Space' — pure unbounded Space — a dark Void that contains no form as such, in its pristine condition, and yet contains everything in principle, or potential. The Druidic teachings refer to It as "the Eternal and Infinite Darkness" — a phrase which encapsulates the idea of the One Space in which is the Holy Trinity: *viz.* Eternity is the Power of God (the Father), Infinity is the Love of God (the Mother), and Darkness is the unmanifest Light or Wisdom of God (the Son.) In this Absolute condition of Being, Darkness fills the Divine Mind, and Divine Matter is void, without form. Hence the second statement in the Hebraic-Christian Bible:

"And the Earth was without form, and void: and Darkness was upon the face of the Deep."[8]

This condition of divine Being, the Absolute, is the highest conception of Reality that is possible to man: beyond that no words or thoughts can possibly suffice, because the Supra-Absolute is beyond any feeling, thought or experience — feeling, thought and experience all being conditions of Being, of which the Absolute is absolutely our limit and the limit of all existence, infinite though that may be. Beyond Infinity nothing can exist as such, and we cannot even express this supra-infinite concept properly, this 'No-Thing' from which all things, all existences, proceed. The best we can do is to use 'negative' words or statements, such as "No-Thing", "Non-Being", "Incomprehensible", "Unknowable", "Negative Existence", and so forth.

But as for the Absolute Being, we do have words to describe It; which words are the most sacred and the most profound of all, because they embody man's highest conceptions of Divinity. I have already mentioned some, but there are others that should be known if we are to comprehend Christianity in all its fullness. The Hebraic-Christian tradition has certain 'God-names' which it uses, each of which carries a vibration or power and a conscious idea or knowledge of all that it represents or symbolises. These God-names are part of the esoteric tradition, but presented exoterically for all to have the chance of grasping their significance if they are capable of it. The names were used by the Chaldeans and the Ancient Egyptians in their Mystery traditions, from whom the Hebraic and Christian teachings were adopted. The Druids also employed the same 'essential' names.

In the Hebraic-Christian tradition, the God-name for the Divine Mind (or Heavenly Father) in *AN*, whilst that for the Divine Mother (or Earthly Mother) is *ANNA*. *AN* is sometimes spelt *ON*, and we derive our word ONE from this mystery name. The very form of the letters, even in English, convey the idea of what they stand for. *ON* signifies the 'O' of Space filled with the potential 'lightning flash' of divine consciousness and creativity, whilst its alternative form, *AN*, says the same thing but in terms of the Holy Trinity (the '*A*' or Triangle) of the Eternal Infinite Darkness. '*O*' or '*A*' actually represents Divine BEING, the Absolute, the initial Desire-to-be from which all else proceeds. The '*N*' is its inherent Wisdom and creative powers that can vibrate Matter into the infinite variety of forms that manifest the divine Being. *ANNA* conveys the idea of *AN*'s polarity, its complementary principle in which is contained the principle of reflection and manifestation — the final '*A*' standing for the ultimate manifestation or revelation of God in living form. The vibratory sound of '*A*' provides, in its first sense, the gasp of wonder or credulous query, whilst in its last sense it conveys the amazed wonder of sudden revelation and comprehension.

AN is the mystery name of the supreme Heavenly Father, who is "Heaven", the Divine Mind, and who is the 'heavenly' polarity of all existence, all being. *ANNA* is the supreme Earthly Mother, who is the "Earth" or Divine Matter, and the 'earthly' polarity of all being. *ANNA* is referred to in exoteric tradition as the Black Virgin or Earth Mother, the Dark Waters or Deep, barren and infertile, but who brings forth the White Virgin, the Virgin Mary. In Christian parable, Anna is the mother of the Virgin Mary. The 'blackness' of Anna refers to the state of darkness, or unconsciousness, of Divine Matter in its Absolute state. The 'virginity' of Anna refers to the pure or pristine condition of that Divine Matter. Her 'barrenness' refers to the fact that the Divine Matter is in a state of passive quiescence or rest, 'unseeded' by Divine Thought because, in the Absolute state, the Divine Mind has not yet begun to think and to create: it remains in a condition of darkness or unconsciousness — the state of Absolute Peace, in which is hidden all potential. Anna's husband is called Joachim, which means 'appointed of the Lord'. He is described as being a high priest of God the Most High, and this is a word and description evolved from the idea that *AN* is the 'Most High' (*i.e.* Heaven) and the 'High Priest' of all Creation. Every subsequent high priest stands in for and represents the manifestation of the Most High in any temple. Heli (Eli) is the alternative name used for Anna's husband which actually means 'God', 'the Most High' or 'Exalted'.

Other words handed down to us, which are now part of our tradition and in everyday useage, are 'Chaos' and 'Hell'. The Greek word, *Chaos*, is their name for the Absolute Being, dark and formless but from which all things proceed. Hell also means the same, being derived from *Hel* or *Hele*, the Eternal Infinite Darkness in which the potential of all things lies hidden. *Hel* is both the Divine Mother and, as *Heli*, the Divine Father. In its more common Hebraic form, *El*, it means 'the Almighty', 'the Omnipotent', the One who knows all and sees all, and in whom all the divine attributes are concentrated.[9] In its original sense, Hell is by no means a place of torment and anguish, or a place of punishment for the wicked. It is the supreme place or condition of Peace — that which we all originally came from and into which we shall all return at the end of time. The Hele stone, such as at Stonehenge, signifies the absolute starting point and ultimate finishing point of all Creation — the real heart-point of any temple. As for 'Hell Fire', it is none other than the 'Fire' of Divine Wisdom that is concealed, or is inherent in, the Mind of the Absolute Being. To be "burnt up in Hell fire" really means to be returned to one's absolute source of being, the Bosom of God, the state of perfect peace and rest. Our forms — our souls and bodies — are indeed "burnt up", but only by the divine fires of Love, because one day we shall have no more need of them. Our essential beings can never be destroyed, but only freed from the limitations of form once we have fulfilled our eternal destiny. Hell fire is eternal, and 'death' in the true sense of the word comes to all of us eventually, as it means the final death or dissolution of all our forms so that we dwell once more in a state of formless, pristine peace. Medieval churchmen seriously misunderstood these sacred meanings, and distorted truths for their own shallow and selfish ends, or just from sheer ignorance, the curse of mankind — a curse which still exists today, and from which none of us are free.

THE VIRGIN MARY AND DIVINE THOUGHT

The Virgin Anna gives birth to (or becomes) the Virgin Mary. Mary (Greek, *Maria*; Hebrew, *Mariam*) is derived from *Mare*, meaning 'the Salt Sea', or 'the Bitter Sea'. *ANNA* is the name for the universal Matter in its Absolute sense, symbolised by the invisible 'Deep' or 'Dark Waters'. *MARIA* however is the name for those self-same 'Waters' that have become impregnated by the light-seed (Spirit) or generative power of God the Most High, the Heavenly Father. Salt is a crystallisation of something that was previously a liquid. It represents the first and fundamental form of life, "born of the Virgin Mary". Salt is the basis of all life forms, and in its structure it exhibits a pattern that expresses the Law or Idea of God in a beautifully simple or pure way. *MARIA* is thus the state of universal Matter when that Matter is vivified (*i.e.* impregnated) with the Spirit or Word of God, whilst *ANNA* is the state of Divine Matter in the Absolute sense, before the creative life force is implanted into it. *MARIA* is in fact the first life-form — Matter made to move (vivified) and organised into form by the Spirit; and, as the crystallisation process proceeds, more and more complex forms, with increasing consciousness and abilities, will be built up, until *MARIA* becomes (or gives birth to)[10] *JESUS* — the final, perfect and fully illumined life-form that manifests the spiritual Idea in its entirety.

But, for *ANNA* to become *MARIA*, the Mind of God must first conceive the Divine Thought of Creation. It must become conscious, creating an Idea and Vision of what God desires, which Idea and Vision can then act as the architectural Plan for building the perfect Temple or life-form out of Matter. Thus the next stage following that of the Absolute is for the Mind of God to become active and think the Divine Thought. The first stage of the Absolute is known as "Emanation", and the second stage of Thought is called "Creation" — the creation of God's Idea in God's Mind. The unconscious Mind becomes conscious: the Darkness is turned to Light in the Heaven of the Divine Father, and the whole Cosmos of form is designed in principle, in thought. In the Hebraic-Christian Bible the rest of *Genesis*, chapter 1, is devoted to this 'Creation', plus the first three verses of chapter 2 (the 'Creation' story running over into the next chapter or stage just as the 'Emanation' story runs into the beginning of chapter 1, the 'Creation' stage).

"And the Spirit of God moved upon the face of the Waters."[11]

God's Mind became active. Its 'Spirit' became active as **Divine Thought**. In early Christian texts the **Holy Spirit** is explained as being the Holy Breath (*i.e.* Divine Love or Life) and the Holy Wisdom. The Holy Breath utters the Wisdom as the **Word of Truth**, radiating it as **Light** into the heavenly Mind, wherein it is reflected as the perfect **Vision** of Truth. The actual process is three-fold, and is described as being a Word of Command (the Will of God), followed by the manifestation of Light as the radiance of Divine Love (which is Blissful Consciousness), followed by the reflection of that Light as the perfect Vision of Truth and Plan of Action (which is Intelligent Consciousness). These three stages are in reality one stage, all three dependant upon and inseperable from each other. They form another expression of the Holy Trinity: this time as Divine Thought[12], whereas before it was Divine Being.[13]

"And God said, 'Let there be Light':
"And there was Light.
"And God saw the Light, that it was Good ...."[14]

The Word of Command is the 'Inner Voice' — "the still small voice" that never ceases to speak to any of us, nor ceases to speak to any part of Creation. It is known in us as the Intuition — the 'Teacher within'. In terms of the Divine Thought it is known as the Word of God (Greek, *Logos*). It conveys and commands the unmanifest Idea to appear, which then does so, radiating into the Mind of God as a limitless Light or blissful Consciousness. This Bliss is then reflected as the perfect Vision or Intelligence and reflected upon, by the Divine Mind, until the Intelligence becomes perfect Understanding. The radiation of Light as pure Bliss, the perfect Wisdom or Intuitive Knowledge of God, is known as the Christ (Greek, *Christos*). The reflection of the Light as the Vision, the perfect Intelligence or Rational Consciousness of God, is known as the 'feminine' counterpart of the 'masculine' Christ, and is called *Sophia* in Greek-Christian tradition. Thus, when the Word (*Logos*, or Will) is uttered, it radiates as Light (*Christos*, Wisdom) and is reflected as the Vision of Truth (*Sophia*, Intelligence). These Three — *Logos*, *Christos* and *Sophia* — are one Divine Thought.

To begin with, the radiating Light is Wisdom, the Bliss or Christ Consciousness of God, and the reflected Light is Intelligence, the Vision or Rational Consciousness of God.[15] But finally, in the last instant, the radiant Light is still Wisdom, changeless, but the reflective Intelligence has become total Understanding or Rational Knowledge. These two instants are important, for they explain the reason for all manifestation of Light in Matter, and the generation and evolution of all life form: which is for the sole purpose of evolving Understanding and producing a manifestation of the whole Being of God. Wisdom is Divine Knowledge, and Understanding is Divine Knowledge: the former is intuitive and the latter is rational. Intelligence itself is not Knowledge, but simply the Knowledge seen (and heard) but yet to be comprehended.

MAN

The mystery of Man lies in the fulfilment of God's Idea. There cannot be a complete reflection of the Christ Light until the Vision itself embodies all that is in the Holy Wisdom. Since the Holy Wisdom contains the idea concerning (i) both radiation and reflection, (ii) the Mind that has the ability to think creatively and perceive its own ideas, (iii) the counterpart of the Mind which is able to receive and manifest the ideas objectively in material form, and (iv) the Absolute Source of them all, this must be a Vision that can become a Divine Being in its own right — an objective 'replica' of God. For, when the Image can independently think, radiating Wisdom as well as reflecting it, then the Understanding of God will be complete and Creation be at an end.

'Man' means 'the Thinker'; that is to say, 'the Mind' — the 'Father' as an Image of the Divine Father. The creative process of Divine Thought culminated with the spiritual creation of Man on the "Sixth Day" of Creation. On this Sixth Day, God completed his perfect Vision of Himself, but only in potential — Man being the culminating point, the intelligent Image or Reflection of the Christ Light in whom all God's ideas and potencies are summarised. In this Image the Heavenly Father may eventually come to see *and know* in all ways His own creative Being.

The Image (*i.e.* Man) must therefore become an active creator in his own right, plus having all other attributes of divinity (for none can be separated from any other), so that at the end he becomes the embodiment of God's Knowledge or Light, Intuitional and Rational. Then will God's Plan be fulfilled and His Mind be glorified in the fully illumined Man,[16] the crowning point of all Creation.

In Hebraic-Christian Caballa, Man is referred to as the 'Kingdom' or 'Bride' — a spiritual Principle and Idea that has to become actualised, first as the Vision or Image in God's Mind, then as the human soul in Matter. For Man to actualise this idea and embody all that God is, thereby fulfilling the Divine Idea, he must involve (*i.e.* incarnate) himself in Matter, generate a body or form that can express this whole Idea, and evolve that form to its glorious perfection. This is the work accomplished on the Seventh Day (the Sabbath) by the incarnating Spirit, whilst the Divine Mind (the Heavenly Father) rests in meditation, transforming Intelligence into Understanding.

It is important to realise that there are three aspects to Divine Thought: namely, the Word (or Will), the radiant Wisdom, and the reflected Intelligence. Then we should realise that the Word itself contains the very Idea of this triune Thought, which Trinity is then radiated as the Light of Divine Wisdom, and then reflected as the Vision of Divine Intelligence. The Word (*Logos*), the Light (*Christos*) and the Vision (*Sophia*) each themselves contain the Word, the Light and the Vision in their own respective ways. Man is associated with *Sophia*, the reflected Intelligence or Vision of Truth, as the summation and crowning principle of all that *Sophia* means; but as *Sophia* he also incorporates the whole triune Thought of God (*Logos*, *Christos* and *Sophia*) as a Reflection or Vision of Truth.

FATHER, SON AND HOLY GHOST

The mystic formula, "**Father, Son and Holy Ghost**" has, like all truths, a real meaning on many levels. In its highest connotation, it signifies the Absolute Trinity of Power, Wisdom and Love — the Divine Father, Son and Mother respectively, who represent (or are) Eternity, Darkness and Infinity. In another connotation it signifies Emanation, Creation and Formation, where the Divine Desire-to-be is the Father of all things, the Divine Thought is the Son that shines in the Heavenly Mind, and the Divine Immanence is the Holy Ghost or Presence of God incarnate in matter. In a further connotation it signifies the *Logos, Christos* and *Sophia* of the Divine Thought, where the Father is the Will, the Son is the radiant Wisdom or Christ Consciousness, and the Holy Ghost is the reflected Intelligence that conceives and gives form to the consciousness of Truth. In terms of Man, the Father is the spirit or archetypal thought of Man, the Son is the spiritual (or illumined) soul of Man, and the Holy Ghost is the procreative or generative vibration embodied in the material forms of Man that evolves them until that perfection is reached wherein the spiritual soul is born.

THE THREE, SEVEN AND TWELVE

The Holy Three are the Trinity, the basis of all life and its manifestation. Inherent in the Love-emanation of God as an essence, in the Mind of God it becomes a creative, active energy. In its activity it has seven principal modes or ways in which it can be active. No aspect of the Trinity can be separated from another, as they are one Unity, but each aspect can be either active or passive. If all three were passive together, there would be no creativity, no thought — and this is in fact the state of Darkness of the divine Mind which preceded the Light. But with the movement of God's Spirit in thought, the three aspects become active in various combinations of activity-passivity — there being seven possible combinations altogether (*i.e.* ABC, the totally passive mode, becomes creative as **A**BC, A**B**C, AB**C**, **AB**C, **A**B**C**, A**BC**, **ABC** — the last being the totally active mode). In terms of the Will of God (the *Logos*), these creative Seven are known as the Seven *Logoi* or Words. In terms of the Bliss Consciousness (the *Christos*), they are known as the Seven Lights, Rays or Voices. In terms of the Intelligent Vision (the *Sophia*) they are known as the Seven Great Archangels or Thought-forms of Light. The Seven Great Archangels give form (in terms of thought) to the Seven Rays of Light, which in turn are the vibrational impulses of the Seven *Logoi* or spiritual Principles. As one great Thought, they are known as the Seven Spirits — the seven creative modes of the one Holy Spirit — the final one summing up all the others.

When incarnate in Matter, the Spirit works with twelve distinct generative powers and generates twelve types of living form with twelve interrelated but distinct functions.

Although the Law of Three, Seven and Twelve is inherent in every aspect of life, nevertheless there is a special relationship of the Three to the realm of Emanation (*i.e.* the expression of divine Love); of the Seven to the realm of Creation (*i.e.* the expression of divine Wisdom); and of the Twelve to the realm of Formation and final Manifestation of Truth (*i.e.* the expression of divine Power). The summation of the numbers 3, 7 and 12 give the mystical number 22 which is associated with the Bride, the final Revelation and Joy of Truth.

BAL, THE WORD OF GOD

The Great Word or Idea of God has, like the divine Mind and Matter, a mystery God-name in the Hebraic-Christian tradition that is little understood today. Usually it is referred to simply as the 'Name'[17]. That name is *BAL*. It is derived from the three 'head' letters of the Hebrew Alphabet (*Aleph*, head of the three 'Mother' letters; *Beth*, head of the seven 'Double' letters; *Lamed*, head of the twelve 'Common' or 'Single' letters), these three 'head' letters standing for the **whole** of the sacred Alphabet. The significance of this is that the Alphabet represents the *Logos* or Word of God — the one Word or Idea from which all other words or ideas are created or derived. *BAL*, and the Alphabet it represents, is the Great Name of God. In the making-up of the sacred Alphabet, with its three divisions giving the sacred numbers or formulae of 3:7:12, is to be found the entirety of Truth. Even in combinations of the three 'head' letters, other names of God as the Creative Thought can be found: *viz. AB*, 'Father'; *BA*, 'Son', *AL*, 'Word'; *BAL*, 'Lord'; *LAB*, 'Heart of Spirit'[18]. *BAL* (sometimes written *BEL*) is found in many sacred traditions and languages, *viz.* Druidic, Arabic, Phoenecian, Greek, as well as Hebrew. In Arabic, Phoenician, Hebrew and Greek the numerological cipher of *BAL* is 33 (B=2, A=1, L=30), which is of prime importance.[19]

But the Alphabet as written down in Ancient Hebrew[20] was one of consonants only, twenty-two in number. *BAL* is derived from and represents these twenty-two consonants.[21] They are all 'feminine', and signify the reflective principle of Light as Divine Intelligence or *Sophia*; hence they can be written down. But the vowels were considered too sacred to write down, and could only be spoken. Said to be seven in number, they represent the radiant aspect of Light, the Divine Wisdom or Bliss Consciousness, the *Christos*. The rays of the Christ Light are reflected in the Heavenly Mind like stars, the radiance only becoming visible and knowable when reflected. Hence the symbolic reason for being able to write down the consonants of the Alphabet, which represent the visible reflections of the otherwise invisible Christ radiance or Word of God (represented by the unwritten vowels).

DIVINE TRANSCENDENCE AND THE IMMANENCE

The Divine Thought (Will, Wisdom and Intelligence) dwells in the Divine Mind and is principally transcendent or 'above', 'beyond', 'uninvolved' in Matter. As part of the Divine Mind it is the heavenly polarity to Matter, the earthly polarity. But the polarities cannot exist without the relationship between them — a relationship that partakes of both polarities. Thus a conscious aspect of the Divine Mind 'descends' or 'incarnates' into Matter, generating a vibrationary aspect of Divine Matter to 'ascend' or 'resurrect' towards the Divine Mind; the two aspects fusing together to form the manifest and conscious relationship of Mind with Matter, an active expression of the Absolute Source itself.

Thus there is, in the Divine Thought, an aspect which is able to incarnate and unite with Matter, generating and evolving the life form that is called the "living soul".[22] As part of the Divine Thought, this aspect is transcendent, but as the incarnating spirit it is immanent in Matter, united with Matter and informing Matter with life, wisdom and intelligence. Its love moves (*i.e.* vivifies) the matter, its wisdom organises the living matter into intelligent life forms, and its power works continuously to transform and transmute those forms until they reach their destined perfection.

This Immanence is Man, an aspect of the Divine Intelligence, *Sophia*. It also exists as an idea in the Divine Wisdom or *Christos*, and as a principle in the Divine Will or *Logos*; but it exists as a thought-form and a generative power in the Intelligence of Divine Thought.

In Hebraic tradition this Immanence is called the *Shekhinah*, whilst in Christian tradition it is known as the Holy Presence, the incarnate aspect of the Holy Ghost or *Sophia*. The Immanence corresponds to the *Shakti* or feminine creative powers that are consorts to the masculine potencies or gods, *Brahma, Vishnu* and *Shiva*. (*Brahma, Vishnu* and *Shiva* are the Vedic names for the *Logos, Christos* and *Sophia* of Graeco-Christian tradition.) As a thought-form of Truth, the Divine Immanence bears within herself the 'seeds' of Christ Consciousness which constitute the *Messiah* or Christ that is "conceived by the Holy Ghost" and carried by her Immanence into incarnation, wherein that Light will eventually be made manifest and known in the pure and illumined human soul (*i.e.* "born of the Virgin Mary") that the *Shekhinah* generates.

When incarnate in Matter, in the form of the living soul she will fulfill the idea of the Bride or Kingdom. She will prepare herself for marriage with the Christ Spirit, her Beloved, by gradually weaving the wedding garment or veil out of matter, purifying and beautifying every form that she makes to clothe herself in, until she is ready for marriage with the Christ — the Christ becoming actively manifest within her bridal form, irradiating her with Light. As the Kingdom, she will allow the Christ to become Sovereign Lord and Master over her whole estate, manifesting his Light, his Glory, from corner to corner. Her Kingdom is her marriage dowry.

As the Immanent Principle, the *Shekhinah* 'falls' from highest Heaven into the abyss or depths of Matter, bearing within her heart the Word and Light of God. She incarnates into the heart of the Earthly Mother, as a spiritual seed sent from the Heavenly Father, impregnating the Virgin Mother with Light and thrilling her from infinity to infinity.

This thrill is a sound, a vibration (*AUM*) which pervades the universe from end to end, expanding, developing, increasing in power whilst building up, sustaining and destroying one living form after another, in endless variety and ever-increasing consciousness and beauty, until at last the spiritual Idea is accomplished and Christ is made manifest in the human form. This form of consciousness, which is the natural living soul evolved to its highest level, is that which is called 'Mary'. The natural living soul of Man begins as Adam-Eve, where Eve is the form and consciousness of Adam, the first personal ego. At this stage Eve, the consciousness, is ignorant; but gradually, through successive experiences, she becomes Mary, the soul who knows good and evil, and who chooses good in that she loves, understands and serves the Truth. At this point the Christ is revealed in her (*i.e.* "born of the Virgin Mary"), whilst Adam, the personal ego, through a series of deaths and rebirths has finally become transmuted to the transpersonal 'I AM' or 'Word made flesh'.

JOHN, THE MAN SENT FROM GOD

The Divine Immanence that works in Matter as the generative Spirit of Nature, also has a mystery name. The name of anything is its own inner nature, light, or creative idea; thus the 'Name' of the *Shekhinah* is her own inherent *Messiah* or Christ Consciousness, which is the perfect embodiment of the transcendent Christ Light. To see and know the inner nature of the *Shekhinah* is to see and know the Christ.

The mystery name of the *Shekhinah* signifies that part of the great Thought of God which incarnates into Matter as Divine Immanence. It signifies the generative, procreative forces of Nature, and its inherent Christ consciousness, which builds and initiates the soul. In its 'feminine' aspect of Intelligence its name is represented by three consonants selected (or 'chosen') from the twenty-two consonants of the Great Name; whilst its 'masculine' aspect of Wisdom is signified by three vowels selected from the seven vowels of the Great Name. These letters, forming the mystery name of Divine Immanence, also sum up the whole Idea of God, so as to manifest that Idea in Matter.

The three consonants of this God-name are *Jod* (*J*)[23], *Heh* (*H*), *Vau* (*V*), chosen from the twelve 'Single' letters. In sacred combination they produce the 'Intelligence' name, *J.H.V.H.*[24], which conceals (and yet reveals) the 'Wisdom' name, *I.O.A.*, composed of three vowels which summarise the whole wisdom of God. The vowel sound '*I*' signifies the very life and being of God and man; the '*O*' signifies the Thought and Consciousness, and the '*A*' signifies the generation and manifestation of this Thought and Being in Matter. These equate to the three stages of divine Emanation, Creation and Formation (or Generation) that culminate in the full manifestation of Divinity. In terms of **Creation** and the Thought of God, '*I*' represents the Word (*Logos*), '*O*' represents the Light (*Christos*), and '*A*' represents the Vision (*Sophia*). In terms of **Formation**, '*I*' inspires desire; '*O*' inspires thought, and '*A*' inspires action; all of which eventually culminate in loving (*I*), understanding (*O*) and service (*A*) when the life-process has perfected the soul.

As for *J.H.V.*, these consonants also express the same truths, in the same respective order, '*J*' being the intelligent reflection of '*I*', '*H*' being the intelligent reflection of '*O*', and '*V*' being the intelligent reflection of '*A*', with the additional '*H*' of *J.H.V.H.* signifying the eventual completion and fulfilment of the Divine Will. In addition, '*J*' is associated with the Divine Ego, and with the creative and generative life force of divine love, the Will of God, that blesses and teaches all things; '*H*' is associated with inspiration and illumination; and '*V*' is associated with the focalisation, unification and synthesis of all things.

The complete mystery name of the Divine Immanence is the combination of the vowels, *I.O.A.*, and the consonants, *J.H.V.H.*, which renders the name *JIHOVAH*. The name has a meaning interpreted as 'He who was, is, and is to come', referring to the generative nature of this divine power in forming the soul and evolving it by means of multitudinous transformations that take place throughout Time. 'He who was, is, and is to come', whilst indicating an eternity of time, also implies the condition of relativity and relative time. *JIHOVAH* is known as the 'God of Israel' in a peculiar sense.[25]

Other consonants, forming other names that mean the same, also conceal and reveal the mystery of *I.O.A*, such as *JONAH, IONA, JOHANNES* or *JOHN*. In that wonderful first chapter of St. John's Gospel is enshrined the secret of the descent of this aspect of the *Logos* into Matter, to generate the natural soul of man, and to become progressively embodied and revealed in the soul until eventually the fully perfected and illumined Christ soul, called *JESUS* (the 'Word made flesh') is produced.

JOHN, a corrupted and popular rendering of *JONAH* or *JIHONAH*, another form of *JIHOVAH*, is the **Name of Man** — the Idea or Light of Man that radiates from the Divine Word and which is the Christ or Messiah. The name 'John' literally means 'the Grace or Gift of God', 'the Truth', 'the Light of God'. It is the name of the Christ Consciousness inherent in spiritual Man.

The first chapter of St. John's Gospel begins with a beautifully condensed statement concerning the Creation of the Divine Thought or Word of God in the Heavenly Mind; followed by an equally succinct statement of the incarnation of spiritual Man, the Divine

Immanence, into Matter, and the final product of that incarnation as the perfectly evolved human soul in whom the Christ Consciousness has become visibly manifest and made known.

> "In the beginning[26] was the Word[27], and the Word was with God, and the Word was God.
> "All things were made by Him; and without Him was not anything made that was made.[28]
> "In him was Life, and the Life was the Light of Men.[29]
> "And the Light shineth in Darkness; and the Darkness comprehended it not.[30]
>
> "There was a Man, sent from God, whose name was John.[31]
> "The same came for a witness, to bear witness of the Light, that all men through him might believe.[32]
> "He was not the Light, but was sent to bear witness of that Light .....[33]
>
> "And the Word was made flesh, and dwelt among us, (and we beheld His glory, the glory as of the Only Begotten of the Father,) full of grace and truth.[34]
> "John bare witness of Him,[35] and cried, saying 'This was He of whom I spake, *He that cometh after me is preferred before me*: for He was before me[36].' "[37]

The work of *JOHN* (or *JIHOVAH*) is to produce a natural living soul that is loving, understanding and helpful, caring for and serving all life as much as a good mother will cherish and nurture her children; and then, at the point where all three conditions are fulfilled in the total self-sacrifice required in serving Truth (which is to do the Will of the Father), to become fully manifest in that soul, resurrecting it into a gloriously radiant form of Light.

In order to accomplish this work the 'feminine' powers of *JIHOVAH*, represented by *JHVH*, 'descend' to the uttermost depths of Matter, to form an earthly polarity to the 'masculine' archetypal wisdom of *JIHOVAH*, represented by *IOA*, which 'rises' to the uttermost heights of Matter to form a heavenly polarity. This polarisation is in imitation of the transcendent polarisations already made, and is what enables the living soul to come into being. *JHVH* becomes the intelligent generative and transforming energies of Nature — the *Shakti* (or *Kundalini*) power. *IOA* becomes the archetypal idea or over-seeing spirit, a hidden light or wisdom that is the incarnate Christ — the three-aspected God that woos and stimulates his bride to rise up and actively unite with him, and to thereby cause his blissful light to be revealed in radiant beauty.

By means of living experience in the cyclic life-process of desiring, thinking and acting, the *JHVH* powers are gradually increased and raised in the evolving natural soul until, when the point of loving, understanding and sacrificial service is reached, the 'Bride' reaches her 'Beloved' *IOA*, and unites with him in marriage. His radiance of Wisdom-light then floods into and through the soul, bestowing upon it a heavenly baptism and converting the reflective soul into a radiant form of light — the perfect image of the Divine Creator. Such a soul can say, "I and my Father are one."[38] The Bride has given birth to (*i.e.* become) the Son — the Son who reveals and makes God known. "No man hath seen God at any time: the only begotten Son — He who is in the bosom of the Father — that One hath revealed Him."[39] "He who hath seen me hath seen the Father."[40]

These two polarities of *IOA* and *JHVH* are symbolically presented in *Genesis*[41] as "the Lord" and "the subtil Serpent" who talk to and influence Adam-Eve in their own respective ways. Adam-Eve, the natural living soul, has to follow the path of generation and transformational experience — cycles of birth, death and rebirth of form — as it is only through such experience that basic desires can be evolved to pure love, thought evolved to a clear understanding of good and evil, and action evolved to that sacrificial point wherein the Will of God is not only known but expertly done.

JOHN AND JESUS: BAPTISM BY WATER AND BAPTISM BY THE HOLY SPIRIT

There are two Baptisms: one of **Water**, and the other of the **Holy Spirit** (or Fire and Breath). The Baptism by Water is of the earth, earthy; and the Baptism by Holy Spirit is of heaven, heavenly. The former is given by **John the Baptist**, and the latter by **Jesus**. Furthermore, the Baptism by Water precedes the Baptism by Holy Spirit, the one having to be undertaken before the other can be bestowed by grace. Similarly, John the Baptist is the forerunner ("the voice crying in the wilderness") of Jesus.

In esoteric tradition both Baptisms are given by **John**, one John being called "**the Baptist**" (Water) and the other John referred to as "**the Beloved**" (Holy Spirit), and in this is revealed the essential mystery of John and Jesus in relation to each other, and of the two Baptisms. "Beloved" is a term for the Holy Spirit, as also is "the Comforter".

The Baptism by Water refers to the generative and transformative process undergone by the mortal soul (*i.e.* the natural soul) in which it is gradually learning to love, understand and serve through its cyclic experiences in matter. 'John the Baptist' is the intelligent and generative aspect of *JIHOVAH*: *viz.* the dragon-serpent *JHVH*, that evolves and guides the natural soul through its "wilderness" of experience until the soul becomes pure in heart, mind and action. The soul will then be a 'Virgin Mary' and ready to give birth to (*i.e.* become) the 'Christ-Child', Jesus. The term 'Water' is given to this Baptism, since the generative and transformative powers of *JHVH* are the 'earthly' or material polarity to the preserving, stimulating and illuminating powers of *IOA*, the 'heavenly' polarity. This Baptism by Water — the transformative experiences of the natural soul — is also known as the Lesser Mysteries (or Degrees) of Initiation, in which the purpose is to produce the pure soul, the 'Virgin Mary'.

The Baptism by Holy Spirit begins at the culmination of the Baptism by Water, the one leading directly into the other. When the Baptism by Water is completed, and the soul has become loving, understanding and of service, then the Baptism by Holy Spirit can and does begin. This second Baptism refers to the 'Grace' of God — the blessing of illumination, the Christhood or Christ consciousness — upon the prepared and virgin soul. The Holy Spirit is the *Logos* — the Fire of Holy Wisdom carried in the Breath of Love. It is the Divine Thought of God, radiant with Light. In particular it refers to the heavenly polarity of the soul, *IOA*, which becomes actively manifest or revealed in the soul, baptising the soul from "above" (from the crown centre) with light or Christ Consciousness. Through the preceeding Baptism by Water, *JHVH* (the *Shakti* or generative power) has been raised up from the root and reached the crown, where it 'marries' with *IOA* and allows *IOA* to flood the soul with light, with divine fire. The result is the transmutation of the reflective soul, Mary, to become the radiant Christ Soul, Jesus. The natural soul, Mary, dies in the fire of the Holy Spirit, and is resurrected as the spiritual soul, Jesus, like the flame arising from the candle.

This Baptism by Holy Spirit — the Christing of the soul — is known as the Greater (or Christ) Mysteries of Initiation, during which the illumined soul grows in strength and

ERRATA

Page 286, Note 158:

The sixth sentence, beginning "The *Complaints*, which followed...", gives an incorrect statement of the cipher signature employed by Francis Bacon. E.D.S.P. does not of course equal 57 in Simple English Cabbala. The sentence should read as follows:

> The *Complaints*, which followed in 1591, used the signature 'Ed.Sp.' - an unusual and pointed abbreviation which renders, in Simple English Cabbala used by Francis Bacon, '9, 33' (*i.e.* E.D.S.P. = 5.4,18.15 = 9, 33), which is the cipher for "I, BACON' (*i.e.* I = 9; BACON = 33), whereas the fuller signature "Ed.Spenser' renders, in the same Simple English Cabbala, '100' (*i.e.* E.D.S.P.E.N.S.E.R = 5.4.18.15.5.13.18.5.17 = 100), which is the cipher for 'FRANCIS BACON' (*i.e.* FRANCIS BACON = 67 + 33 = 100). By using this double cipher signature in such a definitive way, Francis Bacon was able to rule out chance when it came to the matter of others deciphering and recognising his signature and intention.

Page 293, Note 289:

This note should read:

> 289. Francis Bacon, Essay *Of Goodness and Goodness of Nature*.

stature (not in quality, which is perfect, but in power and ability and knowledge), from 'childhood' to 'manhood', from 'resurrection' to 'enthronement' on the right hand (the creative aspect) of God the Father. 'Enthroned' as such, the fully illumined soul has become "like unto God" — a Christ, Son of God, the Light of the Divine Mind. The manifestation of God is complete.

The Baptism by Holy Spirit is also known as the Baptism by John, the Beloved, for it is the Baptism or Illumination of the soul by *IOA*, the spiritual polarity of the incarnate *JIHOVAH*. *IOA* is the incarnate Christ Consciousness or Light of God. It is symbolised by the 'Dove' of white fire which appears above the head and showers its light-blessings or rays down upon the soul. When it manifests, it can be seen (spiritually) in the form of white flames or "tongues of fire" upon the head of the illumined soul, and often these flames can be likened to a dove flying straight down upon the initiate's head. But the real reason for its symbology as a 'Dove' is its original Hebraic word, *Dōhveh*, which means rest, security, strength and plenteousness; in other words, peace — hence the dove is a symbol of peace. True peace is knowledge, or illumination, a state of divine perfection, thus the dove is depicted carrying in its beak an olive branch, or a jewel of fire, or a wafer of light — all symbols of the Christ Illumination that annoints (*i.e.* baptises) the soul with light. The symbol used by the Hebrews for such rest and security, *etc.*, was derived from their principal means of survival — sheep; and so *Dōhveh* was symbolised by the sheep-fold, with the word spelt as *Dohver*. Tied up in this symbology was the idea of perfect sacrifice — the sacrifice of service required to be made by the pure soul in order to attain Christhood — and this was represented by the unblemished male lamb lovingly offered to God, known as the *Agnus Dei*. Hence the use of the term "Lamb of God" to signify the Christ in the Christian teachings. The 'Sheepfold' is the heavenly place in which all the Christs or Sons of God are assembled — the real saints or illumined ones of mankind, who form the true Church of Christ, the 'Heavenly Jerusalem'.[42]

The name *Dohver* or 'Sheep-fold' was also applied to two key constellations in the sky, now called *Ursa Major* ('the Great Bear') and *Ursa Minor* ('the Little Bear') due to a confusion in translation of the word *Dohver* with a similar word *Dōhv* meaning 'bear'. But in our Hyperborean[43] tradition the great Spirit, Arthur, is known as the Bear. Arthur is derived from *Ar Thor*, meaning 'the Thunderer' or Word of God,[44] and in the map of the constellations He is associated with both *Ursa Major* and *Minor*. In older zodiacs it shows Arthur or *Ursa Minor* as a Dove, with the pole star held in its beak (*i.e.* mouth).[45] Our present pole star is this very star, and is called in Arabic *Al Ruccaba*, 'the Turned Upon', the pivot of our present celestial sky and zodiac. This name was used for this star long before it actually became our pole star in the present 26,000 year cycle,[46] the secret of our period always having been known. The Greeks called the star *Kunosoura* (*i.e.* 'Cynosure') or 'Dog's tail', which not only alludes to another mystery (DOG = GOD spelt in reverse, as the "Image"), but is seemingly derived from an older pre-Hellenic word, *An-nas-sur-ra*, meaning 'high in rising', *i.e.* that which is in the heavenly position.[47] The heavenly position corresponds to the crown or spiritual polarity occupied by *IOA*, and it is a fact that the crown centre of the human being and the world is known esoterically as the North Pole, symbolised by the six-pointed Christ Star (the Star of David), whilst the root is called the South Pole. Hence the heavenly map of the stars depicts the Baptism of the World by the Holy Spirit beginning at precisely the time when *Al Ruccaba* of the "Little Dove" is our pole star, which time is now.

It should be noted that when Jesus called James and John "*Boanerges*", an Aramaic term meaning 'sons of Thunder', he was referring to the Word of God and to the fact that these two disciple-initiates were indeed initiates, baptized with the Holy Spirit. Similarly, when

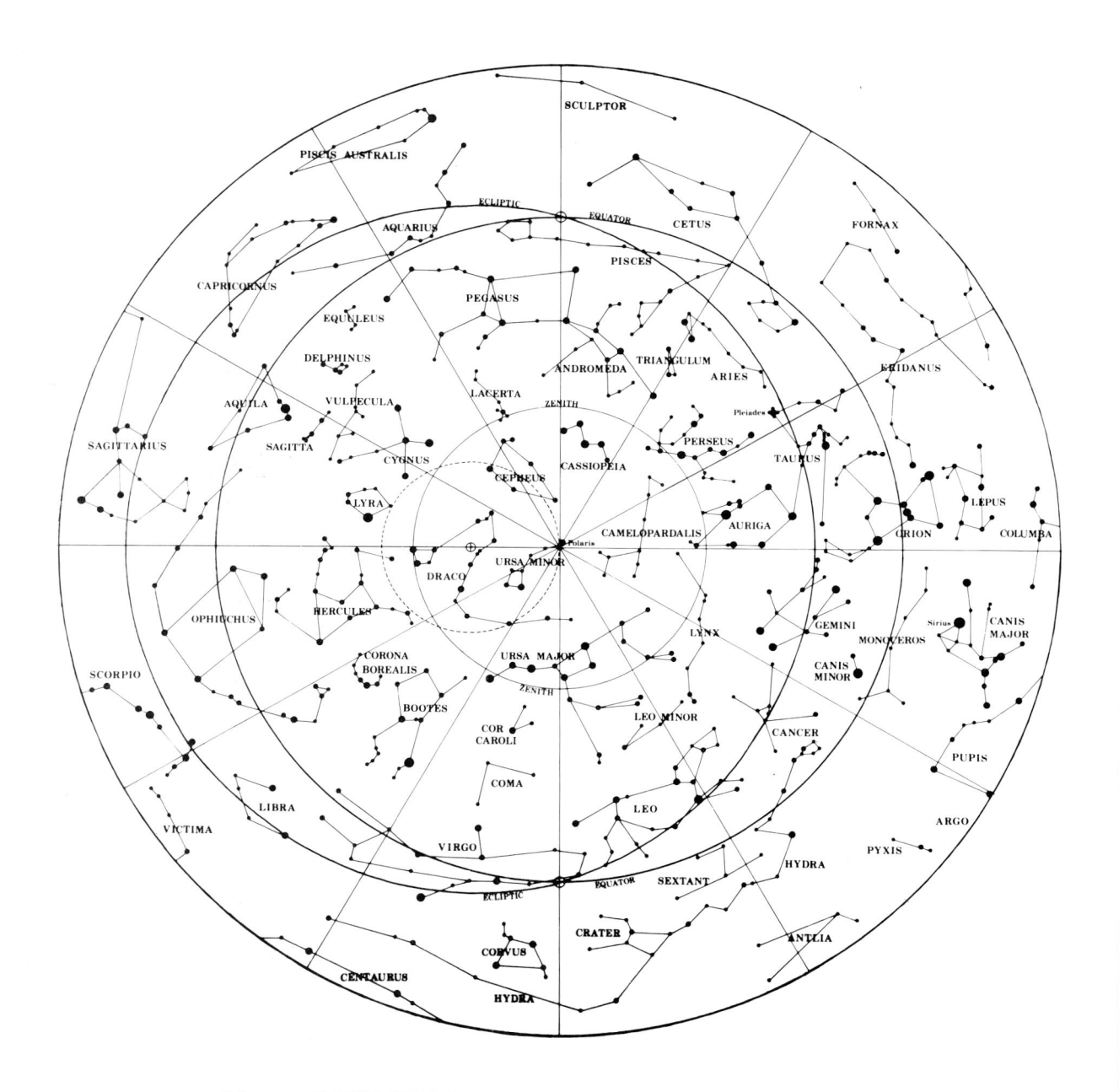

Diagram C: THE CELESTIAL CHART

he called Peter and Andrew "sons of Jonah", he was referring to the same Baptism (by *IOA*, the Dove) and initiateship. There was in fact a Mystery school operating amongst the Essenes at the time of Jesus, which led suitable candidates through the paths of initiation — Lesser and then Greater. When an initiate completed the Lesser Degrees and entered the Greater Degrees, he was referred to as a "son of Jonah", the Dove. The biblical story of Jonah and the Whale is a symbolic story concerning the Christ light (Jonah) incarnate in matter, concealed in the soul (the whale) that swims in the great ocean of life. The release of Jonah from the whale, upon the whale's death, indicates the release of illumination in the soul upon the death of the natural, mortal psyche and the resurrection of the spiritual, immortal soul of man, the "son of Jonah". Another term was also in current useage, but had become misapplied through its popularisation; that is "son of the Torah", where *Torah* is the name for the Law or Word of God, equivalent to the Egyptian *Tot* (or *Thoth*) and the Hyperborean *Thor* (as in *ArThor*). The story that Andrew, and almost certainly the other three principal apostles, were originally disciples of John the Baptist, but were then directed to become disciples of Jesus the Christ, is indicative of the crucial transference from the Lesser to the Greater Mysteries of initiation — the point where those four achieved illumination and became "sons of Jonah", "sons of Thunder".

These four (Peter, John, James and Andrew) were also partners in a fishing concern based at Bethsaida on the **north** shore of the Sea of Galilee. The term 'fishermen' refers, in the Orphic Mysteries, to the fully-fledged initiate who has entered the Greater Degrees of illumination. The word *Orpheus* literally means 'fisherman'. It was a term used also in the Druidic Mysteries, which have a kinship with the Orphic Mysteries. Christianity as we know it was born from a Hebraic womb into a Greek cultural environment. The biblical New Testament is filled with symbols and teachings from both Hebraic and Orphic sources, both of which stem from Ancient Egypt. The 'fisherman' is the grail initiate, who is able to fish in the ocean of life at will and catch the mysteries of God — each fish representing the greatest of mysteries that can be caught, which is man himself; hence the statement by Jesus that these disciples would be "fishers of men". Furthermore, *Beth-saida* (which means 'fisher-home') was the home of these four fishermen, and Peter's house in Bethsaida was made a headquarters of Jesus and his disciples. Not only was Bethsaida on the northern shore of Lake Galilee, but Galilee itself is in the northern portion of the land of Israel, and represents the 'head' of the sacred landscape-temple of Palestine; the most northerly point and crown being the triple-peaked Mount Hermon, called "the Ancient of Days" by the Jews, referring to the Heavenly Father. As the Glory of the Father is revealed through the shining face or "countenance of the Lord" as the corona or halo of light, centred in the forehead, so it is not by chance that the home of Jesus and his "fishermen" disciples should be at Bethsaida in Galilee.

The Baptism by Holy Spirit is from **John the Beloved** (the Christ), or from **Jesus**. Jesus is another term for John, and incorporates the idea of both the Baptist and the Beloved revealed through the perfect Christ soul. *Jesus* actually means 'the illumined soul', 'the Christed soul', 'the soul of light', 'the embodied Light', 'the temple of Light', 'the countenance of the Lord', 'the visible appearance of the Light'. The English form of the word is derived from the Celtic *Jesu*, or *Ies-Hu*, where *Ies* means 'soul' and *Hu* is a word for 'Light'. *HU* specifically means the all-pervading radiant Light or Bliss Consciousness of Christ,[48] the Divine Wisdom that is the radiance of Divine Love. The same word is found throughout the world, often as a basis or root of other words. In Ancient Egyptian tradition, *HU* is directly equated with the 'S', giving *SHU*, the all-pervading Warmth or Light of Love that radiates from the Heart of God like an arrow shot from a bow. Hence the symbol of the arrow was used to signify this radiant Christ Light, whilst the bow represented the Heart. Again, these were symbols used throughout the world — hence the God of Love and Light is represented as an Archer.[49]

The very **sound** of the word *SHU* is like that made by an arrow shot from a bow — a perfect symbol of an all-pervading and powerful influence that has the ability to go straight to the mark. The sound is also like that of the wind, particularly of the breath when it passes through the teeth, hence it represented the Holy Spirit of Breath and Word — the Word being spoken on and by means of the Breath. It is in this mode that the Hebrews used the sacred word, as the consonant *Shin*. The consonant *Shin* stood in for the vowels *IOA* which could not be written down, as they were the **invisible** Spirit. *Shin* represented the manifestation or revelation of IOA in the soul of man. The actual letter *Shin* is written in the form of three flames, signifying the holy dove or tongues of white fire which descend upon the full initiate and baptise him with Christ light.

The Hebraic form of the name Jesus was, therefore, *Jeheshua* (*i.e. J.H.Sh.V.H.*) which is usually rendered Joshua in English translations. Thus the Ancient Hebraic language shows how *J.H.V.H.*, the generative and intelligent spirit that is incarnate as the natural living soul of man, produces the perfect and illumined soul filled with and actively manifesting the radiant Light (Sh) of God. *Jeheshua*, the Hebraic name of Jesus, signifies the Bride who has become perfectly wed to and transmuted by her Beloved, the Christ; being fully illumined by and revealing to others that joyous Spirit and blissful Consciousness by actively embodying it like a radiant sun — a true 'Son' of God.[50]

JESUS AND HIS CHRIST MISSION

Although the Scriptures of the Judaic-Christian Tradition are mainly allegorical, yet they are also historical. This is because all creatures are living out the Laws of God and, when a type of perfection is reached, they become living symbols of the Life and Law of God. For instance, the soul that we know as Jesus, who lived approximately two thousand years ago as Jesus of Nazareth, is historically one of those truly great souls that we call Christs, Sons of God; for such great souls have reached (as far as we are concerned) a divine perfection and are actively manifesting the Christ Light. There are many Master souls connected to our planetary evolution, all of them beautifully illumined beings, but the greatness of the Jesus soul that we call the Christ (soul) for our world surpasses them all. It is said that his soul achieved perfection in worlds other than our own, and that he is a Great Initiate of the star system called Sirius, the evolution of which is far beyond our own. From the great Christ Brotherhood of Sirius he came with two others on a special mission to our solar system, entering it *via* the Sun, its heart, and then incarnating into the coarse environment of our planetry sphere *via* the finer environment of our twin planet, Venus. With these three Sirian Initiates came their circles of disciples, 'picked up' *en route*,[51] souls associated with both the Sun and Venus, in which spheres they had achieved certain levels of initiation, but who needed or chose to undertake such a mission as this one to the planet Earth as part of their next step in initiation.

The mission of the Christ Jesus to our planet is one of overseeing, inspiring, guiding and teaching the younger souls of Earth's humanity throughout their **whole** conscious human evolution on this planet, and helping to provide an evolutionary stimulus at certain required times. By incarnating into the substance and evolutionary chain of events of this planet, this great Teacher and his company of Sages are able to experience and understand the peculiarities of our human and planetary evolution, its pains and joys, and by participating in it to truly love and teach others by example and by close ties of friendship. As well as being exemplars of the perfect life, they act as yeast in the dough of humanity. Such a mission, in which spheres of beauty, harmony and illuminated peace are deliberately left in order to live in and experience the ugliness, inharmony and human injustice of our less evolved world, constitutes a conscious self-sacrifice in the Name of Love. This is part of the process of life: it cannot be avoided, but there are, naturally, various degrees of self-sacrifice and choice depending upon the evolved state of the soul concerned. The sacrifice of the Venusian souls who came with Jesus is great, but not as great as that of the Sun-souls who accompanied Jesus, and theirs is not nearly as great as that of the Sirian Souls. It is difficult for our personalised consciousness to grasp even a little of the nature of these sacrifices, but in moments of intuitive illumination we may gain some insight into these matters.

The Ancient Wisdom Traditions teach us that these mighty souls came to our planet approximately 18 million years ago, living in aetheric bodies[52] and inhabiting the land known subsequently as Hyperborea ('the Land beyond the North Wind') — the area of the planetary landscape associated with and manifesting the heart chakra of our world. Establishing this area as their principal home and temple, they then moved across the world to other key areas in order to act as guides and teachers to the young races of humanity. Young humanity was, at the time of this great incarnation of 'God-men' and 'Sun-men', existing and evolving in areas of the world known as Mu, the Motherland. As a whole they are known historically as the Third Root Race of mankind, a Race of men that, 18 million years ago, had just begun to exist in physical bodies (rather than etheric and astral), to divide sexually into male and female bodies, and to think rationally. They needed, at that dawn of rationality, sexuality and free choice, to be carefully guided, with teachers to answer their questions and to provide allegorical ideas of the life to be aimed for and the experiences to be undertaken. During this period the historic (rather than allegorical) story of Adam and Eve took place, followed by the division of Cain from Abel; the latter representing those souls who followed the teachings of the Christ souls, and the former representing those souls who chose otherwise.

After Mu, the next great epoch of human evolution was in Atlantis, in which humanity is known as the Fourth Root Race. This epoch began about 5 million years ago. The Atlanteans are the *Nephilim* ("giants") of the Bible,[53] their ancestor being Seth, the third 'son' of Adam. During this period the Christ souls and their disciples — the "sons of God"[54] — incarnated totally into the dense substance of the planet, and into its human life cycle, taking on dense physical bodies and involving themselves far more completely with the experiences of younger humanity. Thus "the sons of God saw the daughters of men that they were fair; and they took them wives of all which they chose."[55] Their progeny became "the mighty men, the men of renown",[56] the great initiate-teachers who lived amongst mankind and who established centres of the Ancient Wisdom in Atlantis, to teach the Mysteries of God and lead those who were ready and capable through initiation.

But the overall "wickedness" of the majority of mankind during this Atlantean epoch became so great that it eventually brought destruction upon the people and their lands, in the form of a large asteroid, dislodged from the solar system's asteroid belt by the probable influence of galaxial nebulae, and attracted to the Earth because of the greed of the people for energy and power, acting against the advice of the sages, which increased the Earth's magnetic field and gravitational pull beyond its safe limits in the circumstances. The force of the impact, occurring nearly one million years ago,[57] set up a whole chain of reactions lasting for many thousands of years,[58] destroying the land masses, cities and civilizations of the Atlanteans, and setting the glaciations of the Ice Age in motion. This period of destruction constitutes the historical aspect of Noah's Flood; Noah and his family representing those of humanity (who were to become the 5th Root Race), guided by the initiates and great sages, who took refuge and established "Arks" or landscape temples in areas that could (and of necessity had to) withstand the destruction.

In this Atlantean period the 'Jesus Community' were incarnating and living lives as personal human beings over and over again, including the great soul that we specifically call Jesus. The greatest of the Atlantean incarnations was as Enoch, the full personal Christ manifestation for that Atlantean humanity, in which the profoundest truths, interpretations and examples of the initiatory life were given to humanity which, handed on in tradition, formed the backbone of the Ancient Mystery Tradition of our Fifth Root Race epoch. From the great Atlantean initiates, and particularly Enoch, came the ideas and examples of the Sun-King, the Priest-King, and of the royal dynasties and blood-line. For the Atlanteans and also for the early races of the present Fifth Root Race, the most suitable living example of Truth that these great initiates could give was in the form of a 'Father-Mother' figure — a High Priest and Sovereign Lord ruling his land and guiding his nation of people. For the Third Root Race of Mu, the example had been provided by the Sages existing in aetheric bodies of light and appearing to mankind as 'God-men' or 'Sun-men', clothed in the substance of the Sun.

We are now in the Fifth Root Race epoch, called the Aryan, which was born about one million years ago,[59] just before the time of the asteroid impact, but which has only really come into its own since the last remnants of the Atlanteans were finally swept away in the third great cataclysm that lasted approximately from 80,000 to 13,000 years ago. Noah is the great ancestor of the Fifth Root Race. During this present epoch the 'Jesus Community' have been incarnating over and over again in ways that are increasingly more closely identified with the ordinary man, so that their example can be complete, their virtue and powers being demonstrably attainable by others, and their relationship with ordinary men and women being more of a brother and friend rather than the royal, parental relationship. This has required a gradual 'stepping-down' by the Adepts and Masters, concurrent with the gradual 'stepping-up' of mankind's evolutionary state, until the two may the more nearly meet and 'marry'; so that humanity as a whole might really understand and identify with the initiatory path, and come to realise their potential Christhood.

Just as divine Law governs the workings of Nature and evolution of any life form, so the Law dictates the manner, place and purpose of incarnations. Thus the manner, place and purpose of each individual incarnation of Christ Jesus is predetermined by the divine Will, all according to a beautiful Plan whose operation spans the Ages. The appearance of the Christ Jesus as Jesus of Nazareth, at the beginning of the last Piscean Age, in Palestine, born a Jew, is a unique event which should be seen in the context of time and evolution, as part of a progressive revelation of Truth in the step by step perfecting of humanity on this planet, which in turn has its effect on the evolution of the whole planet, the solar system and beyond.

Although each and every incarnation of man is a unique event, yet each one is but part of a greater whole or series of incarnations that 'stone by stone' build the temple of the spiritual soul. The incarnations of Masters and Adepts, that can reveal God's Law more perfectly and completely than less evolved souls, are always "prefigured" simply because they live in Truth, according to the Law, and that Law is the same in the past as it is now and ever will be. The lives of such souls are thus all essentially similar, differing only in the external circumstances of history, and in their purposes and related powers. The incarnation of Jesus of Nazareth, with his disciples, and John the Baptist, was prefigured by the initiate lives and revelations of Truth which preceded that event; and, in turn, that incarnation prefigures the future great lives and attainment of all mankind. The story of Jesus Christ is thus both historically and allegorically true, being both a specific happening in time related to one great soul and a general happening throughout eternity for all souls of men once they can reach a certain state of purity and enter the true path of initiation.

What the Christ Jesus portrayed for us, in his incarnation as Jesus of Nazareth, was the initiatory path of every man from beginning to end; and to do this he acted out dramatically, in real life situations, the actual symbols and allegorical stories that were used in the Temples and Schools of the Mysteries during preceeding Ages. Helped by his disciples, who all played their parts in this real life Mystery drama to help explain it in detail,[60] Jesus gave the world a living picture book by means of which to see the Christ and the divine processes of life. Although an heir to the throne of Israel, being a prince of the royal line of David, and also a prophet and Essene high priest, yet the older representations of the Sages to mankind in the outward form of king-priests was deliberately sacrificed so that he could become fully known as a brother and friend to all classes of humanity, setting the prime example in that way. Just as Enoch was the grand example of the Christ soul to the Atlanteans, in the form of the regal King-Priest, so Jesus of Nazareth has become the grand example of the Christ soul to the Aryan Root Race, in the form of the humble brother, friend to poor and rich alike. But this was not the first incarnation of this great Christ soul, nor is it his last. He will be with this world, acting as the good Shepherd, until all his sheep (that is, all the souls in his care who are pure in heart) have been gathered in to the sheep-fold.

The sheep-fold is the *Dōhver*, signified by the constellation *Ursa Minor* in the heavens. Whenever the north pole of the world reaches the star *Al Ruccaba* (our present Pole Star), the gate is opened and the sheep (*i.e.* the pure in heart who are ready to climb the ladder of initiation) are gathered in. This is the sheep-fold of our particular planet, and the Christ soul that we call Jesus is our particular Shepherd; but he is also the Shepherd of other sheep belonging to other folds, other planets. The greatness of this particular Christ soul, an Initiate of Sirius, is immense and far beyond our normal human understanding. Certain profound secrets of this great soul are hinted at in the Gospel of St. John, chapter 10:

> "Verily, verily, I say unto you, He that entered not by the door into the sheepfold, but climbeth up some other way, the same is a thief and a robber. But he that entereth in by the door is the shepherd of the sheep. To him the porter openeth; and the sheep hear his voice: and he calleth his own sheep by name, and leadeth them out. And when he putteth forth his own sheep, he goeth before them, and the sheep follow him: for they know his voice. And a stranger will they not follow, but will flee from him, for they know not the voice of the strangers..........
>
> "Verily, verily, I say unto you, I am the door of the sheep. All that ever came before me are thieves and robbers: but the sheep did not hear them. I am the door: by me if any man enter in, he shall be saved, and shall go in and out, and find pasture.
>
> "The thief cometh not, but for to steal, and to kill, and to destroy: I am come that they might have life, and that they might have it more abundantly.
>
> "I am the good Shepherd: the good Shepherd giveth his life for the sheep. But he that is an hireling, and not the shepherd, whose own the sheep are not, seeth the wolf coming, and leaveth the sheep, and fleeth: and the wolf catcheth them, and scattereth the sheep. The hireling fleeth, because he is an hireling, and careth not for the sheep.
>
> "I am the good Shepherd, and know my sheep, and am known of mine. As the Father knoweth me, even so I know the Father: and I lay down my life for the sheep.

"And other sheep I have, which are not of this fold: them also must I bring, and they shall hear my voice; and there shall be one fold, and one Shepherd.

"Therefore doth my Father love me, because I lay down my life, that I might take it again. No man taketh it from me, but I lay it down of myself. I have power to lay it down, and I have power to take it again. This commandment have I received of my Father."[61]

Such is the testimony of the greatest Initiate that our world knows — the Christ soul that embodies and manifests the Christ Spirit, doing the Will of the Father by caring for our world, and other worlds, voluntarily sacrificing his life over and over again in order that the redemption of this world and its solar system might take place. All in this world who reach a purity of heart, purity of motive, become sheep in his sheep-fold, and he henceforth shepherds them through the great initiations, until they achieve Christhood and become Shepherds in their turn.

Nowadays we know Sirius as the Dog Star, the brightest star in the constellation of *Canis Major*, 'the Great Dog'. *Canis Major* has 64 stars[62] altogether, but Sirius is its principal star, marking the mouth and tongue of the Dog. The Egyptians knew this constellation as *Apes*, 'the Head', and symbolised it as the Hawk, sign of Horus,[63] the Christ-soul, who fights the evil Serpent and conquers it.[64] All the stars of this consellation have names which describe the mighty Prince, the bright and shining, glorious Ruler, chief of the Right Hand (of the Father), the Leader, the Prince of Peace. They are all summed up in *Sirius*,[65] 'the Prince', whose Akkadian name was *Kasista*, meaning 'the Leader' and Prince of the heavenly host, Chief of the 64 stars. The Persians called the star *Tistrya*, 'the Chieftain of the East'. In terms of the consellation, it signifies the *Logos* or Word of God, that is spoken from the mouth and which radiates as Light. The three Great Initiates who came from Sirius are the bearers of this three-fold Light of Love.

As the Dog, *Canis Major* signifies Anubis, the Guardian of the Temple and Way-Shower. The name is older than it might seem, the Philistines referring to the Syrian Initiates as *Dogon* or *Dagon*, the same 'God-men' that Babylonian tradition calls *Oannes* or *IOHANNES*, who brought civilization to mankind on our world. It is perhaps not by chance that DOG is the reverse or image of GOD. The star Sirius stands for this whole constellation and its name. The Ancient Egyptians called the star *Sept* (Greek spelling, *Sothis*), and identified it with Isis, with her child Horus seated upon her lap. To the Egyptians it was the principal star of the whole heavens that ruled their calendar and announced their new year, rising on the eastern horizon just before the sun during the months when the sun was in the Sign of Leo,[66] and coinciding with the annual inundation of the Nile which brought life and fertility to the otherwise desert valley. The heavenly Mother and child were further signified by the bow and arrow, sign of the heavenly Archer (*i.e.* Cupid, Eros, Christ), the bow representing the Mother and the arrow symbolising the child of love-light shot from her pure heart. Hence Sirius is known as 'the Archer' or 'Bow Star', and is one of the significances of the sign or 'Bow' in the sky which Noah was shown as the sign of the divine Covenant.[67] Isis and Horus are the same in meaning as Mary and Jesus: only the language is different.

The sign of the Covenant is the sign of the Bow, which is the sign not only of the union of spiritual Fire with the Waters of universal Matter, manifesting and making Light known, but also of the Virgin Mother with her Christ Child 'enthroned' in her lap and shining with the light of love. The Virgin Mother is the 'Water' of universal Matter impregnated with the 'Fire' of the Heavenly Father, who as a result brings forth the "male child" or Christ soul that makes Light manifest and known. But the Virgin Mother and Child is not only the sign (or symbol) of the Divine Creation and Manifestation, but of the means to fulfil this Plan through the initiatory path that mankind can make, with its 'promise' of eventual Christhood to every soul that can become pure in heart.

> "Blessed are the pure in heart, for they shall **see** God."[68]

It is said that the Sirius system is good, pure and perfect; but that our solar system, called 'the Fox', is impure, imperfect. For this reason the good Shepherds come to find and protect their scattered sheep. In order to atone for the Fox's impurity, the good shepherd gives his life: he dies crucified on a 'tree' and is resurrected, acting as a sacrifice and eucharist for us, over and over again. In this way the 'fox' may be rid of sin, of hurtfulness, and be redeemed to a pure and Christed state like that of the 'Dog'.

JESUS AND THE GREAT INITIATIONS

Jesus Christ lived as Jesus of Nazareth in order to show vividly to mankind the Way, Truth and Life of God: and in such a way that every man could have the chance to identify with and hence attempt to follow this great Example. Jesus not only spoke and interpreted the Truth with his voice, but with his very actions; his deeds or works being the principal witness of his state of being.

> Then came the Jews round about Him, and said unto Him, "How long dost thou make us doubt? If thou be the Christ, tell us plainly."
>
> Jesus answered them, "I told you, and ye believed not: the works that I do in my Father's Name, they bear witness of me."[69]
>
> "I am the Way, the Truth, and the Life: no man cometh unto the Father, but by me. If ye had known me, ye should have known my Father also: and from henceforth ye know Him, and have seen Him .... He that hath seen me hath seen the Father .... Believest thou not that I am in the Father, and the Father in me? The words that I speak unto you I speak not of myself: but the Father that dwelleth in me, He doeth the works. Believe me that I am in the Father, and the Father in me: or else believe me for the very works sake."[70]

The Gospel story of the life of Jesus is, as with all scriptures, allegorical and complex. It can be interpreted truly on many different levels and from many viewpoints. Truth is truth, but it transcends human dogma and limited human understanding. It needs to be approached with **compassion** and **humility** — the two great qualities which it is said that the Christ souls are emphasising for us at this period in humanity's evolution:

> I am the Spirit of Humility:
> To every voice I bend the knee,
> And listen well that I might know
> What to do and where to go.

Jesus of Nazareth, a personal incarnation of the Christ soul known as Jesus Christ, fulfilled all natural laws and achieved Christhood *as a new personality* before he began his so-called Christ-mission for that life-time — a Christ-mission that lasted but three years, if the dates are taken as being strictly historical as well as allegorical. During those three years he demonstrated to his disciples, and through them to the whole world, the seven great initiations of life that man can, will and does undertake. Hence, although being already in a state of Christhood when he called his disciples together and began his three-year mission, he yet portrayed the first three initiations of the Lesser Mysteries that precede and give rise to the final four initiations of the Greater Mysteries of Christhood. Although Christed, his work was nevertheless to act out dramatically, for the benefit of others, the pre-Christhood stages of initiation. And, although he was already a *Mahachohan* ('Great Lord') of the 7th degree of Initiation — a fully Christed One — yet his work as Jesus of Nazareth was to show the other degrees of Christhood (4th, 5th and 6th) leading up to the 7th, after he had portrayed the 1st, 2nd and 3rd initiations.

The Gospel story thus relates a historical as well as an allegorical life that was operating and revealing truths on many different levels simultaneously. At one level — the most obvious — Jesus was a child born in the year 7BC to a young Jewish mother who was virginally pure.[71] He was probably named Joseph[72] after his father, with the knowledge that he would one day achieve the name of Jesus, as foretold by the Angel Gabriel to Mary. After thirty years of training he had prepared himself sufficiently so that his **whole** psyche of that incarnation became pure enough to undertake initiation in its fullest possible form. For most of us, only part of our personality ever reaches the purity necessary to take initiation, and so only a part of our psyche can hope to be transmuted into the Christly state of the spiritual soul in each incarnation. Jesus of Nazareth came to show how the **whole** personality, with all its bodies, (mental, emotional, physical), can be made pure and entirely transmuted to form the resurrected spiritual soul.

First Jesus portrayed the **1st Initiation**, related to the *Water* Element and involved with purifying and controlling the emotions, and learning to desire and love truly from the heart. This is referred to as **Baptism**; hence, even though he did not need to in one sense, Jesus submitted to his cousin John's symbolic baptism in the river Jordan, for the benefit of others, as an example of the true path to follow. The baptism in the Jordan was an act of dedication — the beginning of initiation, in which the soul hears the Word of God and sets out to follow its guidance. Afterwards came the Temptations in the Wilderness, testing the true motives of the would-be initiate. This was prefigured in Israel's national history by the Patriarchal period — Abraham and his family leaving behind the pre-initiation life signified by Chaldea, dedicating himself and family to following the initiatory path ahead, and followed by Isaac and Jacob and the testing period in Egypt.

When the portrayal of the 1st Initiation was completed, and Jesus had assembled his circle (or school) of 120 disciples, establishing his home in a spacious house in Capernaum, he then began the portrayal of the **2nd Initiation** by calling the Twelve Apostles, consecrating them and giving them the Word. Then followed the period of careful teaching and development of the understanding. The 2nd Initiation is represented by the *Air* element, and is involved with purifying and controlling the thoughts, becoming aware of and understanding truth. First the Sermon on the Mount was given as the principal initiatory impulse for this 2nd Initiation, corresponding to the *Logos* or speaking of the Word in divine Creation. This was followed by a period of illumination, corresponding to the *Christos* or radiance of the Light of Wisdom, during which Jesus and his disciples embarked upon missionary tours, teaching, healing and feeding the multitudes with natural and spiritual *manna*. The culmination, corresponding to the *Sophia* or vision and understanding of Truth, was marked by the giving of the 'Keys' to Peter, and the Transfiguration on the heights of the triple-peaked Mount Hermon,[73] in which the three senior Apostles saw and understood the perfect Vision of Truth as they beheld it truly 'imaged' in Jesus, Moses and Elias.[74] The three Christ souls, the Great Initiates of Sirius, were beheld; these three

(5) VIRGIN AND CHRIST CHILD, ENTHRONED — southern door, West Porch, Chartres Cathedral. 12th C. sculpture. Virgin and Child in tympanum are surrounded by the 7 Liberal Arts in the archivolts. The two lintels depict the Anunciation, Visitation, Nativity, Annunciation to the shepherds, and the Presentation at the Temple.

7	FIRE		CHRISTED ONE (Sovereign Lord) **ENTHRONEMENT** Full illumination and union with the Christ Spirit through divine love.	Peace Consummation Transformation	GREATER MYSTERIES
6	AIR		GUARDIAN (Priest) **UNIFICATION** Discovery and knowledge of the spiritual Plan, the divine Law or Truth, in order to fulfil it.	Joy Unification	
5	WATER		MASTER (Knight) **ASCENSION** Increasing mastery of the secrets of life. Wielding the sword of illumination and spiritual will.	Promise Dedication	
4	EARTH		ADEPT (Arch Mason) **RESURRECTION** Emergence into a higher life and consciousness. Preparation for a higher function and work.	Rebirth Peace	
3	FIRE		INITIATE (Master Mason) **CRUCIFIXION** Development of latent powers of love; compassion. Complete surrender of personal life and self.	Consummation Transformation	LESSER MYSTERIES
2	AIR		DISCIPLE (Craftsman) **TRANSFIGURATION** Development of purity and harmony of thought. Illumination of the mind.	Joy Unification	
1	WATER		NEOPHYTE (Entered Apprentice) **BAPTISM** Development of peace and harmony in the emotions; tranquility; patience; sympathy.	Promise Dedication	
0	EARTH		CANDIDATE for admission **BIRTH** Preparation. Development of control over physical appetites; determination; courage; responsibility.	Rebirth Peace	

Diagram D: THE LADDER OF INITIATION

embodying and revealing the three-fold Light as its perfect Image. The **Transfiguration** experience sums up and gives its name to the 2nd Initiation. It was prefigured in Israel's national history by the period overseen by Moses, Joshua and the Judges.

The **3rd Initiation**, which is signified by the element *Fire*, is concerned with putting into practice all that one has learnt to love and understand, surrendering one's will totally to the Will of God, and thereby learning to serve God and all God's creatures according to the Will of God, which is perfect Love. It requires self-sacrifice, and so Jesus portrayed this 3rd Initiation of **Crucifixion** by enacting it out in real life in terms of the ancient symbols used to describe this initiatory event. After entering Jerusalem in glory as king, he was literally and willingly crucified on a tree that had been made in the form of the cross used in the Mysteries — the Romans having adopted traditional symbols used to teach the Mysteries and perverted them into a horrific means of execution. This 3rd Initiation is prefigured in Israel's history by the Kings, David and Solomon, whose glory was followed by the conquest, break-up and scattering of the Twelve Tribes that made up the nation of Israel — just as the Twelve Apostles scattered when Jesus was taken for trial and execution.

The lower self or personality of Jesus of Nazareth suffered and died on the cross in the culmination of this 3rd Initiation, and was buried in a tomb. For three nights he 'descended' into Hell, the bosom of the Divine Mother, the pristine state of Chaos or Formlessness. His first, natural form was dissolved and on the third day he arose, reborn in his new spiritual form, as *Jesus*, the Christed soul. This was his demonstration of the **4th Initiation** called **Resurrection**, which is the start of the Greater Christ Mysteries of Initiation. The 4th degree is the degree of the Adept. In terms of Israel's national history, this initiation was not prefigured, as the 'Jesus Event' was the actual resurrection of the new Israel from the crucifixion and death of the old Israel.

From then on Jesus demonstrated the higher initiations to his disciples, and hence to the world by report. Progressing systematically through the 4th Initiation, teaching his disciples every step of the way, he then portrayed the **5th Initiation**, called **Ascension**. This is the degree of the Master soul, the preceding 4th degree of the Adept simply being preparatory to the Higher Mysteries — a state of childhood. In the 5th Initiation the Christ-child becomes a Christ-man. The demonstration of the 4th Initiation took place in the garden of Joseph of Arimathea's house, outside the tomb. As a 'child', Jesus was too 'young' or delicate to be "touched" by Mary Magdalene[75], as he had not yet reached his maturity (*i.e.* his Ascension) when he would be sent out into the outer world. When Jesus came to the reassembled disciples and was "touched" by doubting Thomas,[76] it marked the start of his demonstration of the 5th Initiation. From then on he carefully instructed his disciples, revealing to them the Higher Mysteries and empowering them with spiritual authority to carry out their subsequent work in the world. The Ascension from the summit of the Mount of Olives in the sight of his disciples marked the culmination of that 5th Initiation and the start of the 6th Initiation.

The Ascension from the Mount of Olives — the Mount of Christhood[77] — that stands to the east of Jerusalem, took the Master Jesus into the sphere of Unification with the Christ Light, his individual form of light dissolving into a more universal form of light, disappearing as such from the sight of his disciples. The "cloud" that "received him out of their sight"[78] was a cloud of radiance, which is nevertheless a veil or 'cloud' that conceals the Absolute. The **6th Initiation** is **Unification** with the Christ Glory. It is the degree of the *Chohan* or Lord of Light, who oversees the direction and operations of the particular 'Ray' of Light that he has become at-one with.

The association of the Mount of Olives with the eastern horizon, as the eastern marker or '*Hele* stone' for the rising Sun at the Equinoxes, immediately associates Jesus with the rays of that Dawn Light — the rays which were allowed to stream into the innermost sanctuary, the Holy of Holies of Solomon's Temple, on the Day of Atonement. Once a year, on one day each year (the Autumnal Equinox), the high priest was allowed to enter the Holy of Holies, offering himself as a pure and living sacrifice (*i.e. Agnus Dei*) on behalf of his people; and in that innermost sanctum, facing the east and bathed in the Dawn Light, he would become at-one with that Light. The *Hele* stone or eastern Mount of Olives marked the Gateway and the presence of the Gatekeeper of Light — the Great angel of the rising Sun and the Lord of its Dawn Light. The Dawn Light is the "bright and morning Star", associated with Venus which regularly rises on the eastern horizon before the Sun, acting as its herald; with the Sun itself, when it rises on the eastern horizon; and with Sirius, when it rises once a year, in the Summer, on the eastern horizon. All three signify the Morning Star, in various degrees of glory, and all three are associated with the Christ Jesus in his descent to our world in order to help bring it Light.

Beyond the 6th Initiation lies the **7th Initiation**, known as **Enthronement**. Just as the 5th and 6th Initiations were dealing with the higher, 'spiritual' counterparts of the 1st and 2nd 'natural' Initiations, so the 7th Initiation is the Christly counterpart of the 3rd Initiation. Just as in the 3rd Initiation the purified natural psyche attains dominion as a 'king' over himself, and then sacrifices all that he personally is in charitable service, so the fully Christed soul of the 7th Initiation — a *Mahachohan* or Great Lord — reaches his fullest glory, which is to "sit on the right hand of the Father in Heaven", enthroned as the Christ-King, King of Glory, King of Peace, which immediately leads to his being sent forth again into the darkness of the world as a good Shepherd in order to bring it more life, more illumination, more peace. In this manner the good Shepherd lays down his life over and over again for his sheep, both naturally and spiritually, until all the sheep be gathered back into the fold as illumined beings, like the Shepherd. This is also known as the Second Coming — the coming again in glory to judge both the quick and the dead; but this terminology has a specific relationship to the initiatory status of each individual soul — the sheep that the good Shepherd cares for.

The Christ is essentially Spirit — the spiritual Light of God. It is the Divine Idea or Holy Wisdom that fills the Mind of God — the radiant Son of the Father. The Father is the Divine Mind, the Son is the Bliss Consciousness of that Mind, and the two are essentially One. When Jesus is referring to his Father in Heaven, he is referring to the Christ-Spirit and the Mind of God in which the Spirit dwells. "I and my Father are One."[79] This is a statement referring (a) to the Christ Spirit and the Divine Mind of God, which are One Holy Spirit; and (b) to the Christed soul, Jesus, and the Holy Spirit, the Spirit being the 'Father' to the soul, and the Christ soul being the 'Son' of the Spirit. When Jesus speaks in such terms, with words that are not of himself,[80] but it is the 'Father' in him which speaks, then both meanings are intended and true. For the Christ soul is manifesting the Christ, and is acting as the Christ, being at one with its 'Father' as a "life-giving spirit"[81], a spiritual soul.

The First Coming of Christ relates to the good Shepherd or Christ soul who comes to care for his flock and guide them through initiation, until they too become Christed and good Shepherds in their turn. Christhood is achieved after the 3rd Initiation, and so the external teacher or good Shepherd role is applicable in connection with the natural souls of mankind who are undergoing the Lesser Mysteries of Initiation. For them the Christ appears in the form of the external teacher or *guru*, who can manifest the Christ principle for them to learn from and be guided by. But when the 3rd Initiation is passed, the Christ within the soul becomes active and made known, spiritualising the initiate soul. This is the real Second

(6) CHRIST TRIUMPHANT, KING OF GLORY — central door, West Porch, Chartres Cathedral. 12th C. sculpture. The Christ in his Second Coming is surrounded by the four Kerubim, and by the 24 Elders plus angels in the archivolts. The lintel depicts the 12 Apostles seated, with Enoch and Elijah standing at either end.

Coming of Christ — the appearance of the Teacher within. **The Second Coming of Christ**, in glory, relates to the King of Light, the Christ Spirit, and means true Illumination or Christhood. Henceforth the soul is guided through the Greater Mysteries by his own 'Father', directly, for he is now his own master: he is at one with the Father. His Christ Spirit rules as King within, is manifested through the soul (the Kingdom), and judges the "quick" and the "dead" (the spiritually minded and the carnally minded) as the soul takes up his role as the good Shepherd to others.

The 7th Initiation is linked to the Second Coming, as the fully Christed soul becomes totally identified with and identical to the Christ Spirit in all the works that the Spirit can and does perform. Jesus demonstrated both the 7th Initiation and the Second Coming for his chief disciples, whom he had guided through the Lesser Mysteries in tune with and as part of his own demonstration of the first three initiations. His baptism had been their baptism, his transfiguration had been their transfiguration, his crucifixion had been their crucifixion. When the Holy Spirit, the 'Comforter', came upon them at Pentecost, they entered into the Greater Mysteries of Initiation as Adepts. They achieved the Second Coming, and became Teachers and Shepherds to others,[82] as true Witnesses or Manifestors of the Christ.

> "And I will pray the Father, and He shall give you another Comforter, that He may abide with you for ever; even the Spirit of Truth; whom the world cannot receive, because it seeth Him not, neither knoweth Him: but ye know Him; for He dwelleth with you, and shall be in you."[83]

> "But the Comforter, which is the Holy Ghost, whom the Father will send in my name, He shall teach you all things, and bring all things to your remembrance, whatsoever I have said unto you."[84]

> "But when the Comforter is come, whom I will send unto you from the Father, even the Spirit of Truth, which proceedeth from the Father, He shall testify of me; **and ye also shall bear witness**, because ye have been with me from the beginning."[85]

> And when the day of Pentecost was fully come, they were all with one accord in one place. And suddenly there came a Sound as of a rushing mighty wind, and it filled all the house where they were sitting. And there appeared unto them cloven tongues like as of Fire, and it sat upon each of them. And they were all filled with the Holy Ghost, and began to speak with other tongues, as the Spirit gave them utterance.[86]

THE BIRTH OF THE CHRIST CHILD, JESUS

Interpreted at another level, the story of the Birth of Jesus is, although related to the story of his natural birth, a story recounted by Jesus to his disciples of his own spiritual birth. The Virgin Mary is his own natural, pure psyche, that has undertaken the first three initiations and gives birth (*i.e.* becomes) the Christ child, Jesus. This spiritual aspect of the soul is born from the heart of the natural self, the most virgin part of the lower self and the place in which the Christ Spirit lies, waiting to be manifested. The manger in which the child is born is the humble, pure heart centre, containing the food for the animals. The animals are the various psychological components of the natural self, essentially being grouped as the twelve creatures of the Zodiac, each relating to specific parts of the body and psyche. They come and adore the child once he is born or manifested. The cave or stable signifies the chest, containing the lungs and heart, organs of life and light, providing protection and nourishment for the animals. Bethlehem, meaning 'the House of Bread', also signifies the heart, in which the Christ, the 'Bread of Life' (*i.e. Manna* from heaven) is born. Bethlehem is near Jerusalem ('the Holy Place of Peace'), both representing the heart chakra; but the heart chakra has two parts — a small, secret focus in which the Word of God is first awakened or born as the Christ emanation of love (represented by Bethlehem), and the large, major focus into which the Christ radiance and powers expand as the Adept grows in strength (represented by Jerusalem).

The fact that Jesus is said to be of the royal line of David, King of Israel, is another way of saying that he is a Prince of Light, heir to the Throne of Light. David, meaning 'the Beloved', signifies the Christ,[87] who plays His essential role as Shepherd and then King. Israel, from the Egyptian word *Is-Ra-El*, means 'the child (or soul) of the Light of God', thus all true Israelites are in fact those who have attained the state of Jesus, the spiritual soul. Such are indeed the "Chosen People", anointed (*i.e.* illumined) by the Christ. Israel is composed of Twelve Tribes, which are the twelve separate but unified manifestations or appearances that make up the Christhood. The Tribe of Judah signifies the royal aspect of Kingship. It is associated with the heart, and its symbol is the Lion, signifying the strength of the heart, which is pure, courageous love or faith — the God-desire that initiates and governs all good things. The Christ is the Lion of Judah, in His regal aspect. The so-called Thirteenth Tribe, of Benjamin and Levi, also associated with the heart, signifies the synthesis or union of all the others in Priesthood — which is the role of the good Shepherd. This is in fact the essence of Christhood — a continual sacrifice or pouring out of Love as Light to the world. The King of Israel, as ruler of them all, is also the High Priest in his sacrificial aspect.

The three Kings, who brought gifts of gold, frankincense and myrrh to the baby Jesus, signify the Holy Trinity and also the three major steps in the process of life which are determined by these Three (*i.e.* Loving, Understanding and Service). As *Magi* ('Wise Men') they represent the three great Christ souls or Shepherds of mankind that came bearing the gifts of the Holy Spirit — gifts which are given to and become embodied in each new-born spiritual soul. Gold is the symbol of divine Love or Desire — the Will of God (which is good-will in man). Frankincense, which purifies and exalts the mind, is symbolic of Wisdom and Understanding. Myrrh, used in the rites of the dead, symbolised incarnation and the path of life which involves suffering, death and resurrection.

A further account of this spiritual birth is embodied at another level of interpretation in the story of the Baptism of Jesus. Although at one level this baptism refers to the 1st Initiation, yet at another level it includes all the symbols of the spiritual birth of the Christ soul. The waters of the river Jordan represent universal matter, and more particularly the substance of the natural psyche, which are acted upon by *JHVH*, or John the Baptist. Baptism in the river Jordan, under the direction of John the Baptist, signifies the path of the Lesser Mysteries — the generative and transformative process which develops a purity of love, understanding and service in the natural soul, preparing the way of the Lord and making His paths straight,[88] evolving the natural soul into a 'Virgin Mary'. This is the process of life that must take place in order to give birth to the spiritual soul, Jesus; and so it is that John the Baptist and his Baptism by Water preceeded Jesus and his Baptism by Holy Spirit.

The personality, Jesus of Nazareth, had to submit to the Baptism of John the Baptist. This was his first evolutionary stage as a natural soul. But at the end of the Baptism he "rose up straightway out of the water; and lo, the heavens were opened unto him, and he saw the Spirit of God descending like a dove, and lighting upon him."[89] Dying at the end of the

(7) **VIRGIN AND CHRIST CHILD, ENTHRONED** — oil painting by Margarito of Arezzo. The National Gallery, London. With scenes from the Nativity and lives of the Saints.

The Virgin is seated on a throne supported by royal lions, emblems of the strength of heart. She herself acts as the throne for the Christ Child. The haloed head of the Christ Child shines at the heart centre of the Virgin Mother, his face and crown of light revealing the secret that was once locked up in the heart of the Virgin. The left hand of the Christ Child holds the scroll of the written law, whilst his right hand is raised in blessing, bestowing the grace or light or the Law. The right hand is associated with giving, with energy or pure consciousness, and is the hand of Mercy, Grace, Compassion. The left hand is associated with receiving, with form or the form of consciousness, and is the hand of Judgment, Severity, Discernment. Around the Virgin and her Child is the nimbus of glory, or Vesica Piscis, 'the Vessel of the Fish', the feminine symbol of receptivity and conceptual powers. This vessel has become the auric flame which shines with the light of the manifest Christ. This flame is known as the *Ros* — the Rose, Bud, or Heavenly Dew — and it is also the Grail Chalice, filled with the light of divine Knowledge and Peace that constitutes the *sang real*, the 'royal blood'.

52

Baptism, the natural soul (the Virgin Mary) became transmuted and resurrected as 'Jesus', the spiritual soul. He was straightway filled with light, the *IOA* or Christhood of the Holy Spirit. And the *Logos* spoke from Heaven, "This is my beloved Son, in whom I am well pleased." Embodying and actively manifesting the Holy Spirit, Jesus himself became as *IOA*, or John the Beloved. Using the symbology of the Orphic Mysteries, Jesus bestows the power of the Holy Spirit upon all those pure souls who complete the 3rd Initiation, fanning the threshed harvest corn so as to separate the wheat from the chaff, burning the chaff in unquenchable fire and taking up the wheat to build it into the Bread of Life.[90]

In this manner the account of the Baptism in the river Jordan incorporates the story of Jesus' birth as a Christ soul, paralleling the earlier account of Jesus' birth in the stable.

THE CRUCIFIED SERPENT : MICHAEL AND LUCIFER : ST. GEORGE AND THE DRAGON

Perhaps the primary truth that Jesus Christ came to teach the world, as Jesus of Nazareth, is the Resurrection of the spiritual soul from the death of the purified natural soul. This is the great Promise for mankind — his hope, his Covenant with God. Every soul who follows the path of goodness is promised this resurrection into eternal life and light. Christianity itself (which means Christhood) depends upon the Resurrection. Christianity, or Christhood — the Higher Degrees of Initiation — begin with the Resurrection. But Resurrection cannot occur without the sacrificial death or Crucifixion of the lower self, the natural soul. Jesus portrayed the Crucifixion and Resurrection in a most dramatic and carefully staged public show, so that mankind should never forget this divine Promise and Covenant, and know that it is real. He staged it on the day of the High Sabbath, the Passover Festival, in which the Paschal Lamb is slain, in commemoration of the events which allowed the nation of Israel to leave Egypt, their land of 'bondage'. This festival is held annually on the day of the full moon immediately following the Spring Equinox, when the Sun is rising in the sign of Aries, the Lamb. It marks the beginning of initiation, when the promise or covenant is given concerning the future glory of mankind.

To stage this dramatically, Jesus allowed himself to suffer a method of death that would portray the ancient Teachings by using their symbols. To do this, Jesus presented himself as the wise Serpent crucified upon the Cross — the symbol of the Redeemer, which is the act of giving one's life for others. The Serpent is the ages-old symbol of the generative Spirit of God incarnate in Matter and producing the soul. When Spirit enters Matter, it moves it with Love and vibrates it into patterns and forms according to the Idea or Plan of the Holy Wisdom. The vibrational form of and movement of energy within the soul, plus the fact that the natural soul undergoes successive transformations, gave rise to using the symbol of the serpent to represent the natural soul — the soul being Matter united with and vivified by the incarnate Spirit.

The incarnate, generative power of the Holy Spirit, which is symbolised by the subtil Serpent, is the Divine Immanence or Presence of God's Spirit in Matter. It is this divine power which begets the living soul and gradually transforms and transmutes it into a perfect expression of manifest divinity. It is the pure Intelligence known as 'Man', which bears the Christ Light like a seed of love-consciousness within itself. 'She' falls into Matter, and into the process of generation and evolution, with the golden light-seed inside her womb, with the potential to give birth or active expression to that light in the course of time. This Divine Immanence or *Shekhinah*, which is spiritual Man (male-female), has the God-name *JIHOVAH*, which we commonly use as John. The description of the Divine Immanence as

(8) CRUCIFIX: the Gnostic Serpent — oil painting ascribed to the Master of San Francesco. The National Gallery, London.

Jesus hangs on the Cross as the crucified Serpent of Wisdom. His haloed head lies at the heart centre of the Cross, revealing its central truth of Love as the 'Countenance of the Lord'. A golden disc of light hovers above his head, in place of the more usual Dove, as symbol of the Holy Spirit. On either side of Jesus are depicted the three Maries with (possibly) John the Beloved, and the centurion who exclaimed, "Truly this man was the Son of God".

the light-bearer is given by two names meaning just that: (1) **Christopher** (*i.e. Christo*, 'of Christ', + *Pher*, 'the Bearer'), and (2) **Lucifer** (*i.e. Lucis*, 'of Light', + *Fer* 'the Bearer'). Both names describe the "Bright and Morning Star", which brings light into the world.

The mystery-name, **Lucifer**, has been grossly misunderstood and its significance horribly distorted and perverted by ignorant and malicious souls. When Jesus said that he "beheld Satan as lightning fall from heaven",[91] he was referring to the Divine Immanance incarnating into Matter in order to produce the living soul. Jesus was stating that he had beheld the beginning and origin of all things. There is only one Light, and that is the Son or Thought of God; and there is only one Heaven from which all that Light comes, and that is the Father or Mind of God.

Lucifer is the archangelic name for Man and for Divine Immanence. It is that part of the *Sophia* or Intelligence of God which incarnates into Matter in order to produce the natural living soul. Complementing the Archangel Lucifer is **Michael**, which means 'He who is like unto God', or 'Radiant Countenance of the Lord'. Michael is the archangelic name for that aspect of *Sophia* which remains transcendent in the Mind of God, uninvolved in Matter, acting as the overseer and director of soul evolution. The Archangel Michael represents the image or thought-form of Truth that is held firm and constant in the Divine Imagination as the guiding vision and 'blue-print' for what the natural living soul is to become. Lucifer is the thought-form of Intelligence itself, in which the Christ Wisdom is ensouled, and which may grow in understanding and knowledge of God through experience; whilst Michael is the thought-form of the radiant Wisdom, the Christ, which will one day be realised and manifested by the evolved soul.

Michael thus remains in Heaven with two-thirds of the heavenly host, whilst Lucifer falls from Heaven into the Abyss of Matter with one-third of the heavenly host. From then on Lucifer becomes Satan, 'the Adversary', because Lucifer-Satan is involved in Matter which is the opposite polarity to Spirit. Spirit is infinite movement or life: Matter is infinite inertia, or death. The spiritual freedom of Lucifer 'dies' as He-She becomes enmeshed in Matter, held in bondage; but through the inspiring Light and powers of Michael, Lucifer is gradually stirred into activity as Satan, the Serpent-Adversary. Until the natural soul becomes pure in heart, or motive, it remains an adversary to the Michaelic Light. Once pure, it becomes the Bride to the Spirit, and 'marries' the Spirit at the moment of crucifixion.

Lucifer, or Satan, is symbolised by the Serpent. A compound symbol was invented, that of the Dragon, based upon the Serpent symbol, which was used to teach the details of evolutionary life and composition of the soul. The Dragon is a solid creature that can walk (or slither) upon the earth. It can swim in water, and has fish scales to emphasise this. It can fly in the air, and has wings for this purpose. It breathes fire. All this teaches that the soul is composed of a physical body (*earth*), an emotional body (*water*), a mental body (*air*), and a spiritual body (*fire*) which can be breathed forth as the essence and transmuted nature of the rest of the Dragon. *Earth* also relates to the stage of pre-initiation, *water* to the 1st Initiation, *air* to the 2nd Initiation, and *fire* to the 3rd Initiation. The royal Dragon is the symbol of the Illumined or Spiritual Soul. It is also known as the Wise Serpent — the Saviour or Redeemer — and as the Doorkeeper, holding the keys[92] to the Sanctuary of Light.

"Satan is the Doorkeeper of the Temple of the King; he standeth in Solomon's Porch; he holdeth the key of the Sanctuary, that no man enter therein, save the anointed having the arcanum of Hermes."[93]

To begin with the Serpent is the Serpent of Ignorance and Vice, signifying all that is unevolved and evil in the soul of man. But, under the steady influence of God's Law, that Serpent (which is the natural living soul) is gradually redeemed from inertia and ignorance, passing through many intermediate stages until it reaches the point of virginity, the beginning of true initiation. Then it is said to be 'pierced' by the angelic spear (or sword, or arrow), rendered harmless, controlled and raised up the shaft of light (which is the Cross) until crucified upon it. The Cross represents the Divine Law. To be crucified, nailed (or bound) on the Cross, is to be so joined to the Law of God that the natural soul conforms ideally to that Law, and undergoes sacrificial death in order to do the Will of the Father. The Crucified Serpent is the Wise or Gnostic Serpent, the Dragon of Knowledge, for in dying thus the soul becomes Christed and knows Truth and actively manifests Truth for others to behold.

"Behold, I send you forth as sheep in the midst of wolves: be ye therefore wise as serpents, and harmless as doves."[94]

The Serpent, in being crucified, becomes wise or illumined, embodying and manifesting the Christ principle, *IOA*, the Dove. The sheep are those pure souls ready to undertake the sacrifice as 'lambs of God', and thus to be the wise serpents and the harmless doves. In achieving Christhood, each sheep becomes a Shepherd, each dragon-serpent becomes as a Dove. Each spiritual soul is one who has become 'like unto God', revealing 'the Countenance of the Lord' or Michaelic aspect of Christ. The face shines with light and joy, and the head is crowned with glory. Each is become a Jesus or Horus, 'the Visible Appearance of Light'. In other words, the spiritual soul of man (*i.e.* the Christed Adept or Master) is manifesting and acting as Michael, the Christ Image of Man. In Christian Tradition such a soul is called Saint George, 'the Gardener' who cares for the world, fulfilling God's Plan for mankind.

St. George is the Red (or Rose) Cross Knight, who manifests the Light of Christ (the Cross) and acts as a Saviour by giving his life, his blood (the redness), for others, and working according to the Michaelic Master-Plan, slaying and redeeming each and every serpent-dragon.

JOHN THE BELOVED AND THE THREE MARIES

At the point of death in the Crucifixion enacted out by Jesus and his disciples, the disciples were looking on from afar. Amongst the disciples, Mary Cleopas (Salome), Mary the Mother of Jesus, and Mary Magdalene are purposely picked out in the Gospel stories.[95] Christian symbology later portrays these three as standing beneath the cross, together with John the Beloved. **The three Maries** represent three stages of the natural soul during its evolution, and correspond to the Triple Goddess idea — maiden, nymph and crone.[96] **Mary Magdalene**, who was redeemed from a life of sin to become a devoted disciple of Jesus, portrays for us the pre-initiation condition of the natural soul. **The Virgin Mary**, mother of

(9) CRUCIFIX — painted altarpiece by Raphael. The National Gallery, London.

This painting is another cabbalistic masterpiece, depicting Jesus crucified upon the Cross of the Law, with the three Maries and John the Beloved at the foot of the Cross, representing the states of being that he has embodied and transcended. Above his right hand is the solar symbol of Mercy, Grace — the radiant outpouring of Light; and above his left hand is the lunar symbol of Judgement. Discernment — the receptivity and reflection of Light. Above Jesus' head is the tablet inscribed in gold with the mystical letters *"I.N.R.I."*, meaning both *"Iesus Nazareus Rex Iudaeorum"* ("Jesus of Nazareth, King of the Jews") and *"Igne Natura Renovatur Integra"* ("Nature is completely renewed by Fire"). It is also said to stand for *"In Nobis Regnat Iesus"* ("Jesus reigns within us"). *I.N.R.I.* is a statement of God's Law, and thus represents the Holy Spirit or Divine Thought, the formula standing in place of the Dove of Fire.

Jesus, portrays for us the succeeding stage, when the 'Magdalene' dedicates her life to God and follows the path of initiation. In initiation (the Lesser Mysteries), the natural soul is made virginal or pure, through and through, all the while bearing the Christ-child within her heart-womb. At the point of sacrifice, the Christ child is born from its mother and the Madonna becomes the Jesus; but in giving birth the natural soul must die. As the old psyche dies, in sacrificial service, so its very substance and consciousness becomes transmuted into the form and consciousness of the spiritual soul, like the phoenix rising from its incense pyre. This is the third stage represented by **Mary Cleopas**, the wise old woman. It is the stage of the Wise (old) Dragon. The three Maries can also signify the first three Initiations.

These three, therefore, are depicted standing at the foot of the cross, signifying the three stages that Jesus' natural soul had passed through and transcended in order to reach Christhood. In iconography, the Crucifixion scene usually shows the Holy Dove of Light descending upon Jesus' head, and his head actually crowned with a halo of glory. This depicts the actual Christing with *IOA*, the Christ Principle and divine Unction. The Gospels signify this by the statement that Jesus cried out and gave up his ghost (*i.e.* surrendered completely his soul and his intelligence), and the veil of the temple was rent in twain, allowing the Light to stream into and illumine the whole temple or soul, transmuting it. The addition of John the Beloved in the scene at the foot of the cross, with the three Maries, symbolised the Christ light, *IOA*, that had up until then been latent, hidden within the heart of the natural soul and waiting to be born.

THE LAST SUPPER

The Last Supper is associated with **the Passover Feast** and the slaying of the unblemished sacrificial ram (the *Agnus Dei*) as an offering to redeem the sins of the nation and the world. Jesus desired to eat the Passover with his disciples in Jerusalem, and this he did. What is not generally realised is that the so-called 'Last Supper' in the upper room was not the actual *Pasch* or Pascal Feast of the Passover in which the pascal lamb is eaten in holy communion and remembrance. The **supper** in the upper room was in fact the Passover Eve Feast, termed a 'Supper', which was a **Feast of Brotherhood** called the *Qadosh*, signifying Holiness, Dedication and Sanctification.

The true **Passover** or **Paschal Feast** was strictly a family affair, which was partaken of whilst standing and consisted of the pascal lamb (roasted by fire), bitter herbs, a bowl of salt water, a mixture of ground cinnamon and nuts, and wafers of unleavened bread. In the centre of the table were placed five cups of wine — four of them making the four corners of a square (or ends of a cross). The fifth cup, standing in the centre, represented Isaiah, the Herald of the Messianic Age. A special formula of words was said as the cups of wine were blessed and shared in turn, relating to God's Promise and the initiatory path leading to Christhood. The Feast was known as **the Feast of Remembrance**. It was eaten in haste, and took place at night, between sunset and sunrise on the 15th day of the lunar month called *Abib*, which was known as the Day of Remembrance. Nothing was to be left of the *Pasch* by the time morning came. On that 15th day, during the day-light hours, a Holy Convocation was held, and for the duration of that day no work was to be done. The *Pasch* was the first feast of the Festival of Unleavened Bread (also known as 'The Eight Days of Passover') which culminated on the 21st day of *Abib* with another solemn Holy Convocation. The Days of Holy Convocation were known as High Days or Great Sabbaths.

The People of Israel marked their calendar according to the moon phases, their months beginning with each new moon (the period of darkness). Like other nations, they began each day of 24 hours with sunset and night-time, followed by the daylight-time; for Darkness preceded Light in Divine Creation. The first month of the year was *Abib* (or *Nisan*), associated with the Spring Equinox. On the 10th day of *Abib* each father of a family was to select and take the sacrifice from his flock — the first-born unblemished male lamb of the year — one for each family. At the same time the mother of the family was to clear out all the leaven from the house (*i.e.* spring-clean thoroughly) and prepare the home for the Paschal Feast. The pascal lamb was then to be kept until the 14th day arrived, the day of the full moon, when the whole community of Israel assembled to slaughter the sacrifice of lambs before sunset, usually about 3.00pm (which was called the 9th hour). That particular day, the 14th of *Abib*, was begun with the *Qadosh* or **Supper of Dedication and Sanctification**, held during the night hours.

It should thus be realised that the Last Supper or Lord's Supper was the *Qadosh*, in which Jesus, together with his Twelve, sanctified the Sacrifice and made themselves ready for the pascal offering. After the meal was finished, Jesus ritually shared the bread amongst them, symbolic of the Bread of Life or Divine Wisdom — the only true nourishment which gives us all life, and which Jesus manifested for us in his resurrected body or form of light. Following the bread (which must come first), the grail cup of wine was shared, symbolic of the Wine of Life, which is Knowledge of God attained and enriched by sacrificial experience of life, contained in the Cup of the Soul. This wine of knowledge was the 'blood' that Jesus poured out for us. '**Remembrance**' is in fact the process of acquiring knowledge; for, as we remember our experiences, light dawns in our minds and hearts, and the knowledge bestows the power to change (or transmute) all things. The importance of the entombment after the death on the cross is that it is a period of remembrance. 'Remembrance' also means putting things together again, and refers principally to the mystic marriage of the soul with the Spirit, and of the *Sophia* with the *Christos*, completing God's whole scheme of Creation. The Bread signifies the Spirit, and the Grail the Soul; but also the Bread of Life is the *Christos*, and the Wine of Life is *Sophia*, fulfilled.

The Paschal Feast was indeed shared between them — only it was Jesus who was the Pascal Lamb (and to a lesser extent his disciples as well), slain on the cross at the ninth hour of the 14th day, with his 'body' and 'blood' shared in the *Pasch* that followed. Jesus arose on the third day: the 14th of *Abib* (the Day of Sanctification) being a Wednesday night-Thursday daytime, the 15th (the Passover and Great Sabbath) being a Thursday night-Friday daytime, the 16th (an ordinary Sabbath) being a Friday night-Saturday daytime, and the 17th (the day of Resurrection) being a Saturday night-Sunday daytime. Hence it was "on the third day", in the morning, that Jesus arose from the dead in his Christhood demonstration. The Apostles were longer 'in the tomb' after their Crucifixion experience (which was every bit as real as Jesus', differing only in magnitude and type), and it was not until the day of Pentecost (50 days after the Passover) that they achieved their Resurrection and Christhood as Adepts.

The whole drama was most carefully staged, even to the extent of selecting Jesus as the Paschal Lamb and setting him apart for sacrifice, this selection being staged in the form of Jesus' triumphal entry into Jerusalem on a white ass (the valued animal used by kings and princes for processions), when he was acclaimed king by the people, on the 10th day of *Abib*, (Sunday daytime). In readiness for this public selection and procession, and his approaching death, at the supper in Bethany, after sunset on the 10th day of *Abib*, Jesus was anointed on his head and feet by Mary of Bethany, his sister, who realised that her eldest brother and *Rabboni* ('beloved teacher') was intending to go to his death in a few days, in order to fulfil his teachings. Intuitively she grasped the situation and anointed him with the precious spikenard ointment that she had carefully and specially reserved for the time of Jesus' death and burial. Acting as a prophetess, under direct inspiration, she anointed her royal brother, a prince of Judah, in the manner of a priest and king, signifying thus her own recognition and selection of Jesus as the rightful heir to the Throne of *Salem* ('Peace') and the Paschal Sacrifice that was to be made for the people — the ancient traditional sacrifice of a priest-king.

Mary of Bethany was not one of the 'Three Maries', but she, Martha and Lazarus were the beloved younger sisters and brother of Jesus. Symbolically, Mary and Martha represented and acted out the contemplative and active natures respectively of the natural soul, whilst their younger brother Lazarus ('he whom God assists') represented the personal ego itself. The events concerning this little family during the weeks leading up to the Passover were highly significant, first there being the raising of Lazarus from the dead, followed by the anointing. These two important events prefigured the great event that Jesus was about to undergo.

THE OLD AND NEW COVENANTS

In truth there is only one Covenant or Law, but there are two aspects to it and two ways to experiencing it. The Covenant is God's Promise, which is His Holy Law or Wisdom[97] — the Divine Idea. It is also known as the Testament and the Commandment of God. The two aspects of the Covenant are referred to as the 'Old' and the 'New' Testament or Commandment. The 'Old' deals with loving the Lord God, and the 'New' deals with loving mankind and all of life. Jesus summed them up in the old rabbinical statement:

> "Hear, O Israel, the Lord thy God is one Lord, and thou shalt love the Lord thy God with all thy heart, and with all thy soul, and with all thy mind, and with all thy strength. This is the first and great commandment. The second is like unto it: Thou shalt love thy neighbour as thyself. There is none other Commandment greater than these. On these two Commandments hang all the laws and prophets."[98]

In the Ten Commandments which Moses received from God and delivered to the people of Israel, five are presented on one tablet (or one side), and five on another tablet (or other side). These two groups of five Commandments each are arranged according to the 'Old' and the 'New', the first five dealing with loving God and the second five dealing with loving mankind.

In the symbolic Eucharist which Jesus shared with his Apostles during the Fellowship Feast, the bread symbolised the 'Old' Covenant and the grail symbolised the 'New' Covenant — the one dealing with the Lord God (*i.e.* the Christ), and the other with mankind.

The New Covenant is peculiarly associated with Jesus, whilst the Old Covenant is associated with the preceding history of Israel. This is because the Old Covenant contains the laws which bring life into being, and are linked with the Lesser Mysteries of Initiation and the First Coming, wherein man looks to find the Teacher or God outside of himself. Whereas the New Covenant contains the laws which bring life into its fruition, and are linked with the Greater Mysteries of Initiation and the Second Coming, wherein man has realised his own divinity and is bestowing the divine grace or compassion on others.

In the Judaic-Christian Scriptures there are five specific moments in history when the covenant is made or renewed. The first moment is immediately after the Flood and destruction of the Atlantean Epoch and 4th Root Race of humanity. Inaugurating the new Aryan Epoch, the covenant is made with Noah, the ancestor of the 5th Root Race, and the Testament is portrayed symbolically as the Bow set in the Cloud; signifying the Divine Love set within its radiance of Light, and also the Light shining within the living form of matter and being revealed thereby — hence the Rainbow, which has profound significance. This making of the Covenant is one that generally deals with the new beginning of the whole race of mankind, and marks the period of Preparation or Pre-Initiation.[99]

The next four moments deal with the nation of Israel, and mark the stages of initiation. The first of these four moments takes place in connection with Abraham, the ancestor of the nation of Israel. The Covenant is made that Abraham's seed shall be innumerable, and that they shall inherit the land of Canaan (the 'Promised Land') and share their blessings with others.[100] The symbolic sign for this Covenant was circumcision, signifying "faith which worketh by love"[101] and the casting off of all that is unclean. This Covenant with Abraham deals with the 1st Initiation, and it covers the Patriarchal period of Israel's history.

The next moment occurs in connection with Moses, in which the Covenant with Abraham is renewed and enlarged upon, in detail, and the Ten Commandments are given.[102] The symbolic signs used were the keeping of the Sabbath, the building of the Tabernacle and Ark of the Covenant, and the Pillars of Fire and Cloud. This Covenant with Moses deals with the 2nd Initiation and covers the period of Moses, Joshua and the Judges in Israel's history.

The third moment that deals with Israel is in connection with King David, in which David is promised a son who shall be the Son of God the Father, and that God's House and Kingdom shall be established for ever.[103] The symbolic sign was the building of Solomon's Temple. This Covenant with David deals with the 3rd Initiation and covers the period of the Kings of Israel, including the conquest and dispersal of the nation.

The final moment is associated with Jesus. Its symbolic sign is the Holy Grail and the Brotherhood Feast (also known as the *Agape* or Love Feast), and Jesus himself. This Covenant is the New Covenant, and deals with the Higher Mysteries of Initiation and Christianity (*i.e.* Illumination). Its operation is summed up in Christ Jesus' words:

> "This is my blood of the New Testament, which is shed for many for the remission of sins."[104]

Diagram E: ISRAEL'S NATIONAL HISTORY

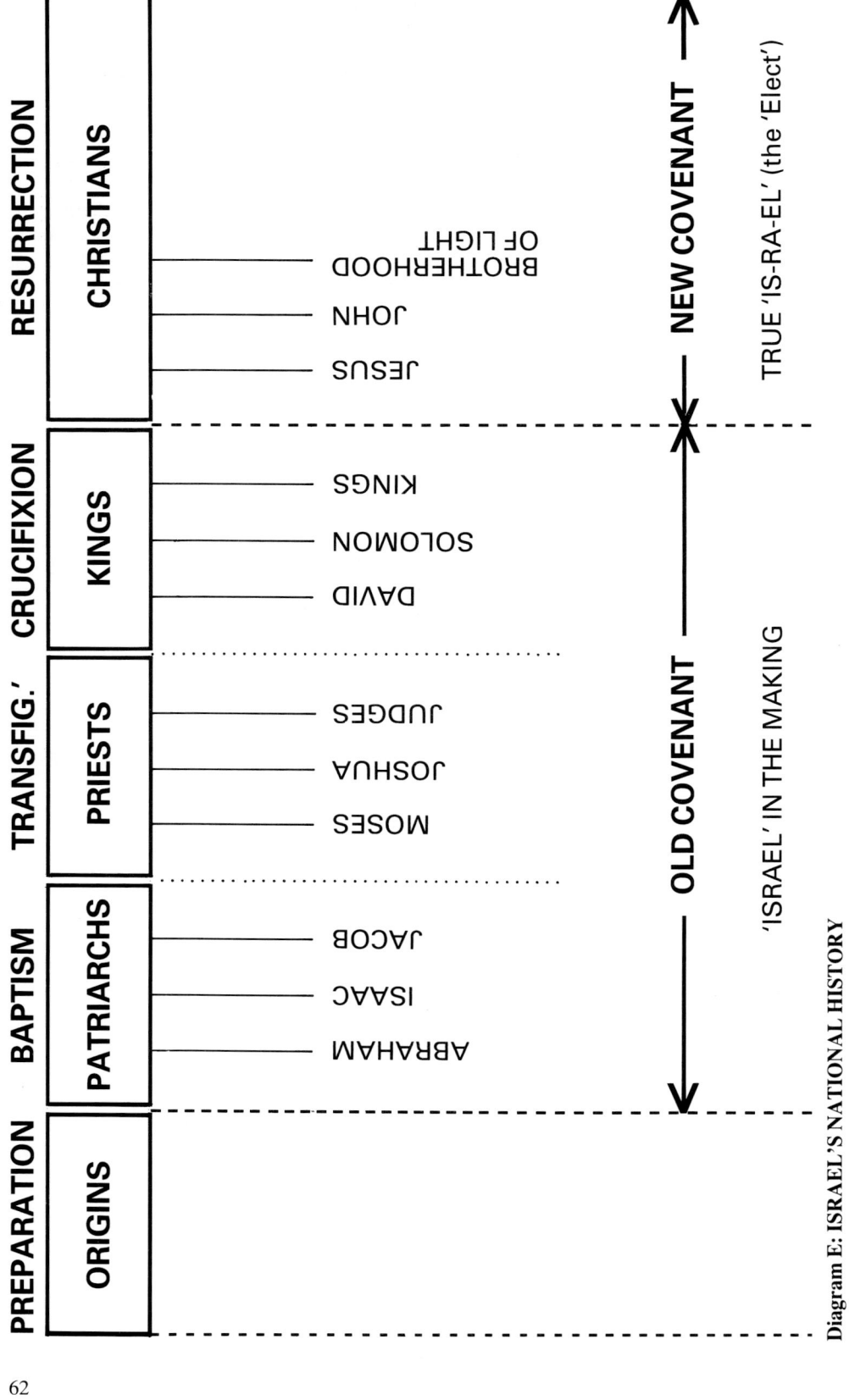

Phase	Group	Figures
PREPARATION	ORIGINS	
BAPTISM	PATRIARCHS	ABRAHAM, ISAAC, JACOB
TRANSFIG.'	PRIESTS	MOSES, JOSHUA, JUDGES
CRUCIFIXION	KINGS	DAVID, SOLOMON, KINGS
RESURRECTION	CHRISTIANS	JESUS, JOHN, BROTHERHOOD OF LIGHT

'ISRAEL' IN THE MAKING — OLD COVENANT (from PATRIARCHS through KINGS)

TRUE 'IS-RA-EL' (the 'Elect') — NEW COVENANT (CHRISTIANS)

"A New Commandment I give unto you, That ye love one another; as I have loved you, that ye also love one another."[105]

"Greater love hath no man that this, that a man lay down his life for his friends."[106]

"Behold, the days come .... when I [the Lord thy God] will make a New Covenant with the House of Israel and with the House of Judah .... I will put My Laws into their mind, and write them in their hearts: and I will be to them a God, and they shall be to Me a people: and they shall not teach every man his neighbour, and every man his brother, saying, 'Know the Lord': for all shall know Me, from the least to the greatest."[107]

PART II

THE VISION AND BIRTH
OF
THE NEW ROSICRUCIANISM

(The Life and Times of Francis Bacon, 1572-1579)

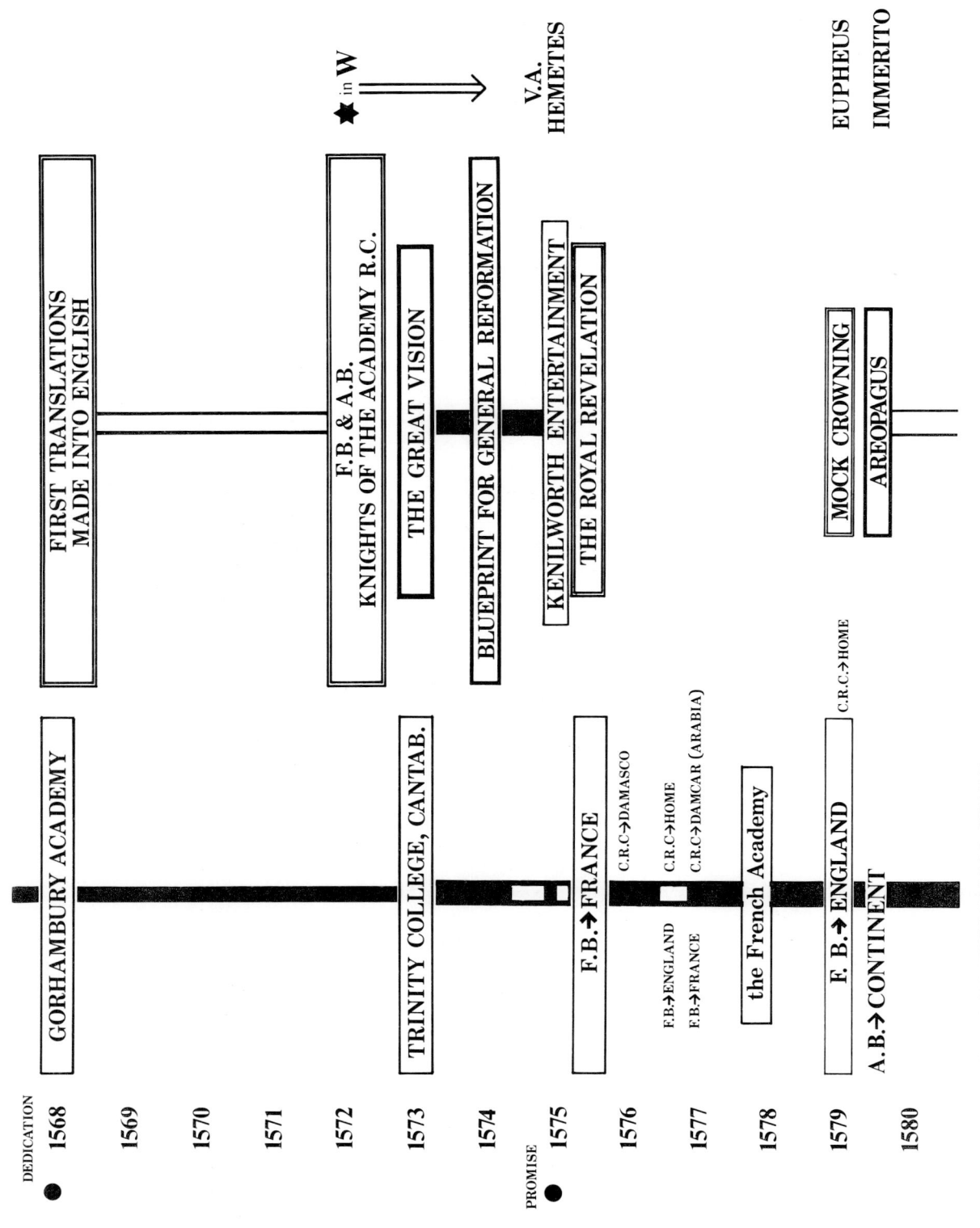

Diagram F: FRANCIS BACON: HISTORICAL CHART

DEDICATION ● **1568** — GORHAMBURY ACADEMY / FIRST TRANSLATIONS MADE INTO ENGLISH

1569

1570

1571

1572 — F.B. & A.B. KNIGHTS OF THE ACADEMY R.C.

1573 — TRINITY COLLEGE, CANTAB. / THE GREAT VISION

1574 — BLUEPRINT FOR GENERAL REFORMATION / KENILWORTH ENTERTAINMENT / THE ROYAL REVELATION (★ in W → V.A. HEMETES)

PROMISE ● **1575** — F.B. → FRANCE
- C.R.C → DAMASCO

1576
- F.B. → ENGLAND
- C.R.C → HOME

1577
- F.B. → FRANCE
- C.R.C → DAMCAR (ARABIA)

1578 — the French Academy

1579 — F. B. → ENGLAND
- C.R.C → HOME
- MOCK CROWNING / AREOPAGUS
- EUPHEUS

1580 — A.B. → CONTINENT
- IMMERITO

Supernova in Cassiopeia

During the 16th century there was a profound belief amongst mankind that a Golden Age was about to be ushered in. With the Renaissance nearing its climax in Europe, great reforming movements were under way in European society, religion and culture. Here a rebirth of some kind was quite literally taking place, and it was genuinely hoped and expected that the new world being born would be 'golden'. Under the Hapsburg Charles V, in the middle of the 16th century, the ideals of the Holy Roman Empire were revived and the old model of a perfect society ruled over by a Sun-king or Holy Emperor (together with a Pope or Supreme Hierophant) was briefly reinstated with great expectations. In Spain the son of Charles V, Philip II, attempted to bring about the same idea with himself as Lord of the World. In France similar aspirations arose when Elizabeth of Austria, grand-daughter of Charles V, married Charles IX of France, both lines claiming descent from Charlemagne the Great, the first Holy Roman Emperor. In the words of Frances A. Yates:

> "Moreover, to sixteenth-century minds saturated in the imperial idea, the union of two great royal lines, both claiming descent from Charlemagne, in the marriage of a *Rex Christianissimus* of France with a daughter and grand-daughter of emperors, was seen as something in the nature of a portent, an event of tremendous significance which might lead to a world religious empire in which religious peace would finally be established."[1]

Others saw it differently, yet were no less inspired by the idea of a rebirth into a Golden Age, presided over by Saturn, the 'genius' of Knowledge and Peace, and by Astraea the Virgin, 'genius' of Purity, Justice, Piety and Plenty. Many looked to a rebirth of Christianity according to its original model, before the 'Roman' takeover and domination of what should have been a free and personal religion. Movements aiming to reform the Church, to return to the first principles of religion and religious life, to study and comprehend the original and true sources of Christian thought and inspiration, spread like wild-fire. Eventually many such reformers broke away completely from the 'Roman' religious hierarchy and dominance, beginning to seek mankind's future hope as lying outside this rigid and, to a large extent, corrupt organisation. Politics soon followed religious thought, which in turn was following the increase in learning, bursting forth from the fetters of ignorance imposed on Christendom during the Mediaeval period by successive popes, cardinals and tyrannical rulers. The invention of the printing press enabled the new gospel — the new learning — to be spread. Renaissance colleges, academies, schools and universities raised the level of education, not only amongst the more wealthy and privileged, but amongst the poorer classes of society too. Growing literacy amongst all classes enabled people to read translations of the Scriptures and the Classics for the first time, as well as commentaries and new thoughts. The Renaissance habit of sending the sons of noblemen and gentry on foreign travels helped to open up minds with new vistas of experience.

Amongst the better educated, moderate and beautiful humanistic thought developed, both mystic and scholarly, surrounded on one side by the mediaeval and conservative dogmas of Roman Catholicism and on the other by new and radical dogmas of various reforming movements. The pendulum was swinging from one side of tyranny and oppression, to another side that was proving to be equally violent and aggressive, in Reformation and Counter-Reformation. Germany, emerging then as a multitude of different states each with their own system of local government, was plunged into civil war as a result of Martin Luther's concern about and direct challenge to the corrupt life, practices and rule of the Roman Catholic hierarchy — a concern which thousands of people shared and a tyranny under which they suffered. The Protestant Reformation spread to England, helping to bring about the formal break with the Roman papacy in Henry VIII's time and the establishment of the Church of England, together with more Puritanical organisations as other teachers, leaders and zealots came to the fore and played their parts. Scotland developed a Presbyterian system of reform, and likewise in the Scandinavian countries revolution and reform spread their wings. In France a hideous series of recurring civil wars commenced in 1562 between the Calvanist Huguenots and the Roman Catholics, the former to find their figurehead in Henri, King of Navarre, the latter to find theirs in the Valois kings of France, even though sometimes the monarchs themselves did not wish these discriminatory roles thrust upon them. Another series of hideous wars was waged in the Netherlands, as an outright rebellion against the increasingly tyrannous Spanish rule, resulting eventually in close alliance with England and the birth of the Dutch Republic in 1579.

In all this turmoil a real renaissance of light — of learning and charitable work — was taking place. A wide underground movement of free thinkers and inspired scholars, artists and craftsmen spread its tentacles wide. It was not in any real sense an organised movement, although certain organisations were formed from time to time as a result of the movement, but rather it was a movement of thought. It was both mercurial and volatile, slipping by silently and discreetly here, and erupting powerfully and publicly there. Out of this free movement were born the real 'giants' of humanity of that era, not all of them by any means well known to common history.

This great change was not confined simply to Europe. Nations world-wide were involved. The Americas, North and South, were discovered anew by the Europeans, and their ethnic cultures were gradually overturned by programmes ranging from peaceful trading and settlements to brutal rapine and overlordship. In Asia the Turks spread their Ottoman Empire through Persia, Kurdistan, Anatolia, Mesopotamia, Syria, Egypt and Arabia, and threatened Europe, ruled by their particular Sun-king, first Selim, then Suleiman ("the Magnificent"). In India the great Mogul Empire was established by Barber, and in Russia the Tsarist autocracy was founded by Ivan III, and the Russian Empire established by his successor Basil III, followed by Ivan IV ("the Terrible"). All these rulers, with their minions and adherants, dreamed the same dream — a World Empire ruled by a World Sovereign, in whom the powers of state and religion are essentially united. It is an old dream — an ever-recurring one, based on an element of truth but so grossly misinterpreted from age to age. The same dream was dreamt in the time of Christ Jesus, who was obliged to spend so much of his time explaining just what this ideal really meant and how it may be truly achieved. Only a few men and women in any generation seem able to comprehend its real truth.

Then, near the height of the furor, a cosmic sign blazed out its message abruptly and unequivocably. In August 1572 a supernova appeared in the constellation of Cassiopeia, and shone brightly for a year and a half until disappearing in March 1574. A supernova is a star which explodes catastrophically, with a sudden liberation of most of its energy and the dissipation of nearly the whole mass of the star into space. Its explosion appears to us about ten thousand times as bright as an ordinary nova. The only supernovae observed in our Galaxy during the past thousand years occurred in A.D. 1006 (in Lupus), 1054 (in Taurus), 1572 (in Cassiopeia), and in 1604 (in Ophiuchus). All reached apparent magnitudes considerably brighter than zero[2], releasing as much energy in one second as our Sun does in sixty years. Imagine a supernova lasting eighteen months! All of them are present-day radio sources. The 1572 supernova appeared brighter than Venus, and was clearly visible in daylight. The fact of its occurrence, and that it appeared in Cassiopeia, the Sign of the Celestial Virgin Queen, was (and still is) of prime significance. A symbolic knowledge or astrological science of the stars has been the concern of sages of all nations and civilisations from the dawn of history. With such knowledge the three Magi were able to know where and when and in what manner the Christ child was born *c.*7 B.C. In 1572 the appearance of the supernova in Cassiopeia was the recognised sign of a new Christ child or Christ appearance being manifested (*i.e.* born) on Earth, as the 'Child' or 'Light' of the 'Virgin'. It was called "the Star of Bethlehem".[3] How truly the divine Idea underlying these cosmic events materialised itself on Earth, we shall see. Michael Srigley comments on it thus:

> "In August, 1572, in the Constellation of Cassiopeia, a new star made a sudden and dramatic appearance. It was as bright as Venus, and to an astonished world taught to believe that the heavens were immutable, it shone as a portent of some convulsive change in the order of the world. To Theodore Beza, the Calvanist Biblical scholar, it heralded the Second Coming, while Paracelcus saw it[4] as 'the sign and harbinger of the approaching revolution: there is nothing concealed which shall not be revealed, for which cause a marvellous being shall come after me, who as yet lives not, and who shall reveal many things'. Tycho Brahe wrote a small tract on the 'new and much admired star' of 1572, which eventually appeared in 1602 dedicated to that royal occultist Rudolph II, Holy Roman Emperor and part model for Shakespeare's Prospero. In it Brahe writes that for him the star marked the entrance of the world into the seventh revolution that would inaugurate the golden age: 'Some great Light is now at hand which shall enlighten and by degrees expell the former darkness.' He goes on to say that the star would be followed by a major conjunction[5] when the actual prophecies of the star of 1572 would be released. Such then was the reaction to the new star which remained visible until 1574; it was the Star of Bethlehem inaugurating the Second Coming and the restoration of the Golden Age."[6]

*   *   *   *

Cassiopeia is a distinctive constellation of five third-magnitude stars that can be seen in the northern hemisphere, and which is circumpolar[7] over Great Britain. The pattern of the five stars appears to us in the shape of the letter 'M', or 'W', depending on which way up it is viewed.[8] Its apparent position in the heavens is not far from Polaris, our North Pole Star, and it transverses the whole zodiacal segment of Pisces when the Zodiac is centred on or near Polaris of Ursa Minor[9], which is the present situation and has been throughout the Age of Pisces. She is also said to preside over the Christmas period, appearing due west of Polaris at midnight for the duration of the Twelve Days of Christmas exactly. In Chaldean astrology she is said to preside over the countries of Syria and Palestine also.

Cassiopeia means 'the Enthroned Lady' or 'Celestial Queen'. She represents the Virgin Queen and Mother of the Christ Child (or Children). Mythologically she is the mother of Andromeda, another constellation in the Piscean sector. Each new star that appears within her basic star pattern signifies the birth of a new Christ impulse or 'Child of Light' that will influence and manifest itself on Earth. Cassiopeia is the 'Woman with Child' of *Revelation*, and was also called by the Druids, *Llys Don*, meaning 'the Court of the Lord'. Each star in her bosom is a 'Child of Don', a Sun-Son of the Supreme Lord. On average, it is said, *Cassiopeia* gives birth to an unusually bright star, or nova, every 300 years, and this approximate periodicity is associated with one of the several legendary cycles of birth, death and rebirth of the Phoenix (and renders 7 stars or Phoenixes per Age). The appearance of a supernova symbolises the birth of a Christ Child in a particularly powerful or 'bright' way. During Francis Bacon's life *two* supernovae appeared in our Galaxy, the 1572-4 one in *Cassiopeia*, and the 1604 one in Ophiuchus ("the Serpent Holder", also known as *Aesculapius*, who holds and masters the Serpent). Both appearances were highly significant and intimately associated with the Bacon-Rosicrucian impulse and work.

The Enthroned Lady or Heavenly Queen is the ancient title of Isis, the virgin mother of Horus, as also of Mary, the virgin mother of Jesus. The hieroglyphic symbol of Isis used by the Ancient Egyptians was that of a throne — Isis being the actual throne or seat upon which the child Horus sits as King of Light (the Sun King). The same symbol is employed in Christian iconography, and Mary is portrayed as the 'throne' or 'seat', with the Christ Child seated in her lap, his head aligned with her heart. It is a very carefully constructed and important symbol, revealing many mysteries.

In the Hebraic cabbala, the Throne is described as the *Merkabah* or Throne of Glory — the perfect dwelling place of the Lord. It is the supreme and pure Tabernacle or Temple in which the Messianic Presence of the Lord God may dwell and be manifest. That is to say, it is the perfect form of divine manifestation, the virgin Soul, Bride to the Spirit. The sublime union of the receptive, responsive Bride with the active, creative Spirit produces (*i.e.* gives birth to) the Messiah as the Christ Child, the living embodiment of Light (which is what the word *Jesus* or *Horus* means). In biblical terms it means that the "living soul", Adam, becomes a "life-giving Spirit", Christ: the former being a receptacle of life, the latter being a creator of life, in the same symbolic sense that a moon may become a sun, or a light-receiving vessel may transform itself into a light-producing star. Out of the cosmic form of Nature are born the shining Sons of Light.

The Temple, synonymous with the Throne, was also used as a hieroglyph for Isis, the Virgin Mother or Soul. When the Temple is so filled with the Presence of God that it is set on fire with divine love and light (which a perfect Temple will be), then it becomes, or is, a pyre of flame — a 'Pyramid'. And, in becoming a pyre of light, it moves, it becomes active. The Throne becomes a Chariot. The motionless Temple becomes a living Flame.

Just as a flame enables light to be manifested as a blazing orb or sun of radiant light, so the perfect Soul enables the Spirit to become fully manifest as the Christ Child. Just as the central orb of light sits enthroned in the midst of the flame, shining its radiance or glory all around, so the Christ Child sits as a King of Glory upon his Throne: Horus sits upon Isis; Jesus sits upon Mary; the Christ Consciousness 'sits' upon man's soul. Then, just as the flame becomes identified with the light, swallowed up by its glory, so the Mother becomes identified with her Child, 'assumed' and 'married', as it were to her Son. Both names, Horus and Jesus, mean 'the Ensouled Light' or 'the Visible Appearance of Light', 'the Countenance' or 'Face of the Lord'. In Christian symbology, Jesus seated upon the lap of his Mother, or Jesus enthroned in the centre of the *Vesica Piscis* (the 'flame-vessel of the

Fish'), mean the same. It is the revelation of the Supreme Mystery or Gnosis; the revelation of Christ, the Son or Light of God; the revelation of the Divine Idea or Divine Consciousness, which is Truth.

The Christ Child is the Sun-King, and his Throne or Mother is the Kingdom of Heaven, the Empire of Peace and Plenty that is bathed in the golden glory of Christ Consciousness. It is, or can be, both individual and universal. Jerusalem (meaning 'the Great Peace') is another name for the Kingdom. Old Jerusalem is the Mother — the psyche; New Jerusalem is the Mother with Child — the 'heavenly' or beautiful Soul that has attained Christ Consciousness.

*   *   *   *

(10) ELIZABETH I, Queen of England (1558-1603), aged 38 years — miniature painted (1572) by Nicholas Hilliard.

At University

In July 1572, just before the supernova burst its message across the heavens, Queen Elizabeth visited Gorhambury again. Probably the main purpose of this visit was to be able to see privately how her eldest son, Francis, was developing, and to discuss his future with Sir Nicholas. From birth Francis had been adopted by Sir Nicholas and Lady Anne Bacon, close friends and confidantes of the Queen, and had been growing up in their capable hands never suspecting his real parentage. Lady Anne was devoted to both children — her real son and her adopted son — and Sir Nicholas, who held the highest office in the land under the Sovereign, was a kind, faithful and wise father. The Queen, however, carefully watched the growth of her 'concealed' child, and had an especial interest in his education (as discussed in F.B.R.T. Journal I-3, *Dedication to the Light*).

Given the best education possible at that time, intensive and wide in its scope, and by the best teachers, Francis's natural genius soon acquired fluency in French, Latin, Dutch, Italian, Greek and Hebrew at an early age. In French and Latin particularly he found it quite natural to think, write and converse. At that time the English language was rough and uncultured, as yet undeveloped and limited in scope, consisting of a multitude of dialects with an extremely limited vocabulary of words usable or agreed upon for a national language of culture and erudition, although, since Chaucer ("the father of English poetry") established the London dialect as the standard literary dialect for subsequent national poets to use, there had been a steady growth in the use of 'London' English for poetic and dramatic uses. Very few books were written in or translated into English, and Latin was still the principal 'literary' language of scholars. Francis also learnt Spanish and Dutch at some stage, and was well versed in the grammar, logic and rhetoric of the various languages, and proficient in music, fencing, riding and the mathematical subjects. In particular his soul was saturated like a sponge with Classical mythology and philosophy, with the scriptures, and with Humanism. He lived in a family of brothers, sisters, cousins, parents, uncles, aunts and grandparents, and their friends, who were at the heart of the English Reformation and who were preparing the ground for the English Renaissance.

> "At twelve his industry was above the capacity and his mind beyond the reach of his contemporaries . . . His memory was fixed and methodical . . . a wonder."[10]

> "His first and childish years were not without some mark of eminence; at which time he was endured with that pregnancy and towardness of wit, as they were presages of that deep and universal Apprehension which was manifest in him afterwards, and caused him to be taken notice of by several persons of worth

and place; and especially by the Queen, who (as I have been informed) delighted much, then, to confer with him, and to prove him with questions: unto whom he delivered himself with that gravity and maturity, above his years, that her Majesty would often term him, The young Lord Keeper."[11]

Francis was a recognised and publicly known 'child-wonder'. Even at a very early age the Queen (and others) nicknamed him, 'baby Soloman'. Many regarded him with great interest — some with fear and trepidation. But Francis was not the only one of the Lord Keeper's wards and children to receive such special attention and education, nor the only one with a good wit (although in sheer genius he was unsurpassed). Anthony, the natural son of Sir Nicholas and Lady Anne (and her only natural child who lived beyond birth), also received the same education alongside Francis, and was said to possess as good a wit as Francis, although not such a proficiency of learning and depth of knowledge.[12] Anthony, born in 1558, was over two years older than Francis. They were brought up together as brothers and became the closest of friends, sharing their dreams and ideals and, for a time, their experiences. For the whole of their lives they remained deeply attached to each other in a true fraternal way, and in letters referred to each other as "my dearest brother", and "Antonie my comforte".

The two brothers, plus another ward of Sir Nicholas, Edward Tyrrell, who seems to have been educated with and been a friend of Anthony and Francis, received a wider education and experience than the categories of subjects named above can imply. York House, by Charing Cross, the London home of Sir Nicholas as Lord Keeper of the Great Seal, was the busy centre of State as well as of cultural affairs — national and international. Sir Nicholas combined the role of Lord Chancellor in his office as Lord Keeper of the Seal, and was not only principal Officer and Counsellor to the Queen, but also the principal Judge or Law Officer of the land, next to the Sovereign. The boys would have seen and heard a great deal that most children hardly even dream of, and not only at York House but also at Court — in London (Whitehall) during the winter, and at other fine palaces, castles and mansions during the summer progresses. Besides affairs of State and royal pageantry, there was a developing form of 'Renaissance' Entertainment involving music, poetry, drama, elaborate and symbolic costumes, mythology, masques and dramas, grand spectacles and processions, and a revival of 'Arthurian' chivalry with regular tournaments on the anniversaries of the Queen's Accession Day. The settings for most of these were in splendid buildings, amongst architecture and art of the best the country could afford, in magnificent countryside, gardens and parks (although the full development of gardens and landscaped parks was yet to come). Then at Gorhambury, the Bacon's country home at St. Albans, the beauty and marvel of nature would have had a real chance to sink into the boys' souls in peace and tranquility. That the boys had keen interests other than the more academic and social subjects is well testified. Their adventurous spirits, their investigations into nature and life in general, and their experiments into realms of knowledge and consciousness other than what was commonly known and accepted sometimes caused much consternation to others, particularly the devout Lady Anne and her household chaplain. Their chaplain had once been admonished by the Archbishop of Canterbury for not subscribing to the new Prayer Book, and in his defence he reported:

> "Sir Nicholas Bacon needs Christian help regarding his youthful retinue, among whom all manner of vices do increase apace, and zeal, virtue, and the true fear of God decrease through lack of due admonition and instruction."

The lives of the boys were indeed full! Intense instruction, study and rigorous discipline occupied a large part of their lives, but could not dampen the delight and wonder that they took in life all around them, or in their sheer vitality and exuberance of youthful spirit that was stirred up in such environments and conditions. Anthony and Francis had many characteristics in common, which no doubt helped to bond them together as such firm life-long friends and real 'brothers of the Spirit'. Both were adventurous, daring and courageous in adversity. Being deeply religious and idealistic, imbued with the Humanistic spirit[13] and a perception of esoteric truths, they were each passionate seekers after truth. They were both naturally kind and open-hearted, and their religious natures — although possessing strong personal sympathies with the principles of the Reformation — were stamped with a liberality far in advance of their age. They created and maintained good friendships with men and women of other religious persuasions, much to the alarm of Lady Anne. They both enjoyed the beauties of life — culture, good food, fine clothes and surroundings, and so forth — but with taste and temperance. In fact the seven cardinal virtues were instilled into them from an early age, as also was cultural taste and reserve: but this did not prevent the boys, from their youth onwards, from demonstrating a liking and preference for gayer, more colourful clothing and surroundings than was the more sober tendency of Sir Nicholas and Lady Anne. Both Anthony and Francis had a good wit, or intelligence, as remarked before, with Francis demonstrating sheer genius from childhood; and, very markedly, both had a melancholic disposition. This melancholy humour[14], as it is called, is that inherent nature which produces the deep thinkers and philanthropic geniuses of mankind. In its worst or unredeemed condition, it produces depression and despair, but in its more positive and redeemed state it gives rise to the great thinkers, poets, philosophers, philanthropists, prophets and religious seers of mankind. As Frances Yates describes it:

> "There was a line of thought through which Saturn and the melancholy temperament might be 'revalued', raised from being the lowest of the four [humours] to become the highest, the humour of great men, great thinkers, prophets and religious seers. To be melancholy was a sign of genius; the 'gifts' of Saturn, the numbering and measuring studies attributed to the melancholic, were to be cultivated as the highest kind of learning which brought man nearest to the divine."[15]

The importance of this melancholy humour cannot be overestimated, especially in regard to the Bacon 'brothers' and the whole Baconian work. In the Humanistic or Neoplatonic teachings of the Renaissance, three principal stages of 'melancholy redeemed' were recognised, the redemption depending upon the infusion of 'black' melancholy with the 'white-hot' fire of heroic frenzy or *furor,* which according to Plato is the source of all inspiration (*i.e.* the Fire or Love-consciousness of God). Such a combination "produces the truly great men; it is the temperament of genius. All outstanding men have been melancholics, heroes like Hercules, philosophers like Empedocles, Plato, and practically all the poets."[16] Frances Yates provides us with an excellent abridgment of a quotation taken from an English translation of Agrippa's manuscript version of *De occulta philosophia* (1510), which denotes the three main stages of 'melancholy redeemed':

> "The *humor melancholicus,* when it takes fire and glows, generates the frenzy (*furor*) which leads us to wisdom and revelation, especially when it is combined with a heavenly influence, above all with that of Saturn . . . Therefore Aristotle says in the *Problemata* that through melancholy some men have become divine beings, foretelling the future like Sybils . . . while others have become poets . . . and he says further that all men who have been distinguished in any branch of knowledge have generally been melancholics.

"Moreover, this *humor melancholicus* has such power that they say it attracts certain demons into our bodies, through whose presence and activity men fall into ecstacies and pronounce many wonderful things . . . This occurs in three different forms corresponding to the threefold capacity of our soul, namely the imagination (*imaginatio*), the rational (*ratio*), and the mental (*mens*). For when set free by the *humor melancholicus,* the soul is fully concentrated in the *imagination*, and it immediately becomes a habitation for the lower demons, from whom it often receives wonderful instruction in the manual arts; thus we see a quite unskilled man suddenly become a painter or an architect, or a quite outstanding master in another art of some kind; if the demons of this species reveal the future to us, they show us matters related to natural catastrophes and disaster, for instance approaching storms, earthquakes, cloud-bursts, or threats of plague, famine and devastation . . . But when the soul is fully concentrated in the *reason*, it becomes the home of the middle demons; thereby it attains knowledge of natural and human things; thus we see a man suddenly become a philosopher, a physician, or an orator; and of future events they show us what concerns the overthrow of kingdoms and the return of epochs, prophesying in the same way that the Sybil prophesied to the Romans . . . But when the soul soars completely to the *intellect* it becomes the home of the higher demons, from whom it learns the secrets of divine matters, as for instance the law of God, the angelic hierarchy, and that which pertains to the knowledge of eternal things and the soul's salvation; of future events they show us for instance approaching prodigies, wonders, a prophet to come, or the emergence of a new religion, just as the sybil prophesied Jesus Christ long before he appeared."[17]

In addition to these qualities which Francis and Anthony had in common, there were also complementary differences and other purely individual characteristics. For a start, Francis's memory, imagination, reason and whole intellectual faculties and powers were breathtakingly immense. His intuitive capacity was highly developed, and he was capable of receiving quite remarkable inspiration and vision. He had a rich humour which involved a jesting nature (it was said that he could never pass by a jest), being at the same time a natural and quick-witted conversationalist and orator. He had an inquisitive nature and, as a child, was rather precocious — which developed into an inherent boldness in his youth. His deep passions tended to be fiery and ebullient, and, coupled with an innocent naivity in his early days regarding the real thoughts and motives of other people, he made many devoted friends but also some deadly enemies. But he learnt quickly from both sweet and bitter experience.

Anthony, though, was notably grave of mien, quiet, assiduous, naturally reserved and careful, secretive by nature (whilst Francis had to learn secrecy). His passion ran deep and silent, hidden from most. He was an unselfish person and, like Francis, generous in nature, but sometimes generous beyond his means (it is said). He greatly admired Francis, for whom he thought no sacrifice was too great for the forwarding of his enterprises. He also seemed to have had a remarkable magnetic power of attacting others to his cause or to be his friends. These were all qualities which were so useful when employed later in his particular role in the Baconian work, and in his service to the Queen and State.

*   *   *   *

When the Queen came to Gorhambury in July, 1572, Anthony was 14 years old (or nearly so) and Francis was but 11 years old. Thomas Howard, the 4th Duke of Norfolk, had just been beheaded (2nd June 1572) as a result finally of the Ridolfi Plot and its discovery (in

September 1571), and through Cecil's persuasive and constant pressure. Sir William Cecil, the Principal Secretary of State, had by now become Baron Burghley of Burghley (in 1571). He acted for reasons of expediency and logical statecraft, but the execution of such a powerful member of the nobility and relative of the royal family gnawed at Elizabeth's conscience for years to come, and she firmly laid the blame for it on Burghley. But it did not prevent her from asking for and following Lord Burghley's advice, nor of advancing him to the office of Lord Treasurer in November 1572 when William Paulett, Marquess of Winchester, died. But when Elizabeth went to Gorhambury that July, she may have needed comfort. In April of that year (1572) the Treaty of Blois had been signed with France, by which England and France each undertook to aid the other in case of attack, and France at last acknowledged the *status quo* in Scotland. At the same time the political stage-show of François, Duc d'Alençon, negotiating for Elizabeth's hand in marriage was under way.

Immediately following the Queen's visit, the supernova exploded on the scene, and certain events occurred which were to determine the course of European history and eventually World history.

On Monday, August 18th, 1572, the Princess Marguerite de Valois, sister of the feeble and neurotic Catholic King of France, Charles IX, and daughter of the powerful and ruthless Queen Mother, Catherine de Medici, was married in great pomp to the Huguenot King of Navarre, Henri de Bourbon. The marriage festivities continued in great style until Thursday, August 21st. It was followed the next day by the attempted assassination of the Admiral of France, Gaspard de Coligny, a Huguenot and Queen Catherine's imminent rival as the principal influencer of Charles IX's mind and actions. The final result of this botched attempt was the ghastly massacre of all the now-indignant Huguenot followers of Henri de Navarre, who had come to Paris as guests for the great wedding. The massacre was begun by order of Charles IX, persuaded by his mother, Catharine de Medici, at midnight on Sunday, 24th August, St. Bartholomew's Day. It was mercilessly and cruelly carried out by the Catholic soldiers and citizens of Paris. On hearing the news, other French cities carried out similar nightmare events. At least five thousand people were murdered — two thousand in Paris and three thousand in the provinces, at a very conservative estimate. It was one of the most notorious events in European history, but the atrocity helped to unite the Huguenots and other Protestants, and gave them increased determination to fight for their cause. The outbreak of the fourth war of religion in France inevitably followed, and in England had far-reaching effects on the mind and policies of Queen Elizabeth and her people, as also on Henri de Navarre (who was one day to become Henri IV of France). Sir Nicholas Bacon even supported a Bill in Parliament for the expulsion of all Frenchmen resident in England.

In March 1573, the supernova still shining brightly, Queen Elizabeth was entertained at Gorhambury. The purpose of this visit was almost certainly linked with the arrangement that had been made for Francis, together with Anthony Bacon and Edward Tyrell, to go up to Cambridge University that Easter. The College chosen was not St. Bennet's College where Sir Nicholas had been educated, but Trinity College which had been founded by the Queen's father, Henry VIII, and visited by Elizabeth and Leicester in 1564. The Master of Trinity was then Dr. John Whitgift, one of the Queen's private chaplains, whom she admired, later to be made Archbishop of Canterbury. Whitgift's reputation as a favourite at Court, the wide recognition of his services to the State Church, and his abilities as a strong administrator, enabled him to rule the College practically as a dictator. He had been appointed Master of the College in 1567, was made Bishop of Worcester in 1577, and Archbishop of Canterbury in 1583. In after years Lady Anne Bacon described Whitgift as "the destruction of our Church, for he loved his own glory more than the glory of Christ."

"At the ordinary years of ripeness for the university; or rather, something earlier; he was sent by his father to Trinity College in Cambridge, to be educated and bred under the tuition of Doctor John White-Gift, then Master of the College, afterwards the renowned Archbishop of Canterbury, a prelate of the first magnitude for sanctity, learning, patience and humility: under whom he was observed to have been more than an ordinary proficient in the several Arts and Sciences."[18]

On April 6th, 1573,[19] Anthony Bacon, Francis and Edward Tyrell matriculated (*i.e.* entered) Trinity College, Cambridge, as Fellow Commoners. With them went a servant, named Griffith. The three boys (Anthony was then 14 or 15 years old, Francis was 12 years old) were lodged under the Master's roof, sharing rooms, and they were placed under the direct charge of Whitgift. As privileged Fellow Commoners they each wore a special cap and gown, and dined at the Fellows' Table for 'commons'.[20] Breakfasts, however, were taken in their own rooms. Whilst there they made good friends with a popular young professor named Gabriel Harvey, who tutored them in rhetoric and poetry. Another Fellow at Trinity at that time was Philemon Holland,[21] translator of Pliny's *Natural History*.

Neither Anthony nor Francis possessed a strong bodily constitution that could stand up well to the vicissitudes of University life, and quite soon they had periods of sickness. Anthony, when only 2 years old, suffered a dangerous fever, and was continually subject to rheumatic disorders, inheriting from his father a tendency to gout and other infirmities. Throughout his life he was lame. In particular Anthony developed severe eye trouble at the age of 14, stating in later years that he had nearly lost the sight of both eyes when he was 14 years old. Whether this occurred before or after he entered Trinity College is not recorded, but we do know that the trouble persisted for some years, for a letter written by Francis in July 1574 mentions Anthony being delayed in London (presumably at York House) "by reason of sore eyes". No explanation was ever given, and they may not have known the cause, but it could have been due to severe eye strain and a deficiency of Vitamin A through prolonged and intensive study, bad lighting conditions and an insufficiently well-balanced diet or inability to properly absorb his dietary needs. As for Francis, he was delicate or sensitive all his life, and suffered from sleeplessness, "vapours" and "clouds of melancholy". He soon adopted a method for stilling his 'melancholic' and very active mind, by consuming a glass or two of home-brewed ale just before retiring to bed. I am not sure, from what he wrote, as to whether this method worked well or not for him, but as he continued in its practice I can only assume it did! After leaving University Francis's health seems to have rapidly improved, as later in his life he speaks of having had good health in his youth.

The only records known to exist of the Bacon brothers at Trinity are the account books that were kept so meticulously by John Whitgift. We have no records of their studies or the progress they made, although from Whitgift's ledgers we can readily infer that in the first six months the boys were reading Titius Livius (*i.e.* Livy's *History of Rome*), Caesar's *Commentaries,* Homer's *Iliad,* the Olynthiac Orations of Demonsthenes, the works of Aristotle and Plato, the *Orations* and *Rhetoric* of Marcus Tullius Cicero, and the *Life of Zeno* by Diogenes Laertus in Greek and Latin. We also know that they paid a seven-day visit to Redgrave during these first six months, hiring horses to do so. Redgrave was Sir Nicholas Bacon's seat in Suffolk which was lived in by the eldest son of his first marriage, Nicholas and his family. (The fine house was bequeathed to Nicholas in the Will of his father.) At Redgrave they met "Lord North", who would have been at that time Roger, the second Lord North (b.1530 - d.1600). He was the elder brother of Sir Thomas North, who was then translating Plutarch's *Lives* (publ. 1579). Lord North was Knight of the Shire of Cambridge, and High Steward of the Town of Cambridge. In 1596 he was made Treasurer of the Household, and was known to have been a patron of players. In June 1574 he was sent to France as Ambassador Extraordinary to the Court of Henri III, to congratulate the new King of France on his accession to the throne. He resided in Paris with Dr. Valentine Dale, who was the appointed English Ambassador at that time, and returned to England in November 1574.

The three boys (Anthony, Francis and Edward Tyrell) entered Cambridge just at the height of excitement at the University concerning the hot-tempered Dr. Caius, Master of Gonville and Caius College. Dr. John Caius (originally 'Kays', but his name was latinised shortly after he first went up to Cambridge as a student) was born at Norwich on October 6th, 1510. He was admitted as a student to what was then Gonville Hall, Cambridge. In 1533 he visited Italy, studying under Montanus and Andreas Vesalius at Padua University, and obtained a medical degree there. He then spent some time travelling Europe, principally in France, Italy and Germany, looking for original texts of Galen and Hippocrates. Returned to England he practised medicine in London, where he was for several years president of the College of Physicians. In 1557 he enlarged the foundation of his old college at Cambridge, renaming it Gonville and Caius, endowing it generously. In January 1559 he accepted the Mastership of the College. He was a Catholic, which was not easy for him in a period of strong anti-Catholic feeling that followed the St. Bartholomew's Day Massacre. He also had an extraordinary dislike or antipathy to Welshmen, and the ordinances of his College even went so far as to expressly exclude Welshmen from the privileges of fellowship. His temper was notably fiery.

> "Caius' relations with the society over which he ruled at Cambridge were less happy. Lying, as he did, under the suspicion of aiming at a restoration of the Catholic doctrine, he was an object of dislike to the majority of the fellows, and could with difficulty maintain his authority. He retaliated vigorously on the malcontents. He not only involved them in law-suits which emptied their slender purses, but visited them with personal castigations, and even incarcerated them in the stocks. Expulsions were frequent, not less than twenty of the fellows, according to the statement of one of their number, having suffered this extreme penalty."[22]

During these first three months (April, May, June, 1573) the choleric and unfortunate professor and the students "furiously raged together". Feelings ran so high, and the action developed so much, that the Bacons' uncle, Lord Burghley, was called upon to quell the riots and adjudicate the rights and wrongs of the case. These were the last months of Dr. Caius's life, for he died in July. The Cambridge professor was used by the Bacon brothers as a model for the hot-tempered Dr. Caius, whose pet aversion is a Welsh parson, Sir Hugh Evans, in the Shakespeare Play, *The Merry Wives of Windsor*.

*   *   *   *

It should not be thought that, in Queen Elizabeth's time, the universities in England or even on the continent were illustrious seats of learning. Far from it. Oxford and Cambridge Universities had deteriorated considerably, with little money to pay for teachers and few subjects taught — the main emphasis being on turning out an educated clergy. Few sons of the more educated and noble families were sent to university. With endless and useless argument, dispute, sloth and lechery, scholarly ignorance and arrogance abounded. The great majority of college students were "ragged clerks", labourer's sons and such like, with little previous education. As a contemporary wrote:

> "It was thought good enough for a nobleman's sons to wind their horn, carry their hawk fair and leave study and learning to mean people."[23]

The historian, Mullinger, tells us that:

> "Intermingling with a certain small minority of scions of noble houses and country squires we find the sons of poor parsons, yeomen, husbandmen, tailors, shoemakers, carpenters, innkeepers, tallow chandlers, bakers, vintners, blacksmiths, curriers, ostlers, labourers, and others, whose humble origin may be inferred from the fact that they are described merely as 'plebeians'."[24]

Except for a small minority, the students preferred not to learn — even if they could obtain a good lecturer or tutor, which was rare — but to spend their time on other matters:

> "It was only when some lecturer of more than ordinary reputation, like Albericus, appeared, that his fame, and perhaps the novelty of the subject, attracted more than one or two listeners . . . We learn, on authority which can hardly be called in question, that the schools still usually presented the same deserted aspect as in the days when Walter Haddon and Dr. Caius uttered their pathetic remonstrances and laments, that to ignore the ordinary lectures of the professors had become by this time a tradition in the college."[25]

Contemporary observers depict the universities as "abodes of discontent and brawling". A professor of that time characterised the average student in England and on the Continent as one who:

> "cares nought for wisdom, for acquirements, for the studies which dignify human life, for the Churches' weal or for politics. He is all for buffooneries, idleness, loitering, drinking, lechery, boxing, wounding, killing."[26]

Walter Travers, a Fellow of Trinity, describes the colleges as:

> " . . . the haunts of drones, the abodes of sloth and luxury [*i.e.* lasciviousness], monasteries whose inmates yawn and snore, rather than colleges of students, trees not merely sterile but diffusing a deadly miasma all around."[27]

Robert Greene, who was a student at St. John's College, Cambridge, from 1575 to 1578, and who then led a dissolute life abroad in Italy and Spain for about five years before returning to Clare Hall, Cambridge, to take a Master of Arts degree in 1583 and to marry, commented:

> "At the University of Cambridge I light among wags as lewd as myself, with whom I consumed the flower of my youth."

In one year (1570) the students of Trinity College consumed 2,250 barrels of beer. That's quite a feat when one considers that Trinity was but one college in a university numbering less than 1500 students! As for the tutors, Henry Peacham (who was at Trinity College, Cambridge, in 1593-5) averred:

"Whereas they make one schollar, they marre ten."[28]

Peacham went on to describe one 'country' tutor as whipping his boys on a cold morning "for no other purpose than to get himself a heate."[29] Whilst Giordano Bruno, who visited Oxford in 1582-4, commented that the pedantry of its scholars, their ignorance and arrogance, conjoined with the rudeness of their demeanor, would have tried the patience of Job.[30]

The full Arts course at Oxford or Cambridge, leading to a Master's degree, took seven years, but many students left after the four-year Bachelor's degree. Some practically minded middle-class students often chose to leave the university course even earlier and spend instead a year or more at one of the Inns of Court in London, thus gaining some knowledge of the law as well as of Latin and oratory. Latin then was a 'living' language — the principal language of scholars — and the main substance of university learning was to debate certain Classical philosophers or historians and numerous commentaries upon Aristotlean thought, in Latin. Hebrew and Greek had come to be neglected and despised, with few to teach it at the Universities, and other subjects were barely attempted.

"The Universities did little or nothing to instruct in natural philosophy, either for the want of men to teach, or the means to pay them."[31]

"The University giveth degrees and honours to the unlearned, and the Church is filled with ignorant ministers, being for the most part poor scholars."[32]

The actual requirements for the attainment of Master of Arts had become limited to the keeping of one or two 'Acts' and the composition of a single declaration. An 'Act' was a logistic dispute. If successful, the disputer was then said to "commence in Arts", the ceremony at which he was so admitted being called 'the Commencement'. If the candidate went on to a higher degree, he was said to "proceed". Theology and Aristotlean-style disputation were the easiest means to attain distinction, and the universities had to come to be "regarded as little more than seminaries for the education of the Clergy of the Established Church."[33]

Francis Bacon had many sober comments to make upon university life, with feeling, as well as of the state of learning generally in his time:

"In the Universities they learn nothing but to believe: first that others know that which they know not; and after themselves know that which they know not. They are like becalmed ships; they never move but by the winds of other men's truth and have no oars of their own to steer withall."

"I found myself amidst men of sharp wits, with abundance of leisure, shut up in the cells of a few authors, as their persons were shut up in College; and who, knowing little of either nature or time, do spin cobwebs of learning of no substance or profit."

It is hardly surprising that the three of them were soon to leave: Edward Tyrell in 1574, before a year was out, and Anthony and Francis by the end of 1575. But the time was hardly wasted, for their depressing experience at University now opened their eyes wide to a deep-seated problem besetting the States of Christendom in particular, and the whole world in general. Furthermore, it stirred up in them a powerful determination to do something about it.

"Capacity [*jugement*] and memory were never in any man to such a degree as in this one: so that in a very short time he made himself conversant with all the knowledge he could acquire at College."[34]

"Whilst he was commorant in the University, about 16 years of age (as his Lordship hath been pleased to impart unto myself), he first fell into the dislike of the philosophy of Aristotle: not for the worthlessness of the Author, to whom he would ever ascribe all high attributes, but for the unfruitfulness of the way, being a philosophy (as his Lordship used to say) only strong for disputations and contentions, but barren of the production of works for the benefit of the life of Man. In which mind he continued to his dying day."[35]

The Vision of Universal Reformation and Enlightenment

The first four terms at Cambridge, from Easter 1573 to the Summer 1574, whilst the 'Star of Bethlehem' still blazed away (disappearing from sight in March 1574), were critical to Francis's life and the birth of his great work for mankind. Already deeply versed in the Christian faith, in the principles of Humanistic thought, in the ideals of the Reformation, in the esoteric teachings of the Ancient (or Ageless) Wisdom: already wooed by the beauty of God's life or truth hidden in Nature and becoming manifest through her in the course of time; already inspired by Sir Nicholas's example and ideas for the advancement of learning; already imbued with philanthropic desires and ideals, he was now fired with an idea of what needed to be done — and what he could do — to procure the good of the whole wide world.

In August 1574 the plague broke out and Cambridge was closed to students for a full eight months, until the end of March 1575. We have no clear record of what Anthony and Francis did during this time, but the seed idea had now been sown and, by the time they both returned to Trinity College for the Easter term, 1575, the idea had grown into a distinct project to which the two brothers pledged themselves. They requested to be allowed to leave University, it having nothing more to teach them, and it would seem from Francis's own notes and those of his early biographers that he at least (and possibly Anthony) had put in a further request to be allowed to travel abroad, for wise and far-seeing reasons:

> "Capacity and memory were never in any man to such a degree as in this one: so that in a very short time he made himself conversant with all the knowledge he could acquire at College. And although he was then considered capable of undertaking the most important affairs yet, so that he should not fall into the usual fault of young men of his kind (who by hasty ambition often bring to the management of great affairs a mind still full of the crudities of the school), M. Bacon himself wished to acquire that knowledge which in former times made Ulysses so commendable, and earned for him the name of Wise; by the study of the manners of many different nations. I wish to state that he employed some years of his youth in travel, in order to polish his mind and to mould his opinion by intercourse with all kinds of foreigners. France, Italy, and Spain, as the most civilized nations of the whole world, were those whither his desire for knowledge carried him."[36]

James Spedding, the famous 19th century biographer of Francis Bacon, begins his *Letters and Life of Francis Bacon* with an eloquent and beautiful summary of the silent but vital months in Francis's life:

"It seemed that toward the end of the sixteenth century, men neither knew nor desired to know more than was to be learned from Aristotle; a strange thing at any time; more strange than ever just then when the heavens themselves seemed to be taking up the argument on their behalf, and by suddenly lighting up within the region of 'the unchangeable and incorruptible,' and presently extinguishing a fixed star as bright as Jupiter, to be protesting by signs and wonders against the cardinal doctrine of the Aristotelian philosophy.

"It was then that a thought struck him, the date of which deserves to be recorded, not for anything extraordinary in the thought itself, which had perhaps occurred to others before him, but for its influence upon his after life. If our study of nature be thus barren, he thought, our method of study must be wrong; might not a better method be found? The suggestion was simple and obvious, and the singularity was in the way he took hold of it. With most men such a thought would have come and gone in a passing regret . . . . But with him the gift of seeing in prophetic vision what might be and ought to be was united with the practical talent of devising means and handling minute details. He could at once imagine like a poet, and execute like a clerk of the works. Upon the conviction, This may be done, followed at once the question, How may it be done? Upon that question answered, followed the resolution to try to do it.

"Of the degrees by which the suggestion ripened into a project, the project into an undertaking, and the undertaking unfolded itself into distinct proportions and the full grandeur of its total dimensions, I can say nothing. But that the thought first occurred to him during his residence at Cambridge, therefore before he had completed his fifteenth year, we know upon the best authority — his own statement to Dr. Rawley. I believe it ought to be regarded as the most important event of his life; the event which had a greater influence than any other upon his character and future course. From that moment there was awakened within his breast the appetite which cannot be satiated, and the passion which cannot commit excess. From that moment he had a vocation which employed and stimulated all the energies of his mind, gave a value to every vacant interval of time, an interest and significance to every random thought and casual accession of knowledge; an object to live for, as wide as humanity, as immortal as the human race; an idea to live in, vast and lofty enough to fill the soul forever with religious and heroic aspirations. From that moment, though still subject to interruptions, disappointments, errors, regrets, he never could be without either work, or hope, or consolation."[37]

Throughout his whole life Francis remained true and totally dedicated to the initial idea and great vision that he received during his university days. Humbly and devotedly he always attributed it as a direct inspiration and revelation from God. For ever after he faithfully laboured to bring this vision to full birth in a difficult and antagonistic world, helped by his brother Anthony and a few others who were gradually drawn into the project. Francis gave the project several titles from time to time, such as: "The Greatest Birth of Time", "The Great Renewal of the Empire of Man over the Universe", "The Advancement and Proficience of Learning", "The Renewal of all Arts and Sciences", "The Great Instauration", "The Great Renewal of Learning", "The Universal and General

Reformation of the Whole Wide World". He himself he saw quite clearly (like an Elias or John the Baptist) as a *"buccinator novi temporis"*, "herald of a new time" — the time of the "last ages":

> "The beginning is from God: for the business which is in hand, having the character of good so strongly impressed upon it, appears manifestly to proceed from God, who is the author of good, and the Father of Lights. Now in divine operations even the smallest beginnings lead of a certainty to their end. And as it was said of spiritual things, 'The kingdom of God cometh not with observation,' so is it in all the greater works of Divine Providence; everything glides on smoothly and noiselessly, and the work is fairly going on before men are aware that it has begun. Nor should the prophecy of Daniel be forgotten, touching the last ages of the world:- 'Many shall go to and fro, and knowledge shall be increased;' clearly intimating that the thorough passage of the world (which now by so many distant voyages seems to be accomplished, or in course of accomplishment), and the advancement of the sciences, are destined by fate, that is, by Divine Providence, to meet in the same age."[38]

Let us continue in Francis's own words, written later in his life when rich experience gained from his travels and contact with the peoples of all Europe had tried, tested and proved for him the validity of his initial inspiration, argument and new method:

> "Again, in the customs and institutions of schools, academies, colleges, and similar bodies destined for the abode of learned men and the cultivation of learning, everything is found adverse to the progress of science. For the lectures and exercises there are so ordered, that to think or speculate on anything out of the common way can hardly occur to any man. And if one or two have the boldness to use any liberty of judgment, they must undertake the task all by themselves; they can have no advantage from the company of others. And if they can endure this also, they will find their industry and largeness of mind no slight hindrance to their fortune. For the studies of men in these places are confined and as it were imprisoned in the writings of certain authors, from whom if any man dissent he is straightway arraigned as a turbulent person and an innovator. But surely there is a great distinction between matters of state and the arts; for the danger from new motion and from new light is not the same. In matters of state a change even for the better is distrusted, because it unsettles what is established; these things resting on authority, consent, fame, opinion, not on demonstration. But arts and sciences should be like mines, where the noise of new works and further advances is heard on every side. But though the matter be so according to right reason, it is not so acted on in practice; and the points above mentioned in the administration and government of learning put a severe restraint upon the advancement of the sciences."[39]

> "Those who have handled sciences have been either men of experiment or men of dogmas. The men of experiment are like the ant; they only collect and use: the reasoners resemble spiders, who make cobwebs out of their own substance. But the bee takes a middle course; it gathers material from the flowers of the garden and of the field, but transforms and digests it by a power of its own. Not unlike this is the true business of philosophy: for it neither relies solely or chiefly on the powers of the mind, nor does it take the matter which it gathers from natural history and mechanical experiments and lay it up in the memory whole, as it finds it; but lays it up in the understanding altered and digested. Therefore

from a closer and purer league between these two faculties, the experimental and the rational (such as has never yet been made) much may be hoped."[40]

"We have as yet no natural philosophy that is pure; all is tainted and corrupted; in Aristotle's school by logic; in Plato's by natural theology; in the second school of Platonists, such as Proclus and others, by mathematics, which ought only to give definiteness to natural philosophy, not to generate or give it birth. From a natural philosophy pure and unmixed, better things are to be expected."[41]

"No one has yet been found so firm of mind and purpose as resolutely to compel himself to sweep away all theories and common notions, and to apply the understanding, thus made fair and even, to a fresh examination of particulars. Thus it happens that human knowledge, as we have it, is a mere medley and ill-digested mass, made up of much credulity and much accident, and also of the childish notions which we at first imbibed."[42]

Francis saw the need — the desperate need — for a complete revolution in learning, at least in the Western world of Christendom, for man is destined (through his own efforts) to know God, to become God-conscious or Life-conscious, which is a state of light or illumination. If, however, man remains in ignorance, apathy and corruptness, he is doomed of his own free-will to remain in a state of darkness, disease and death. To realise love is the starting point, but to feel love is not enough, for man has to become *conscious* of love in order to develop love and use it positively and wisely as the life force which conquers all death. Who would dare take the next gigantic step for mankind, to point out the errors of mankind's method of learning, his deficiencies and weaknesses, and not only to propose a new and better method but to actually begin the new way forward in a real, true and earnest way? The young Francis took up the challenge, offering all that he was, and was endowed with, in the service of humanity — as a servant of mankind and of God Who is the Life and Light in all things:

"Accounting myself born for the uses of mankind, and judging the case of the commonweal to be one of those things which are of public right, and like water or air lie open to all; I sought what might be the most advantage to men, and deliberated what I was most fitted for by nature. I discovered that nothing is of such estimation towards the human race, as the invention and earnest of new things and arts, by which man's life is adorned. For I perceived that, even in old times among rude men, the inventors and teachers of things rude were consecrated and chosen into the number of the gods; and I noted that the deeds of heroes who built cities, or were legislators, or exercised just authority, or subdued unjust dominations, were circumscribed by the narrowness of places and times. But the invention of things, though it be a matter of less pomp, I esteemed more adapted for universality and eternity. Yet above all, if any bring forth no particular invention, though of much utility, but kindleth a light in nature, which from the very beginning illuminates the regions of things, which lie contiguous to things already invented, afterwards being elevated lays open and brings to view all the abstrusest things; he seems to me a propogator of the empire of man over the universe, a defender of liberty, a conqueror of necessities. But I found myself constructed more for the contemplations of truth than for aught else, as having a mind sufficiently mobile for recognizing (what is most of all) the similitude of things, and sufficiently fixed and intent for observing the subtleties of differences, and possessing love of investigation, patience in doubting, pleasure in meditating, delay in asserting, facility in returning to wisdom, and neither affecting novelty, nor admiring antiquity, and hating all imposture. Wherefore I judged my nature to have a kind of familiarity and relationship with truth."[43]

Truth is the very essence of what Francis sought, and what he (in company with all the great sages of the world) knew that men should seek. If we do not seek truth, think truth, live truth, then we are nowhere near attaining our divine inheritance, the kingdom of God, which is the kingdom or state of truth, peace and all-goodness. Worse still, if we insist on imposing our own deluded concepts and wild imaginations of what we think is truth, or what *we* want truth to be, upon each other, upon society and upon nature, then not only are we not rectifying the original sin of man, by deliberately ignoring truth, but we are in fact compounding the original sin many times over and gradually forfeiting any human powers that we may have left to us. Such a mortifying existence could lead mankind (although Francis does not actually state this) on a devolutionary path, till he ends up being no better — and perhaps a good deal worse — than an animal, losing his intellectual powers that set him higher than the other creatures of nature. The Creative Word that brought man into being commanded man to behold Creation[44], to behold the Truth as it manifests itself in Nature, the Work of God. The very word 'man' means 'the thinker'. Man is created to see, to think, and thus come to understand and know Truth. Revelation is man's ultimate goal. In his beautiful language, Francis expresses this deeply felt knowledge in the introduction to his *Natural and Experimental History for the Foundation of Philosophy:*

> "Men are to be admonished, nay urged and entreated as they value their fortunes, to be lowly of mind and search for knowledge in the greater world, and to throw aside all thought of philosophy, or at least to expect but little and poor fruit from it, until an approved and careful Natural and Experimental History be prepared and constructed. For to what purpose are these brain-creations and idle displays of power? In ancient times there were philosophical doctrines in plenty; doctrines of Pythagorus, Philolaus, Xenophanes, Heraclitus, Empedocles, Parmenides, Anaxagoras, Leucippus, Democritus, Plato, Aristotle, Zeno, and others. All these invented systems of the universe, each according to his own fancy, like so many arguments of plays; and those their inventions they recited and published; whereof some were more elegant and probable, others harsh and unlikely. Nor in our age, though by reason of the institutions of schools and colleges wits are more restrained, has the practice entirely ceased; for Patricius, Telesius, Brunus, Severinus the Dane, Gilbert the Englishman, and Campanella have come upon the stage with fresh stories, neither honoured by approbation nor elegant in argument. Are we then to wonder at this, as if there would not be innumerable sects and opinions of this kind in all ages? There is not and never will be an end or limit to this; one catches at one thing, another at another; each has his favourite fancy; pure and open light there is none; every one philosophises out of the cells of his own imagination, as out of Plato's cave; the higher wits with more acuteness and felicity, the duller, less happily but with equal pertinacity. And now of late by the regulation of some learned and (as things now are) excellent men (the former variety and licence having I suppose become wearisome), the sciences are confined to certain and prescribed authors, and thus restrained are imposed upon the old and instilled into the young; so that now (to use the sarcasm of Cicero concerning Caesar's year)[45], the constellation of Lyra rises by edict, and authority is taken for truth, not truth for authority. Which kind of institution and discipline is excellent for present use, but precludes all prospect of improvement. For we copy the sin of our first parents while we suffer for it.

They wished to be like God, but their posterity wish to be even greater. For we create worlds, we direct and domineer over nature, we will have it that all things are as in our folly we think they should be, not as seems fittest to the Divine Wisdom, or as they are found to be in fact; and I know not whether we more distort the facts of nature or our own wits; but we clearly impress the stamp of our own image on the creatures and works of God, instead of carefully examining and recognising in them the stamp of the Creator himself. Wherefore our dominion over creatures is a second time forfeited, not undeservedly; and whereas after the fall of man some power over the resistance of creatures was still left to him — the power of subduing and managing them by true and solid arts — yet this too through our insolence, and because we desire to be like God and to follow the dictates of our own reason, we in great part lose. If therefore there be any humility towards the Creator, any reverence for or disposition to magnify His works, any charity for man and anxiety to relieve his sorrows and necessities, any love of truth in nature, any hatred of darkness, any desire for the purification of the understanding, we must entreat men again and again to discard, or at least set apart for a while, these volatile and preposterous philosophies, which have preferred theses to hypotheses, led experience captive, and triumphed over the works of God; and to approach with humility and veneration to unroll the volume of Creation, to linger and meditate therein, and with minds washed clean from opinions to study it in purity and integrity. For this is that sound and language which went forth into all lands[46], and did not incur the confusion of Babel; this should men study to be perfect in, and becoming again as little children condescend to take the alphabet of it into their hands, and spare no pains to search and unravel the interpretation thereof, but pursue it strenuously and persevere even unto death."[47]

In his Dedication to the King, that introduces his Latin version of *The Advancement of Learning*[48], Francis summarily identifies the three main Grail vessels or objects pertaining to the advancement of human learning or philosophy: places of learning, books of learning[49], and learned persons. Cabbalistically, the Holy Grail essence is Knowledge of Truth. Francis envisages the truly 'golden age' when men (*i.e.* men and women[50]) of learning and illumination will form a world-wide "noble and generous brotherhood", as 'sons of God' according to that paternity which is called "the Father of Lights". This illumined Brotherhood that Francis envisions and works for is none other than the Fellowship of the Holy Grail, the mythical yet real Rosicrucian Fraternity, the true Church of Christ. The following are extracts from his Dedication:

"The works or acts which pertain to the advancement of learning are conversant about three objects; the places of learning, the books of learning, and the persons of the learned. For as water, whether it be the dew of Heaven or the springs of the earth, easily scatters and loses itself in the ground, except it be collected into some receptacle where it may by union and consort comfort and sustain itself (and for that cause the industry of man has devised aqueducts, cisterns, and pools, and likewise beautified them with various ornaments, for magnificence and state as well as for use and necessity); so this excellent liquor of knowledge, whether it descend from divine inspiration or spring from human sense, would soon perish and vanish into oblivion, if it were not preserved in books, traditions, and conferences; and especially in places appointed for such matters, as universities, colleges, and schools, where it may have both a fixed habitation and means and opportunity of increasing and collecting itself.

"And first, the works which concern the places of learning are four; buildings, endowments with revenues, grants of franchises and privileges, and institutions and ordinances of government; all tending (for the most part) to retirement and quietness of life, and a release from cares and trouble; like the stations which Virgil prescribes for the hiving of honey bees.

> First for thy bees a quiet station find,
> And lodge them under cover of the wind.[51]

"The principal works touching books are two; first, libraries, which are as the shrines wherein all the relics of the ancient saints full of true virtue are preserved. Secondly, new editions of authors, with more correct impressions, more faithful translations, more profitable commentaries, more diligent annotations, and the like.

"The works pertaining to the persons of the learned (besides the advancement and countenancing of them in general) are likewise two. The remuneration and designation of lecturers in arts already extant and invented; and the remuneration and appointment of writers and inquirers concerning those parts of learning not yet sufficiently laboured or prosecuted.

"These are summarily the works and acts wherein the merits of many excellent princes and other illustrious personages towards learning have been manifested. As for the particular commemoration or any one who has deserved well of literature, I call to mind what Cicero said when, on his return from exile, he gave general thanks; "It is hard to remember all, ungrateful to pass by any."[52] Let us rather (after the advice of Scripture) look forward to that part of the race which is still to be run, than look back to that which has been passed.

"First therefore, among so many noble foundations of colleges in Europe, I find it strange that they are all dedicated to professions, and none left free to the study of arts and sciences at large. For if men judge that learning should be referred to use and action, they judge well; but it is easy in this to fall into the error pointed at in the ancient fable; in which the other parts of the body found fault with the stomach, because it neither performed the office of motion as the limbs do, nor of sense, as the head does; but yet not withstanding it is the stomach which digests and distributes the aliment to all the rest. So if any man think that Philosophy and Universality are idle and unprofitable studies, he does not consider that all arts and professions are from thence supplied with sap and strength. And this I take to be a great cause, which has so long hindered the more flourishing progress of learning; because these fundamental knowledges have been studied but in passage, and not drunk deeper of. For if you will have a tree bear more fruit than it is used to do, it is not anything you can do to the boughs, but it is the stirring of the earth, and putting richer mould about the roots, that must work it. Neither is it to be forgotten that this dedication of colleges and societies to the use only of professory learning has not only been inimical to the growth of sciences, but has also been prejudicial to states and governments. For hence it proceeds that princes when they have to choose men for business of state find a wonderful dearth of able men around them; because there is no collegiate education designed for these purposes, where men naturally so disposed and affected might (besides other arts) give themselves especially to histories, modern languages, books of policy and civil discourse; whereby they might come better prepared and instructed to offices of state.

"And because founders of Colleges do plant, and founders of Lectures do water, I must next speak of the deficiencies which I find in public lectures; wherein I especially disapprove of the smallness of the salary assigned to lecturers in arts and professions, particularly amongst ourselves. For it is very necessary to the progression of sciences that lecturers in every sort be of the most able and sufficient men; as those who are ordained not for transitory use, but for keeping up the race and succession of knowledge from age to age. This cannot be, except their condition and endowment be such that the most eminent professors may be well contented and willing to spend their whole life in that function and attendance, without caring for practice. And therefore if you will have sciences flourish, you must observe David's military law; which was, "That those who stayed with the baggage should have equal part with those who were in the action;"[53] else will the baggage be ill attended. So lecturers in sciences are as it were the keepers and guardians of the whole store and provision of learning, whence the active and militant part of the sciences is furnished; and therefore they ought to have equal entertainment and profit with men of active life. Otherwise if the fathers in sciences be not amply and handsomely maintained, it will come to pass, as Virgil says of horses, —

Et partrum invalidi referent jejunia nati:[54]

the poor keeping of parents will be seen in the weakliness of the children."[55]

"It is a general custom (and yet I hold it to be an error) that scholars come too soon and too unripe to the study of logic and rhetoric, arts fitter for graduates than children and novices; for these two rightly taken are the gravest of sciences, being the arts of arts, the one for judgment, the other for ornament; besides they give the rule and direction how both to set forth and illustrate the subject matter. And therefore for minds empty and ignorant (and which have not yet gathered what Cicero calls 'stuff' or 'furniture', that is matter and variety) to begin with those arts (as if one should learn to weigh or to measure or to paint the wind), works but this effect, that the virtue and faculty of those arts (which are great and universal) are almost made contemptible, and either degenerate into childish sophistry and ridiculous affectation, or at least lose not a little of their reputation. And further, the premature and untimely learning of these arts has drawn on, by consequence, the superficial and unprofitable teaching and handling of them, — a manner of teaching suited to the capacity of children. Another instance of an error which has long prevailed in universities is this: that they make too great and mischievous a divorce between invention and memory. For most of the speeches there are either entirely premeditate, and delivered in preconceived words, where nothing is left to invention; or merely extempore, where little is left to memory; whereas in common life and action there is little use of either of these separately, but rather of intermixtures of them; that is of notes or commentaries and extempore speech; and thus the exercise fits not the practice, nor the image the life. But it must ever be observed as a rule in exercises, that they be made to represent in everything (as near as may be) the real actions of life; for otherwise they will pervert the motions and faculties of the mind, and not prepare them. The truth whereof appears clearly enough when scholars come to the practice of their professions, or other offices of civil life; which when they set into, this want I speak of is soon found out by themselves, but still sooner by others. But this part, touching the amendment of the Institutions and Orders of Universities, I will conclude with a sentence taken from one of Caesar's letters to Oppius and Balbus; 'How this may be done, some means occur to me, and many may be found; I beg you therefore to take these matters into consideration.'[56]

"Another defect which I note ascends a little higher than the preceding. For as the progress of learning consists not a little in the wise ordering and institutions of each several university; so it would be yet much more advanced if there were a closer connexion and relationship between all the different universities of Europe than now there is. For we see there are many orders and societies which, though they be divided under distant sovereignties and territories, yet enter into and maintain among themselves a kind of contract and fraternity, insomuch that they have governors (both provincial and general) whom they all obey. And surely as nature creates brotherhood in families, and arts mechanical contract brotherhoods in societies, and the anointment of God superinduces a brotherhood in kings and bishops, and vows and regulations make a brotherhood in religious orders; so in like manner there cannot but be a noble and generous brotherhood contracted among men by learning and illumination, seeing that God himself is called 'the Father of Lights.'[57]"[58]

In his *Introduction to the Great Instauration,* "Francis of Verulam reasoned thus with himself, and judged it to be for the interest of the present and future generations that they should be made acquainted with his thoughts":

"Being convinced that the human intellect makes its own difficulties, not using the true helps which are at man's disposal sound and judiciously; whence follows manifold ignorance of things, and by reason of that ignorance mischiefs innumerable; he thought all trial should be made, whether that commerce between the mind of man and the nature of things, which is more precious than anything on earth, or at least than anything that is of the earth, might by any means be restored to its perfect and original condition, or if that may not be, yet reduced to a better condition than that in which it now is. Now that the errors which have hitherto prevailed, and which will prevail for ever, should (if the mind be left to go its own way), either by the natural force of the understanding or by help of the aids and instruments of Logic, one by one correct themselves, was a thing not to be hoped for: because the primary notions of things which the mind readily and passively imbibes, stores up, and accumulates (and it is from them that all the rest flow) are false, confused, and overhastily abstracted from the facts; nor are the secondary and subsequent notions less arbitary and inconstant; whence it follows that the entire fabric of human reason which we employ in the inquisition of nature, is badly put together and built up, and like some magnificent structure without any foundation. For while men are occupied in admiring and applauding the false powers of the mind, they pass by and throw away those true powers, which, if it be supplied with the proper aids and can itself be content to wait upon nature instead of vainly affecting to overrule her, are within its reach. There was but one course left, therefore, — to try the whole thing anew upon a better plan, and to commence a total reconstruction of sciences, arts, and all human knowledge, raised upon the proper foundations. And this, though in the project and undertaking it may seem a thing infinite and beyond the powers of man, yet when it comes to be dealt with it will be found sound and sober, more so than what has been done hitherto. For of this there is some issue; whereas in what is now done in the matter of science there is only a whirling round about, and perpetual agitation, ending where it began. And although he was well aware how solitary an enterprise it is, and how hard a thing to win faith and credit for, nevertheless he was resolved not to abandon either it or himself; not to be deterred from trying and entering upon that one path which is alone open to the human mind. For better it is to make a beginning of that which may lead to something, than to engage in a perpetual struggle and pursuit in courses which have no exit. And certainly the two ways of contemplation are much like those two ways of action, so much celebrated, in this — that the one, arduous and difficult in the beginning, leads out at last into the open country; while the other, seeming at first sight easy and free from obstruction, leads to pathless and precipitous places."[59]

Between the new philosophy, which Francis is beginning or regenerating in the world (and which the world is still a long way from understanding fully, let alone practising in its complete sense), and the old philosophy, there is a whole world of difference. To differentiate what is meant by the old and the new philosophy (like the Old and the New Jerusalem) Francis employs two descriptive titles, *Anticipation of the Mind* and *Interpretation of Nature* respectively, which should be self-explanatory to most thinking minds. The Interpretation of Nature by which we discover God's hidden laws, or Truth, is the path of the initiates, the "true sons of knowledge", sons of Light. Nature, or God's Work, is but the Veil which simultaneously conceals and reveals the formless vibratory essence that is Truth. Without the Veil of Nature we could not discover or see Truth, yet Nature is not Truth itself. Nature is but the *maya*[60] or thought-illusion created as a dream in the Imagination of God, seemingly real and tangible to us since we are living in it as part of the picture and with a measure of free will and creativity. The development of our conscious knowledge depends on the dream-picture, the world of imagination; yet all that we deem solid and substantial is but polarised energy held together by the energising Will of God, patterned and directed according to the Divine Idea. How many souls yet comprehend this truth, taught to us century after century by the illumined souls of East and West? The 'Baconian Light' is teaching it quietly, and gently leading mankind on a large scale to discover the truth steadily, surely, progressively and scientifically. The 'Revolution' is proceeding apace, step by step. One day Science and Religion will be remarried. Science will become religious, and Religion will be scientific once more. Ignorance and superstition will be forever banished to the realms of Yesterday.

"Be it remembered then that I am far from wishing to interfere with the philosophy which now flourishes, or with any other philosophy more correct and complete than this which has been or may hereafter be propounded. For I do not object to the use of this received philosophy, or others like it, for supplying matter for disputations or ornaments for discourses, — for the professor's lecture and for the business of life. Nay more, I declare openly that for these uses the philosophy which I bring forward will not be much available. It does not lie in the way. It cannot be caught up in passage. It does not flatter the understanding by conformity with preconceived notions. Nor will it come down to the apprehension of the vulgar except by its utility and effects.

"Let there be therefore (and may it be for the benefit of both) two streams and two dispensations of knowledge; and in like manner two tribes or kindreds of students of philosophy — tribes not hostile or alien to each other, but bound together by mutual services; — let there in short be one method for the cultivation, another for the invention, of knowledge.

"And for those who prefer the former, either from hurry or from considerations of business or for want of mental power to take in and embrace the other (which must needs be most men's case), I wish that they may succeed to their desire in what they are about, and obtain what they are pursuing. But if any man there be who, not content to rest in and use the knowledge which has already been discovered, aspires to penetrate further; to overcome, not an adversary in argument, but nature in action; to seek, not pretty and probable conjectures, but certain and demonstrable knowledge; — I invite all such to join themselves, as true sons of knowledge, with me, that passing by the outer courts of nature, which numbers have trodden, we may find a way at length into her inner chambers. And to make my meaning clearer and to familiarise the thing by giving it a name, I have chosen to call one of these methods or ways *Anticipation of the Mind*, the other *Interpretation of Nature*.

"Moreover I have one request to make. I have on my own part made it my care and study that the things which I shall propound should not only be true, but should also be presented to men's minds, how strangely soever preoccupied and obstructed, in a manner not harsh or unpleasant. It is but reasonable however (especially in so great a restoration of learning and knowledge) that I should claim of men one favour in return; which is this; If any one would form an opinion or judgment either out of his own observation, or out of the crowd of authorities, or out of the forms of demonstration (which have now acquired a sanction like that of judicial laws), concerning these speculations of mine, let him not hope that he can do it in passage or by the by; but let him examine the thing thoroughly; let him make some little trial for himself of the way which I describe and lay out; let him familiarise his thoughts with that subtlety of nature to which experience bears witness; let him correct by seasonable patience and due delay the depraved and deep-rooted habits of his mind; and when all this is done and he has begun to be his own master, let him (if he will) use his own judgement."[61]

Nowadays the Western world is being permeated with Yoga teachings and techniques from the East, but in Bacon's day a true and soundly-based spiritual science was no longer known or practised, and the religious and philosophical attempts to discover and practise truth (which is what Christ and Christianity means) that were made by the courageous few had little effect on the life of the vast majority. Francis Bacon was inspired with a method and scheme for remedying the stagnant and polluted condition of humanity on a world-wide scale, and he had the ability to begin the great philanthropic process which is in motion now and will move inexorably on step by step towards its goal in many centuries hence. It is only because of the increasing revolution in thinking and learning, that has taken place since Francis Bacon set it in motion, that the Western world — and indeed more than just the Western world — is able to consider seriously and receive with joy on a massive scale such an advanced spiritual-science as Yoga, and to begin to develop its own spiritual sciences. *Yoga* means 'union' — conscious union — with God. This is the precise goal of the Great Instauration and, just as all the different types or paths of Yoga can be combined to form one supreme or complete Yoga, so the scheme of the Great Instauration combines, in its various stages, everything relevant to attain the fullest knowledge of or conscious union with God, who is Life and Truth. The Great Instauration *is* Yoga, precisely calibrated and designed to lead *all* mankind up the ladder of attainment, one step at a time, carefully and surely. Eventually the "sons of science" or "sons of Wisdom" (science means 'knowledge', and wis-dom means 'knowledge of the Lord'), who are thus "sons of Yoga", will be living world-wide on this planet, loving truth, understanding truth and serving truth. Only by feeling, thinking and practising (or living) truth does one achieve real KNOWLEDGE of truth.[62] Knowledge is more than just understanding something. Knowledge depends on experience: it is a state of BEING. This is the purpose and goal of all Ancient Wisdom traditions. This is the purpose and goal of Francis's Great Instauration. The philosophy that Francis Bacon gave to the Western world began the great Occidental revolution in thinking and learning, combining thought with action for the purpose of producing a useful and good result: yet the full extent and essence of his philosophy has barely been glimpsed yet by mankind, and the New Method that he gave by which the Great Instauration could be achieved has not yet been fully understood, let alone practised! This is yet to happen; and it will only happen when the purpose of science is seen to be to search for Truth, the divine Good, and to practise it. Meanwhile a foundation has been laid.[63] But the true Temple — the Temple of Wisdom — has yet to be properly built.

> "I confess that I have as vast contemplative ends as I have moderate civil ends; for I have taken *all* knowledge to be my province. This, whether it be curiosity or vain-glory, or, if one may take it favourably, *philanthropia,* is so fixed in my mind that it cannot be removed."[64]

> "The essential form of knowledge . . . is nothing but a representation of truth: for the truth of being and the truth of knowing are one, differing no more than the direct beam and the beam reflected."[65]

> "The mind is the man, and the knowledge is the mind. A man is but what he knoweth."[66]

> "For he (Solomon) saith expressly, the glory of God is to conceal a thing, but the glory of a king is to find it out[67]; as if according to that innocent and affectionate play of children, the Divine Majesty took delight to hide his works, to the end to have them found out; and as if kings could not obtain greater honour than to be God's play-fellows in that game; specially considering the great command they have of wits and means, whereby the investigation of all things may be perfected."[68]

> "Truth, which only doth judge itself, teacheth that the enquiry of truth, which is the love-making or wooing of it, the knowledge of truth, which is the presence of it, and the belief of truth, which is the enjoying of it, is the sovereign good of human nature."[69]

> "For Knowledges are as pyramids, whereof History is the basis. So of Natural Philosophy, the basis is Natural History; the stage next the basis is Physique; the stage next the vertical point is Metaphysique. As for the vertical point[70] . . . the Summary Law of Nature, we know not whether man's enquiry can attain unto it."[71]

> "But we are not dedicating or building any Capitol or Pyramid to human Pride, but found a holy Temple in the human Intellect, on the model of the Universe . . . For whatever is worthy of Existence is worthy of Knowledge — which is the Image (or Echo) of Existence."[72]

> "For we are founding in the human Intellect a true pattern of the Universe; such as it is actually found to be, not such as any one's own reason may have suggested it to him. But this cannot be completed, save after a most diligent

dissection and anatomy of the world. We declare that those foolish models, and as it were apish imitations of worlds, which men's fancies have erected in their systems of Philosophy, are to be scattered to the winds. And so let men know, as we said above, what a difference there is between the Phantoms of the human Mind, and the Ideas of the Divine Mind. The former are nothing but fanciful abstractions; the latter are the true signatures of the Creator upon creatures, as they are impressed and limited in matter by real and exquisite lines. And so the chief things of all are, in this kind, Truth and Usefulness; and effects themselves are to be accounted of more worth, in so far as they are pledges of truth, than as they give the comforts of life."[73]

The purpose of knowledge is to discover Truth. Man is the Thinker, and ultimately the Knower — knowing Truth or Good, which is God. God, the Truth, is the Principle of Goodness. The manifestation of Truth or Good is Goodness, which is also perfect Usefulness, called 'Charity' in the Christian gospels. Good, the Principle of Goodness, is the All-Wise Love. This is the Truth that man seeks to know. The Orphic teachings, which underlie Christianity and Humanism, state that:

"All existence comes from one immeasurable Good Principle, the One Cause."[74]

"God is revealed as an eternally abiding Good, an ever-flowing fountain of Truth and Law, an omnipotent Unity, an omniscient Reality."[75]

"Love is the eldest and noblest and mightiest of the gods, and the chiefest Author and Giver of virtue in life and of happiness after death."[76]

"God is the Truth and Reality which sustains the universe. This Truth animates all things; it is the Spiritual Principle in all of life."[77]

The young Francis achieved a profound revelation of Truth whilst at University, and he dedicated his life to the attainment of further and further knowledge of this Truth, which he also realised was inextricably linked with helping others to attain this state of consciousness and being also. He recognised, as few do, that active labour in service to all life was a fundamental part of becoming divinely conscious; that contemplation and action must go hand in hand — the first to direct the second, and the second to perfect the first. Together they form a living entity.

Truth, which is Love, is All-Wise. Again as the Orphics taught, Love is the luminous principle of Light; Light (or Wisdom) is the radiance of Love. This, in Christian terminology, is what is meant by Christ. Christ is Truth — the principle of Divine Consciousness.[78] Christ is Love — the principle of Life. To know Truth is to become divinely conscious and in a state of perfect love. This Love is the life force or power that moves matter. It is the Divine Emotion or Passion, the emotive force of the Universe. This is Love, which is Truth; and it is eternally All-Wise. Its Wisdom orders matter and imprints itself on matter like the seal imprints itself on molten wax, creating the form that manifests the Divine Love-Wisdom; and it is Goodness or Charity which is the manifest form of Truth. Let us continue again in Francis's words:

"Usefulness is what makes gods and men great."

"Goodness the habit, answers to the Theological virtue Charity. This of all virtues and dignities of the mind is the greatest, being the character of the Deity."

"The perfection of the human form consists in approaching the Divine or Angelic Nature."[79]

"A man doth vainly boast of loving God whom he never saw if he love not his Brother whom he hath seen . . . "[80]

"But this is that which will dignify and exalt knowledge: if contemplation and action be more nearly and straitly conjoined and united together than they have been; a conjuction like unto that of the highest planets, Saturn, the planet of rest and contemplation, and Jupiter, the planet of civil society and action."[81]

"The Understanding of Man and his Will are twins by birth as it were, for the purity of illumination and the liberty of will began together. Nor is there in the Universal Nature of things so intimate a sympathy as that of Truth and Goodness."[82]

"Let men know . . . what a difference there is between the phantoms of the human mind, and the Ideas of the Divine Mind. The former are nothing but fanciful abstractions; the latter are the true signatures of the Creator upon creatures, as they are impressed and limited in matter by real and exquisite lines. And so the chief things of all are, in this kind, TRUTH and USEFULNESS; and effects themselves are to be accounted of more worth in so far as they are pledges of truth, than as they give the comforts of life."[83]

"The unlearned man knows not what it is to descend into himself, or to call himself to account . . . the good parts he hath he will learn to show to the full, and use them dexteriously, but not much to increase them: the faults he hath he will learn how to hide and colour them, but not much to amend them: like an ill mower, that mows on still and never whets his scythe. Whereas with the learned man it fares otherwise, that he doth ever intermix the correction and amendment of his mind with the use and employment thereof. Nay, further and in sum, certain is it that *Veritas* and *Bonitas* differ but as the seal and the print: for **Truth prints Goodness . . .** "[84]

"Let no man upon a weak conceit of sobriety or an ill-applied moderation think or maintain that a man can search too far, or be too well studied in the Book of God's Word, or in the Book of God's Works — Divinity or Philosophy. But rather, let men endeavour an endless progress or proficience in both; only let men beware that they apply both to CHARITY, and not to swelling; to USE, and not to ostentation; and again that they do not unwisely mingle or confound those learnings together."[85]

" . . . 'The spirit of man is as the lamp of God, wherewith he searcheth the inwardness of all secrets.'[86] If then such be the capacity and receipt of the mind of man, it is manifest that there is no danger at all in the proportion or quantity of knowledge, how large soever, lest it should make it swell or out-compass itself; no, but it is merely the quality of knowledge, which, be it in quantity more or less, hath in it some nature of venom or malignity, and some effects of that venom, which is ventosity or swelling. This corrective spice, the mixture whereof maketh Knowledge so sovereign, is Charity, which the Apostle immediately addeth to the former clause: for so he saith, 'Knowledge bloweth up, but **Charity buildeth up'**[87]; not unlike that which he delivereth in another place; 'If I spake', saith he, 'with the tongues of men and angels, and had not charity, it were but as a tinkling cymbal';[88] not but that it is an excellent thing to speak with the tongues of men and angels, but because, if it be severed from charity, and not referred to the good of men and mankind, it hath rather a sounding and unworthy glory, than a meriting and substantial virtue.'"[89]

" . . . charity, which is excellently called the bond of perfection[90], because it comprehendeth and fasteneth all virtues together . . . so certainly, if a man's mind be truly inflamed with charity, it doth work him suddenly into a greater perfection than all the doctrine of morality can do, which is but a sophist in comparison of the other. Nay, further, as Xenophon observed truly, that all other affections, though they raise the mind, yet they do it by distorting and uncomeliness of ecstasies or excesses: but only love doth exalt the mind, and nevertheless at the same instant doth settle and compose it; so in all other excellencies, though they advance nature, yet they are subject to excess; only charity admitteth no excess. For so we see, aspiring to be like God in power, the angels transgressed and fell; 'I will ascend above the heights of the clouds; I will be like the Most High'[91]: by aspiring to be like God in knowledge, man transgressed and fell; 'ye shall be as gods, knowing good and evil'[92]; but by aspiring to a similitude of God in goodness or love, neither man nor angel ever transgressed, or shall transgress. For unto that imitation we are called; 'Love your enemies, do good to them which hate you, bless them that curse you, and pray for them which despitefully use you, and persecute you; that ye may be the children of your Father which is in heaven: for he maketh his sun to rise on the evil and on the good, and sendeth rain on the just and the unjust'[93]."[94]

Christianity has derived its name from *Christos,* a Greek word for 'Divine Consciousness' which is also Divine Love, as Divine Consciousness is a state of Love. This *is* Divine Being. "I AM the way, the Truth and the Life"[95] is a simple, beautiful statement of the *Way* of service, the *Truth* of understanding, and the *Life* of loving. To love, to understand, to serve — these three are one Beingness, and they correspond to the Christian gospel terms of faith, hope and charity respectively. The Orphics had yet another word for the Christ, and this was *Eros*, the Omniscient Love that motivates (*i.e.* animates) and gives form to matter. The Romans used their Latin word, *Cupid.* Francis is able to convey the most esoteric of Christian teachings to mankind in a manner that is veiled by the use of either very ancient or entirely new words that are able to transcend time and human 'religious' reaction. Because of his deliberate 'veiling' of the esoteric truths behind Christianity, Francis was able to both avoid being burnt as a heretic and also to imitate the Divine Majesty by creating a game of hide-and-seek as an essential part of his New Method. The idea is so simple and so delightful — yet it has fooled the vast majority of students for four centuries! Other terms that Francis used to describe Christ, the Truth, are *"the Summary Law of Nature"* and *"the work which God worketh from the beginning to the end."*

"The principles, fountains, causes, and forms of motions, that is, the appetites and passions of every kind of matter, are the proper objects of philosophy . . ."[96]

" . . . It was not the pure knowledge and nature of universality, a knowledge by the light whereof man did give names unto other creatures in paradise, as they were brought before him, according unto their proprieties, which gave the occassion to the fall; but it was the proud knowledge of good and evil, with an intent in man to give law unto himself, and to depend no more upon God's commandments, which was the form of the temptation. Neither is it any quantity of knowledge, how great soever, that can make the mind of man to swell; for nothing can fill, much less extend the soul of man, but God and the contemplation of God; and therefore Salomon, speaking of the two principal senses of inquisition, the eye and the ear, affirmeth that the eye is never satisfied with seeing, nor the ear with hearing; and if there be no fulness, then is the continent greater than the content: so of knowledge itself, and the mind of man, whereto the senses are but reporters, he defineth likewise in these words . . . : 'God hath made all things beautiful, or decent, in the true return of their seasons: Also he hath placed the world in man's heart, yet cannot man find out the work which God worketh from the beginning to the end': declaring not obscurely, that God hath framed the mind of man as a mirror or glass, capable of the image of the universal world, and joyful to receive light; and not only delighted in beholding the variety of things and vicissitude of times, but raised also to find out and discern the ordinances and decrees, which throughout all those changes are infallibly observed."[97]

"The accounts given by the poets of Cupid, or Love, are not properly applicable to the same person; yet the discrepancy is such that one may see where the confusion is and where the similitude, and reject the one and receive the other.

"They say then that Love was the most ancient of all the gods; the most ancient therefore of all things whatever, except Chaos, which is said to have been coeval with him; and Chaos is never distinguished by the ancients with divine honour or the name of a god. This Love is introduced without any parent at all; only, that some say he was an egg of Night. And himself out of Chaos begot all things, the gods included. The attributes which are assigned to him are in number four; he is always an infant; he is blind; he is naked; he is an archer. There was also another Love, the youngest of all the gods, son of Venus, to whom the attributes of the elder are transferred, and whom in a way they suit.

"The fable relates to the cradle and infancy of nature, and pierces deep. This Love I understand to be the appetite or instinct of primal matter; or to speak more plainly, the *natural motion of the atom;* which is indeed the original and unique force that constitutes and fashions all things out of matter. Now this is entirely without parent; that is, without cause. For the cause is as it were parent of the effect; and of this virtue there can be no cause in nature (God always excepted): there being nothing before it, therefore no efficient; nor anything more original in nature, therefore neither kind nor form. Whatever it be therefore, it is a thing positive and inexplicable. And even if it were possible to know the method and process of it, yet to know it by way of cause is not possible; it being, next to God, the cause of causes — itself without cause. That the method even of its operation should ever be brought within the range and comprehension of human inquiry, is hardly perhaps to be hoped; with good reason therefore it is represented as an egg hatched by night. Such certainly is the judgment of the sacred philosopher, when he says, *He hath made all things beautiful according to their seasons; also he hath submitted the world to man's inquiry, yet so that man cannot find out the work which God worketh from the beginning to the end.* For the summary law of nature, that impulse of desire impressed by God upon the primary particles of matter which makes them come together, and which by repetition and multiplication produces all the variety of nature, is a thing which mortal thought may glance at, but can hardly take in.

"Now the philosophy of the Greeks, which in investigating the material principles of things is careful and acute, in inquiring the principles of motion, wherein lies all vigour of operation, is negligent and languid; and on the point now in question seems to be altogether blind and babbling; for that opinion of the Peripatetics which refers the original impulse of matter to privation, is little more than words — a name for the thing rather than a description of it. And those who refer it to God, though they are quite right in that, yet they ascend by a leap and not by steps. For beyond all doubt there is a single and summary law in which nature centres and which is subject and subordinate to God; the same in fact which in the text just quoted is meant by the words, *The work which God worketh from the beginning to the end.*[98]"[99]

"The stories told by the ancients concerning Cupid, or Love, cannot all apply to the same person; and indeed they themselves make mention of two Cupids, very widely differing from one another; one being said to be the oldest, the other the youngest of the gods. It is of the elder that I am now going to speak. They say then that this Love was the most ancient of all the gods, and therefore of all things else, except Chaos, which they hold to be coeval with him. He is without any parent of his own; but himself united with Chaos begat the gods and all things. By some however it is reported that he came of an egg that was laid by Nox. Various attributes are assigned to him: as that he is always an infant, blind, naked, winged, and an archer. But his principal and peculiar power is exercised in uniting bodies; the keys likewise of the air, earth, and sea were entrusted to him. Another younger Cupid, the son of Venus, is also spoken of, to whom the attributes of the elder are transferred, and many added of his own.

"This fable, with the following one respecting Coelum, seems to set forth in the small compass of a parable a doctrine concerning the principles of things and the origins of the world, not differing in much from the philosophy which Democritus held, excepting that it appears to be somewhat more severe, sober, and pure . . .

"This Chaos then, which was contemporary with Cupid, signified the rude mass or congregation of matter. But matter itself, and the force and nature thereof, the principles of things in short, were shadowed in Cupid himself. He is introduced without a parent, that is to say, without a cause; for the cause is as the parent of the effect; and it is a familiar and almost continual figure of speech to denote cause and effect as parent and child. Now of this primary matter and the proper virtue and action thereof there can be no cause in nature (for we always except God), for nothing was before it. Therefore there was no efficient cause of it, nor anything more original in nature; consequently neither genus nor form. Wherefore whatsoever this matter and its power and operation be, it is a thing positive and inexplicable, and must be taken absolutely as it is found, and not to be judged by any previous conception. For if the manner could be known, yet it cannot be known by cause, seeing that next to God it is the cause of causes, itself only without a cause. For there is a true and certain limit of causes in nature; and it is as unskilful and superficial a part to require or imagine a cause when we come to the ultimate force and positive law of nature, as not to look for a cause in things subordinate. And hence Cupid is represented by the ancient sages in the parable as without a parent, that is to say, without a cause, — an observation of no small significance; nay, I know not whether it be not the greatest thing of all. For nothing has corrupted philosophy so much as this seeking after the parents of Cupid; that is, that philosophers have not taken the principles of things as they are found in nature, and accepted them as a positive doctrine, resting on the faith of experience; but they have rather deduced them from the laws of disputation, the petty conclusions of logic and mathematics, common motions, and such wanderings of the mind beyond the limits of nature. Therefore a philosopher should be continually reminding himself that Cupid has no parents, lest his understanding turn aside to unrealities; because the human mind runs off in these universal conceptions, abuses both itself and the nature of things, and struggling towards that which is far off, falls back on that which is close at hand. For since the mind, by reason of its narrowness, is commonly most moved by things of familiar occurrence and which may enter and strike it directly and at once, it comes to pass that when it has advanced to those things which are most universal in experience, and yet cannot be content to rest in them, that then, as if striving after things still more original, it turns to those by which itself has been most affected or ensnared, and fancies these to be more causative and demonstrative than those universals themselves.

"It has been said that the primitive essence, force and desire of things has no cause. How it proceeded, having no cause, is now to be considered. Now the manner is itself also very obscure; and of this we are warned by the parable, where Cupid is elegantly feigned to come of an egg which was laid by Nox. Certainly the divine philosopher declares that 'God hath made everything beautiful in its season, also he hath given the world to their disputes; yet so that man cannot find out the work that God worketh from the beginning to the end.'[100] For the summary law of being and nature, which penetrates and runs through the vicissitudes of things (the same which is described in the phrase, 'the work which God worketh from the beginning to the end'), that is, the force implanted by God in these first particles, from the multiplication whereof all the variety of things proceeds and is made up, is a thing which the thoughts of man may offer at but can hardly take in. . .

"But one who philosophises rightly and in order, should dissect nature and not abstract her (but they who will not dissect are obliged to abstract): and must by all means consider the first matter as united to the first form, and likewise to the first principle of motion as it is found. For the abstraction of motion also has begotten an infinite number of fancies about souls, lives, and the like; as if these were not satisfied by matter and form, but depended on principles of their own. But these three are by no means to be separated, only distinguished; and matter (whatever it is) must be held to be so adorned, furnished, and formed, that all virtue, essence, action, and natural motion, may be the consequence and emanation thereof."[101]

*　　*　　*　　*

I hope that I have been able to convey something to you, the reader, of this immense and glorious idea that Francis had conceived before he was 15 years old and which he then put into effect, knowing that if the vision was true then it must reach its eventual conclusion and fulfillment, even if it took many generations to do so. Before moving on to the next chapter I shall conclude this one with a further quotation from the first chapter of Francis's *A Description of the Intellectual Globe*. In this he shows clearly how he perceives the three main divisions of human learning, corresponding to the three faculties of the understanding (*i.e.* the capacity of our mind or soul). It is quite clear that Francis must have perceived these three divisions from the start; then, having proved the truth of them to his own satisfaction by experiment, he employed the knowledge and the fact of their existence in his philosophical scheme. If you study them carefully you will see that Francis is adopting the three-fold division of the soul expounded by the Humanistic, Neo-Platonic teachers, which we have already mentioned in connection with the Melancholy humour; *viz.* Memory (corresponding to the Intellect), Reason and Imagination. For those who do not understand why Memory and Intellect are synonymous, and comprise the highest faculty of the soul, then it is worth pointing out, as indeed Francis and others before him did, that all knowledge is but a remembrance — a re-membering of the Divine Consciousness that was divided and, as far as man is concerned, lost with the Fall of man. The 'Last Supper' is a "feast of Remembrance". This Memory is connected with the very spine of man, and with his actions or life experiences.

Although Imagination, Reason and Memory are faculties of the human understanding, yet they have a correspondence with the three faculties of Being — Love, Understanding and Service (or Desire, Thought and Action) respectively. Francis used all three in his life's work to launch the Great Instauration. That is to say, he dealt with (and embodied) Love, Understanding and Service as the overall framework or 'Law' of his scheme. Then, in respect of Understanding, he dealt with (and used) the three faculties of Imagination, Reason and Memory; and to do this he employed the corresponding parts of human learning — Poesy, Philosophy and History. In addition he played the divine game of 'hide-and-seek' and deliberately veiled (or masked) a large proportion of his work, openly revealing only a small part to the public view. How expressly necessary this veiling or secrecy would prove to be was soon to be thrust home upon the 14-year-old Francis in a devastating way, circumscribing with bands of iron the rest of his life. Seen from the vantage point of time, this was clearly the wise operation of Divine Law acting upon the human destiny of Francis, ensuring that he carried out the Plan to minute perfection; but to Francis at the time it must have been a difficult burden to bear.

> "I adopt that division of human learning which corresponds to the three faculties of the understanding. Its parts therefore are three; History, Poesy, and Philosophy. History is referred to the Memory; poesy to the Imagination; philosophy to the Reason. And by poesy here I mean nothing else than feigned history. History is properly concerned with individuals; the impressions whereof are the first and most ancient guests of the human mind, and are as the primary material of knowledge. With these individuals and this material the human mind perpetually exercises itself, and sometimes sports. For as all knowledge is the exercise and work of the mind, so poesy may be regarded as its sport. In philosophy the mind is bound to things; in poesy it is released from that bond, and wanders forth, and feigns what it pleases. That this is so any one may see, who seeks ever so simply and without subtlety into the origins of intellectual impressions. For the images of individuals are received by the sense and fixed in the memory. They pass into the memory whole, just as they present themselves. Then the mind recalls and reviews them, and (which is its proper office) compounds and divides the parts of which they consist. For the several individuals have something in common with one another, and again something different and manifold. Now this composition and division is either according to

the pleasure of the mind, or according to the nature of things as it exists in fact. If it be according to the pleasure of the mind, and these parts are arbitrarily transposed into the likeness of some individual, it is the work of the imagination; which not being bound by any law and necessity of nature or matter, may join things which are never found together in nature and separate things which in nature are never found apart; being nevertheless confined therein to these primary parts of individuals. For of things that have been in no part objects of the sense, there can be no imagination, not even a dream. If on the other hand these same parts of individuals are compounded and divided according to the evidence of things, and as they really show themselves in nature, or at least appear to each man's comprehension to show themselves, this is the office of reason; and all business of this kind is assigned to reason. And hence it is evident that from these three fountains flow these three emanations, History, Poesy, and Philosophy; and that there cannot be other or more than these. For under philosophy I include all arts and sciences, and in a word whatever has been from the occurrence of individual objects collected and digested by the mind into general notions. Nor do I think that there is need of any other division than this for Theology. For the information of revelation and of sense differ no doubt both in matter and in the manner of entrance and conveyance; but yet the human spirit is one and the same; and it is but as if different liquors were poured through different funnels into one and the same vessel. Therefore I say that Theology itself likewise consists either of sacred history, or of divine precepts and doctrines, as a kind of perennial philosophy. And that part which seems to fall outside this division (that is, prophecy) is itself a species of history, with the prerogative of divinity wherein times are joined together, that the narrative may precede the fact; and the manner of delivery, both of prophecies by means of visions and of divine doctrine by parables, partakes of poesy."[102]

Early Elizabethan Chivalry and Pageantry

During the winter of 1574, on the 16th December, Sir Nicolas Bacon signed an agreement with James Paget of Grove Place, Southampton, in which Paget's daughter, Dowsabell, was betrothed in marriage to Anthony. James Paget was the son of a wealthy London merchant. He was also an alderman and, at one time, Sheriff of the City of London. He was thrice married, each wife being an heiress and 'well connected'. Dowsabell was the only child by his second wife. When the agreement was made Dowsabell was 14 or 15 years old, Anthony was 16. The marriage was planned to take place in May 1575. The young couple were to have an allowance of £75 for the next three years, and Sir Nicholas promised £100 *per annum* plus various leases, together with a further lump sump of £500. James Paget on his part offered an estate of £314 per annum after his death and that of his wife. Paget was so certain and proud of his future son-in-law that he went so far as to place a heraldic shield, combining the arms of Cooke of Gidea Hall and Bacon of Redgrave, in the gallery of his house, Grove Place, alongside the heraldic shields of other distinguished families to whom he was allied by marriage.

The marriage never took place. No explanation is recorded either for it being contracted in the first place or for it miscarrying, but the most probable likelihood is that Anthony and Dowsabell met, took more than a passing fancy to each other, and that it seemed pleasing and serious enough to the respective parents to warrant drawing up a marriage agreement. There is no evidence that Sir Nicholas ever forced his sons into marriages that they did not themselves agree to, and it is patently obvious that neither Anthony nor Dowsabell was ever held to the agreement. The agreement miscarried probably as a result of the young lovers falling out of love, or realising as the fateful day approached that their fancy for each other and their plans for the future did not really stretch far enough for marriage with each other. The other possibility is that Francis's great vision had by now been worked out as a project with the help of Anthony, or was in the process of being worked out, and therefore Anthony was too fully committed to this grand philanthropic adventure to consider settling down with a wife, family and fairly mundane job. All the time and energy of both brothers was needed if the project was to get off the ground. In any case, whatever the reason, the marriage was a non-event.

By the Spring, 1575, the plague had died down and both Anthony and Francis returned to Cambridge for the Easter term. This was to be the last term at university for Francis. When they came down for the summer vacation they were involved in Court pageantry of some importance, that took place during the Summer Progress of the Queen and her

Court. As university fledglings and as sons of the Lord Keeper, both youths were almost certainly included in the Royal Progress. Francis definitely was, and it would appear that not only did he write down a record of his experiences at these and other pageants that he witnessed, but that he may have contributed to them in a major way.

The pageantry of the Elizabethan Court was extremely important in the life of those who governed and guided the course of England's destiny, and in the building up of the theological, political and cultural position of Protestant England. One of the principal factors in the establishment of the Church of England had been the claim not only to a Church foundation independant to that of Rome, but also of its precedence (in terms of time) to the Roman Church. The roots of the Reformation in England lie as far back as William the Conqueror, who replied to the insistent autocracy of the Roman Pope:

> "Fealty I have never willed to do nor will I do it now. I have never promised, nor do I find that my predecessors did it to yours."[103]

The actual English Reformation itself was begun by Wycliffe, 136 years earlier than Luther, with the assistance of the great reformer Edward III, who took the first steps towards the abolition of papal jurisdiction over the British national Church. Edward III, with King Arthur as his ideal, carried out Richard I's cherished scheme for refounding the old British 'Order of St. George and the Round Table'. King Edward established St. George as the patron saint of England and set up 'the Order of St. George and the Garter', having for its purpose "Good Fellowship". Arthur had been of direct lineage of the earlier British Kings, and a direct descendant of Joseph of Arimathea's daughter, Anna, who had married a Welsh prince of that royal Silurian line. (Joseph of Arimathea, under instruction from Jesus's Apostles and appointed by St. Philip as Apostle to Britain, had set up the first 'Christian' Church in Britain in A.D. 37). From this marriage were born a son who became Beli the Great and a daughter, Pernadim, who married the Cornish King, Llyr Llandaff (King Lear). From these were descended the British (and Christian) High Kings, Cunobelinus (Cymbeline) who was also known as Bran the Blessed, Caradoc (Caractacus), Cyllinus, Coel (King Cole), Lucius (the Great Luminary), Cadwallader, the Princess Helena who married the Emperor Constantius and whose son was the Emperor Constantine the Great, Ambrosius Aurelianus (Prince of the Sanctuary), King Arthur, and the whole royal House of Tudor. Caradoc and his children were those who were captured and taken to Rome by the Emperor Claudius (A.D. 52), where they were given their freedom and a palace to live in, and established the first organised Christian Church in Rome at their *Palatium Britannicum* which also became the most fashionable and cultural centre in Rome, known as the "Home of the Apostles". St. Paul lived at this home for eight years. In A.D. 45 the British Arviragus (High King) Caradoc had been offered and accepted the hand in marriage of Venus Julia, daughter of the Roman Emperor Claudius, in an attempt to come to terms. The youngest daughter of Caradoc and Venus Julia was actually 'adopted' as a daughter by the Emperor Claudius and given the name Claudia. She married a Roman senator, Rufus Pudens, a relative of St. Paul. The second, priestly son of Caradoc, Prince Linus[104], was consecrated as the first Bishop of Rome by St. Paul in A.D. 58. St. Peter, who had first visited Rome in A.D. 44, and who periodically returned to stay with his British friends in Rome, affirmed this fact:

> "The First Christian Church above ground in Rome was the Palace of the British. The First Christian Bishop, was a Briton, Linus, son of a Royal King, personally appointed by St. Paul, A.D. 58."[105]

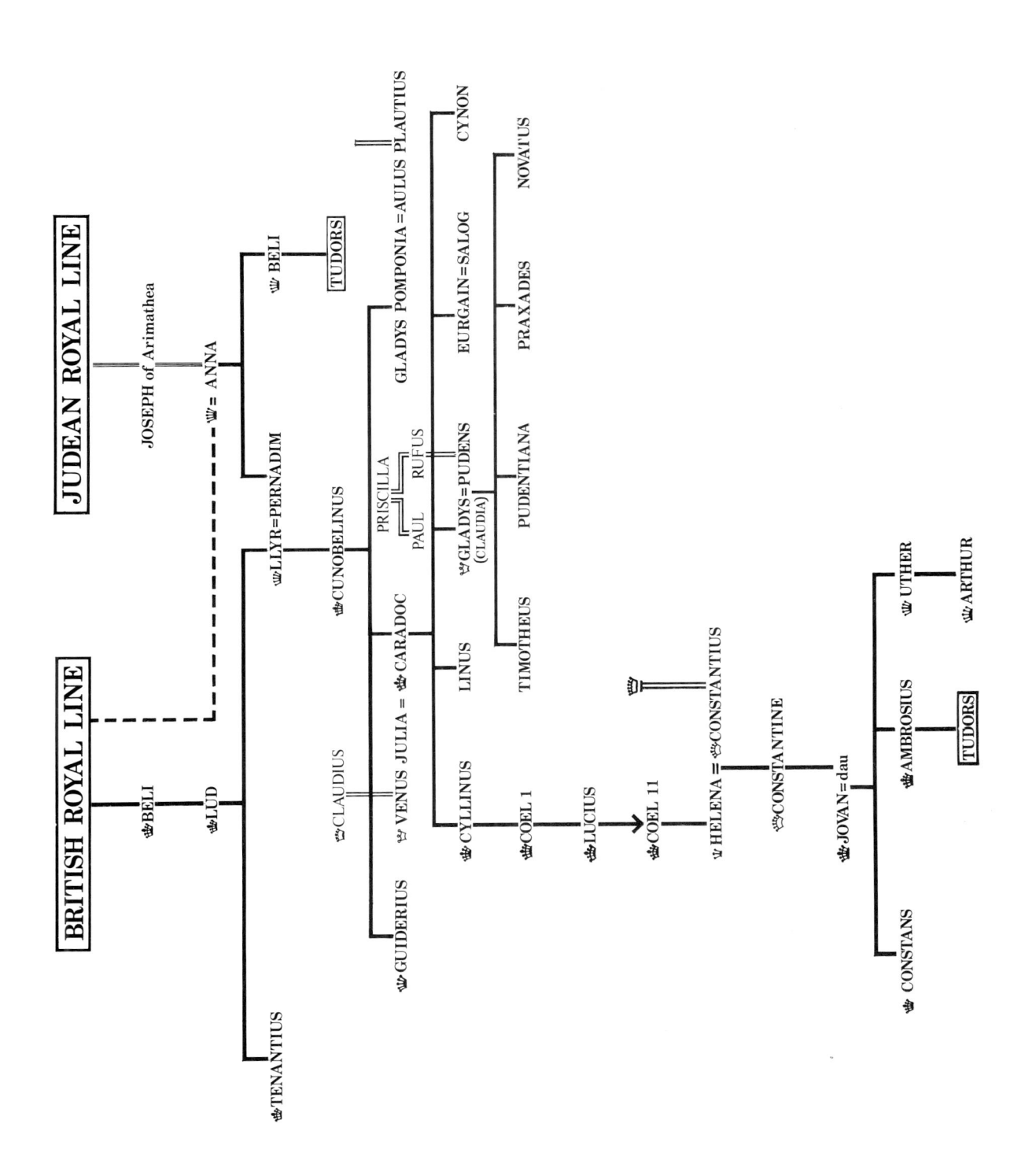

Diagram G: THE ROYAL LINE OF BRITISH KINGS — HOUSE OF JUDAH

> "Concerning those Bishops who have been ordained in our lifetime, we make known to you that they are these; Of Antoich, Eudius, ordained by me, Peter; Of the Church of Rome, Linus, brother of Claudia, was first ordained by Paul, and after Linus's death, Clemens, the second ordained by me, Peter."[106]

Clemens, or Clemens Romanus, the second Bishop of Rome, ordained by St. Peter, was the St. Clement who was one of the original band of Twelve sent to Britain with Joseph of Arimathea by St. Philip in A.D. 36. It should be noted that Apostles are not Bishops: their role and jurisdiction is of a much higher order than Bishops. Apostles travel the world, with no 'fixed' abode, and are responsible for the teaching and welfare of people on a large scale. They have the 'Christly' consciousness, power and authority to recognise and consecrate other Apostles if needed for particular purposes, such as St. Joseph, Apostle to Britain (consecrated by St. Philip), and St. Paul, Apostle to the Gentiles[107] of Asia Minor and Aegea (consecrated by St. Peter), and to ordain Bishops, appointing them jurisdiction over a certain locality and its people: *i.e.* the Bishop of Rome is responsible for and has jurisdiction over the place and people of Rome. Despite later Roman Catholic claims, neither St. Peter nor St. Paul were ever Bishops of Rome; their work and responsibility was far greater than that. Neither were they responsible for the way the Church of Rome went when it was corrupted by power politics in the 6th century onwards.[108] Except for Claudia, all of her royal family living in Rome plus thousands of other Christians were martyred during the persecutions instigated by the Emperor Nero, and the early Christian Church in Rome was virtually wiped out.

However, Caradoc returned to Britain with some of his sons and his father, Bran the Blessed, well before the Nero persecutions began. Caradoc was a good friend of Joseph of Arimathea, and in A.D. 36 had given Joseph the land of Glastonbury on which to establish his principal Christian centre. Before he became the 'Arviragus' or High King of the Britons, Caradoc was one of the three 'Princes' and king of the Silures of South-West Britain. He was elected 'Pendragon' (Military Leader) of the Britons and led the British armies against the Romans during their mammoth invasion of Britain (A.D. 42-52) in which the Romans hoped to stamp out the seat of Druidism and Christianity. Caradoc's father was the High King Cunobelinus (Cymbeline), who eventually abdicated his throne in favour of Caradoc during the military campaign, so that he could devote himself entirely to the Christian faith as an Arch-druid (Arch-bishop). Cunobelinus became known as 'Bran the Blessed' ('the Blessed King'). Neither Cunobelinus nor Caradoc were unknown to the Romans, both having spent some time in Rome at the Imperial Court in previous years. They were highly cultured and judicial. The Emperor Claudius so respected his valiant 'enemy' that he gave his daughter, Venus Julia, to Caradoc in marriage.[109]

Caradoc was the first to bear the title "Defender of the Faith", given to him by St. Joseph of Arimathea together with the 'Long Cross' to display on his coat of arms. The 'Long Cross' was the 'Red' or 'Rose Cross', displayed on a white (or silver) ground, and it was used as the badge of the elected Christian King. Later in history it became associated with St. George, when Constantine the Great used it to commemorate his friend and martyr, St. George of Lydda, whom Constantine made "the Champion Knight of Christendom". Under Constantine the Red Cross became the public symbol of Christianity everywhere. (The Crucifix was not adopted until the end of the 7th century.) All Christian Kings of Britain adopted the Red Cross as their symbol, together with the title "Defender of the Faith" — a title which echoed the old Druidic motto and war-cry, "The Truth against the World". Since the time of Caradoc the Red Cross flag has always been the Christian flag of the British Church. With the association of St. George's name with the flag, Britain adopted the name and the saint as its own in Richard I's time[110], the name 'St. George's Cross' being an alternative or substitute name for the 'Red Cross' or 'Rose Cross'.

These matters are important when dealing with the motives and beliefs of the British people, the establishment of the Christian Church in the world, and the causes of the Reformation. Druidism accepted Christianity with ease, seeing in it a natural fulfilment of their own teachings. The Druidic colleges, vibrant centres of religion and learning, soon became Christian colleges that sent out missionaries all over the known world, founding Bishoprics in numerous cities of importance. It was these missions and their teachings that alarmed Rome and precipitated the massive Roman invasions of Britain in an attempt to destroy what the Romans saw as the main centre of Christian teaching and evangelism. The Romans were unable to defeat the Britons entirely or stop the Christian missions, despite decades of war and persecution (A.D. 43-86), and eventually a treaty was concluded that incorporated the British as allies of the Roman Empire, recognising most of the British native freedoms and kingly prerogatives. The peace that ensued lasted until the Diocletian persecution, c. A.D. 300. In A.D. 137, St. Timotheus, son of Claudia Pudens, returned to Britain to baptize his nephew, the High King Lleurug Mawr (Lucius) in the Chalice Well at Glastonbury, consecrating him in turn as "Defender of the Faith".[111] A new wave of Christian fervour and evangelism began in Britain with this public act of consecration, and in A.D. 155, at the National Council held at Caer Winton (Winchester),[112] with the consent of the people and by Act of the Gorsedd ('Parliament'), Lleurug Mawr officially proclaimed Christianity and the Christian Chruch to be the national Faith and Church of Britain. Because of his exemplary life and outstanding achievements he became known as Lleuver Mawr, 'the Great Light', which in Latin translates as 'Lucius'. He was also named 'the Most Religious King' — a title held by every British sovereign since. Lucius had his coinage stamped with the sign of the Cross. Not until the time of his descendant, Constantine the Great, was Christianity proclaimed the national Faith of any other country besides Britain.

> "Christianity was privately confessed elsewhere, but the first nation that proclaimed it as their religion, and called itself Christian, after the name of Christ, was Britain."[113]

St. Augustine, sent over to Britain on a mission to the unconverted Saxons by the Bishop of Rome in A.D. 597, at the time when the new title of Supreme Pontiff was being foisted onto the Roman Bishop, tried to coerce the Bishops of the British Church into accepting the authority of the Bishop of Rome. Their reply to Augustine was:

> "We have nothing to do with Rome. We know nothing of the Bishop of Rome in his new character of Pope. We are the British Church, the Archbishop of which is accountable to God alone, having no superior on earth."

Francis Bacon, writing later on this subject, said:

> "The Britons told Augustine they would not be subject to him, nor let him pervert the ancient laws of their Church. This was their resolution, and they were as good as their word, for they maintained the liberty of their Church five hundred years after this time, and were the last of all Churches in Europe that gave up their power to the Roman Beast, and in the person of Henry VIII, that came of their blood by Owen Tudor, the first that took that power away again."[114]

When Augustine came, the Anglo-Saxons occupied a large part of Britain. Augustine and his monks of the Roman Church evangelised the Anglo-Saxons from the south-east, whilst the missionaries of the British Church were mostly effective in the western and northern Anglo-Saxon kingdoms. The Anglo-Saxon Kings, however, tended to prefer the arguments of Augustine to those of the British Bishops, and through the Anglo-Saxons (*i.e.* the English) the Roman Catholic Church and its Papacy became established in England. Augustine even moved the Principal Archbishop's See, that had been set up in Caer Troia (London) at St. Peter's, Cornhill, by King Lucius, to Dorobernia (Canterbury), where it still remains.[115]

A century before Augustine, the Angles, Saxons and Jutes were still striving to conquer and colonise Britain, in the wake of its dissolution into an unorganised multitude of small kingdoms ruled by petty tyrants — a mere shadow of the original British-Druidic State of Britain with its hierarchy of 40 kingdoms and tribes ruled by 40 *Rigs* (uncrowned Kings) and Chief Druids, over whom ruled 3 *Cunos* (crowned Princes) and Arch-Druids, all of whom were overlorded by the elected *Arviragus* (the supreme crowned King, or High King) who combined the offices of priest and king within his one being. At the end of the 6th century a royal prince, Ambrosius Aurelianus (A.D. 516-570), said to be a grandson of Constantine the Great, stepped in and was elected *Arviragus*. He managed to largely unite the warring *Rigs* and restore to some extent the ailing Christian Church which was suffering under the petty tyranny, the invasions and the Pelagian heresy, yet which was still the centre of learning. He defeated Hengist and the invading tribes sufficiently to inaugurate 40 years or so of relative peace, which were kept peaceful due to the efforts of the elected *Pendragon* or War-leader, Arthur. It is not known whether Arthur was actually a king (*i.e. Rig),* but it seems the most likely if a Pendragon was going to be able to command other kings in a war emergency. In fact he was probably a *Rig* by right of birth, then created a *Cunos* or crowned Prince by the High King. In legend he is said to have been born the son of Uthyr Pendragon through the magic of Merlin. Merlin, or *Myrrdin,* meaning "Wise Man", was a title given to Ambrosius (*i.e.* Merlin Ambrosius), who was also known as "the Prince of the Sanctuary" because of his holiness and wisdom. His brother was Uthyr (Uther) Pendragon, and it is quite possible and in keeping with Ancient British practice that Uthyr's son Arthur was adopted by Ambrosius, raised and initiated by him into the Mysteries, groomed for princehood, became a *Rig* on his father's death and finally crowned as a *Cunos* before being elected *Pendragon* in his father's footsteps. The other possibility is that Uthyr Pendragon was himself 'Arthur', and that the story of his magical birth is simply an initiation story concerning the birth of his 'higher self' or true soul, his 'Golden child' that was given the Romanised name, Arthur.

Arthur (or Uther) Pendragon founded an Order of Chivalry based on the traditional pattern adopted by Jesus and his disciples — the Zodiacal Round Table (*i.e.* Christ, the 'Sun', surrounded by 12 Apostles or 'Signs'). Each Christian community was established on this same pattern, each Apostle representing the Christ and presiding over one or more 'Round Tables' of disciples, with a Bishop as chief or head of each Twelve (and being one of those Twelve). With the ideal and example of St. George brought into the Christian symbology, it became possible to translate the priestly office to that of a knight for those of the fighting, chivalrous ilk. Thus the Holy Grail of the priest-priestess could become, at another level of expression, the Holy Grail of the knights and their ladies. This was no new idea, but a revival of ancient traditions and systems of initiation. Arthur called his Order of Chivalry, "The Order or Society of St. George and the Round Table", and he adopted (or was allowed to adopt) the Cross of St. George (*i.e.* the Rose Cross) as his personal banner. Windsor Castle[116] was his principal residence and the centre of the Order — an Order established to protect and defend the Faith, proclaiming "The Truth against the World". Arthur also revived the ancient British tradition of 'tournament', the knights jousting against each other in the setting of fine pageantry, and the chivalric 'quest' for adventure and accomplishment of a good deed.

Even with the 'Romanisation' of England through the spread of Roman Catholicism, the original British Church never died out, and it guarded its mysteries — the esoteric mysteries of Christianity — against the perversion of the Roman Catholic hierarchy until such time as the Roman dominance could be shaken off and the real Truth revealed publicly. Richard I vowed to refound the old British 'Arthurian' Order, and he promised to make any knight who could successfully scale the walls of Jerusalem during the Crusades a 'Companion of St. George'. His Knight Companions were to be distinguished by wearing light-blue thongs like a garter around their legs, bound tightly on or just below the knee, to remind the knight never to flee. This insignia was previously used by the Druids, and its deeper meaning was to represent unity amongst a fellowship who never fled from living, proclaiming or defending the Truth. (The British Pendragon was described as an "elected man of the garter, leader of your ranks".) Richard instituted the battle-cry, "For St. George", esoterically meaning the same as the Druidic battle-cry. His knights carried his personal banner with his device of the Red Cross on a White Field — the flag which has now become the British Ensign. Richard altered England's Patron Saint from being the gentle, pious and learned Edward the Confessor, to be St. George — a fighting, chivalrous "Defender of the Faith", more in keeping with the mood of his day and more possible of attainment by the many.

Richard I died before he could fittingly found his Order of the Round Table, but successive Kings of England implanted the ideal more firmly into the young English nation. Richard II ordered that every soldier should be distinguished by wearing the Red Cross of St. George, and from his time until the 16th century the Red Cross was borne as a badge over the armour of each English soldier. Edward I renewed the ancient tournaments which had fallen into abeyance since Arthur's time and, with 100 knights at the Round Table gathering at Kenilworth Castle, he revived the Arthurian glories.

Edward III made the next big step, and he actually managed to carry out Richard I's cherished scheme. He refounded 'The Order of St. George and the Round Table', renamed 'The Most Noble Order of St. George and the Garter', as "a society, fellowship and college of knights", having for its purpose "Good Fellowship". The Knights of this Order were also known as "The Knights of the Blue Garter". The Order originally included ladies, who were known as *"dames de la confraternitè de St. George"*. There were 26 Knight Companions, including the King and his Queen (*i.e.* 13 x 2 in total), epitomising the Christ presiding over his Zodiac in such a way as to show the male-female aspects of each part. They were to show fidelity and friendliness towards another. The light-blue garter represented Amity and Courage; the gold collar, having 26 garter links encircling 26 roses of Sharon tied together with gold knots, represented the bond of Faith, Peace and Unity; the purple robe indicated that the Knight was the equal of Kings because of his courage, piety and devotion to Truth. The Patrons of the Order were the most Holy Trinity, the Virgin and St. George. The Royal Chapel of St. Edward the Confessor in Windsor Castle was consecrated on 6th August, 1348, as the Chapel of the Knights of the Order of St. George and the Garter, and rededicated to the Virgin Mary, St. George and St. Edward. King Edward further obtained a papal Bull from Pope Clement VI (in 1351) declaring the Chapel free of the jurisdiction of the Archbishop of Canterbury and the Bishop of Salisbury (in whose diocese Windsor then lay). The Sovereign, who was Head of the Order, and the Bishop of Winchester, the Prelate, nominated the Dean and the 12 Canons of the Chapel, with appeal to the visitor, the Lord Chancellor. The Royal Chapel and its College was the first religious foundation in England free of the control of Abbot, Prior, Archbishop or Bishop, and it was in this sense the very foundation-stone of the Reformation, being a "peculiar" and "royal free chapel".

But the complete freeing of the English national Church from papal jurisdiction and the Roman Catholic hierarchy was only able to take place 200 years later, during Henry VIII's reign. When the Tudors finally came to the Throne of England they brought with them the original British Church, together with all the symbology, chivalry, pageantry and history associated with it. It is absolutely crucial to an understanding of the Reformation, the Tudors and Francis Bacon's own scheme, to realise that the religious and chivalrous elements of British Christianity are inextricably woven together, and that the ancient idea of a 'High King' who is head of both Church and State, as a Priest-King, is an essential part of the whole religious-political structure. Moreover, the Arthurian-type chivalry and pageantry formed a vital part of it all, with its deeply significant 'heraldic' symbology and the profound learning which lay behind it — a true British heritage revived, or restored, by the Tudors.

But the restoration of the learning, the symbolism and the pageantry took time to accomplish, and it was not until the entrance of Francis Bacon that they all truly began to be restored in any fullness. As we have seen (in the previous chapter) learning was still in a bad state in Elizabeth I's reign — yet it had advanced considerably from the ignorance of the 14th century when even the Bible could not be read by the vast majority of both clergy and laity. But, with the invention of printing and the efforts made at translations, the Bible began to become accessible to all people, which in turn caused the public to query the doctrines and style of life of the Roman Catholic clergy. On Elizabeth's accession, when the Archbishop of York, Nicholas Heath, (who refused to take the Oath of Supremacy), exhorted Elizabeth to follow the Pope, she replied bravely:

> "I will answer you in the words of Joshua. As Joshua said of himself and his, 'I and my realm will serve the Lord.' My sister[119] could not bind the realm, nor bind those who should come after her, to submit to a usurped authority. I take those who maintain here the Bishop of Rome and his ambitious pretensions to be enemies to God and to me."[118]

The Reformation was established, though not perfected, during the reign of Elizabeth. The yoke of Rome, unwittingly accepted and imposed on all England by King Oswy in A.D. 664, was at last thrown aside and the "cleansing of the Sanctuary" had begun.

Edward III had established St. George's Day (April 23rd) as the day of the Feast of the Order of the Garter at Windsor, in 1344. In 1415 St. George's Day was made a major public double Feast and was ordained to be observed like Christmas Day. Both Henry VII and Henry VIII were zealous supporters of these traditions and largely revived them, but under Edward VI certain statutes were promulgated severing the link between the Saint and the Noble Order of the Garter. The Feast was transferred to Whitsun, the original Feastday of Arthur's Knights. Mary I restored the connection, and the festival continued to be observed on St. George's Day. Elizabeth I, however, was advised to order (in 1567) the discontinuance of the public festival, reserving it only for the Order of the Garter, as it was thought to be incompatible with the reformed religion. Later on, Elizabeth had a substitute day for the public 'Arthurian' festivities — her Accession Day Anniversary on November 17th each year, together with the Twelve Days of Christmas festivities for the Court, both festivals celebrating the Queen as the Virgin or Fairy Queen, Bearer of the Holy Grail, "Defender of the Faith, *Etc.*"[119] But St. George's Day continued to be kept as a Feast for the Knights of the Garter, first at Windsor, then at Whitehall (from 1567), and the ceremonies developed into a grand public spectacle by the 1590's. Any Knight who had to be absent from Court on that day was bound to hold a corresponding service and ceremony on St. George's Day wherever he might be, and thus many thousands came to behold the Garter spectacle.

The Queen and her Knights, as a legend realised (or Arthur and the Round Table returned), was the major focal point and ideology of the whole reign. It only remained for Francis, as the Queen's natural son, fathered on her by her chief Knight, to complete the story of the Holy Grail — which he did.[120]

To be a Knight of the Garter was deemed to be the highest honour possible, and only those of the highest rank and position were admitted to the Order. It was greatly sought after not only by high-ranking Englishmen but also by foreign princes and dukes, and it came to form an important part of major political alliances. It even united, temporarily, Catholics and Protestants in Garter services. Knighthood was never bestowed lightly. The symbolic image of St. George slaying the Dragon became universally accepted by both Catholics and Protestants as a sacred emblem or hieroglyph, as it was at the time of Constantine the Great. Although interpreted in detail in many different ways, yet the general meaning was unmistakable and unanimously recognised: "Good conquers Evil".

The imagery and symbology of the Order of the Garter promotes the ancient idea of a Supreme or High King, who is not only a king ('a prince of State') but also a high priest ('a prince of the Sanctuary'), embodying both religious and secular rulership in one being. This is the ancient ideal of all great traditions and cultures; and we find it, for instance, not only in the Arviragus of Britain, but also in the Pharoah of Egypt, the Inca of South America, the Emperor or Sun King of both Near and Far East, the King of Israel (David and Solomon), the Augustus Caesar of Rome, and in Melchizedek, the Great King and High Priest of Righteousness and Peace — an epithet for the Christ. The very essence of Christianity is that the Christ is both Supreme King and High Priest all in one, and that all true Christians are called to become the same, rising by degrees to the exalted state of Christhood, of the Order of Melchizedek. Around this Great King are arrayed, like a zodiac, his principal officers. Futhermore, this Zodiacal pattern is extended through four 'manifestations of being', represented in society by the four divisions or 'castes', *i.e.* (1) the priestly or spiritually enlightened, (2) the chivalrous or warrior-ruler, (3) the farmer and business man, and (4) the peasant or labourer caste.[121] With Jesus and his Apostles this Order was manifested in its highest caste — the 'head' of the Universal Man — and the early Christian disciples such as Joseph of Arimathea repeated the pattern within the priestly context. With Arthur, the Pendragon, the Order was manifested in the second highest caste, that of the warrior or knight; and it is from this 'caste' that the Knights of the Round Table are derived, with St. George, the Rose Cross Knight, acting as the ideal of the Knight who serves his King in his aspect of "Defender of the Faith".

In Constantine the Great, first Emperor of the Holy Roman Empire and descendant of both the Roman and British royal lines, this full ideal was revived and extended throughout Christendom. In his grand- or great-grandchildren, Ambrosius Aurelianus and Arthur, the ideal was again revived and used in the face of harsh opposition. In the descendants of these famous 'Emperors' — the Tudors — the ancient ideal was again restored, as an essential and integral part of the Restoration. Thus, immediately Elizabeth Tudor came to the Throne of England she was declared the heir and successor of Constantine and likened to him, defending the true Faith and overthrowing the Antichrist — the Antichrist being in this case the usurpation by the Roman Bishop or Pope of a power and authority not his own, together with the falsified history and the appalling abuse of the usurped powers by the papal authority and hierarchy which has had such terrible and far-reaching effects for Christianity as a whole. The symbol of St. George and the Dragon represented for many this apocalyptic act of Reformation and Revelation of the Truth.

One of the key writings of Elizabeth's reign in connection with this fundamental principle of the Reformation and the 'Divine Right of Kings' to rule over both Church and State was the

(11) INITIAL 'C' — woodcut, from first page of John Foxe's *Acts and Monuments* (1563).

Apology for the Church of England, written in 1560 in Latin by Sir Nicholas Bacon's good friend, John Jewel, Bishop of Salisbury, and translated into English by Lady Anne Bacon for general publication in 1564. Jewel was the official apologist for the Church of England and he followed up his *Apology* with the *Defence* (1567) of the *Apology of the Church of England* in which he provided more detail and support of his official thesis, referring readers to Marsilo of Padua, Ockham, Dante[122], Petrarch[123], Pico della Mirandola[124], Baccace, Mantuan, Valla, Chaucer, Raymond of Toulouse, Wyckiff, and many others. In John Foxe's *Acts and Monuments* (popularly known as 'Foxe's Book of Martyrs'), of which the first English edition was published in 1563, the analogy of Elizabeth with Constantine was still further emphasised[125] and a 'universal history' of humanity provided, set within the framework of divine revelation and cabbalistic numerology — a history culminating with the triumph of Light in the accession of Elizabeth, restorer of the pure and true religion.

Not only was Elizabeth seen as the successor of Constantine and other Priest-Kings, but also as Astraea, the Virgin Queen, whose return to the world of mortals is destined to overthrow vice and misery, and usher in a golden age of virtue, justice, peace and plenty. Such a golden age is traditionally an age ruled by Saturn, the god-name given to the divine principle of Peace, Knowledge and Justice. Cabbalistically Saturn has two aspects — male and female, solar and lunar — the one being creative and the other being receptive and formative. The 'male' aspect is the *Logos* of creative Wisdom, the Invisible Light of God, whilst the 'female' aspect is the *Sophia* of receptive and reflective Intelligence that can receive, reflect upon, visualise and understand the radiant Idea or holy Wisdom. The comprehension and eventual knowledge of the creative Idea is the light that is born or reflected in the Intelligence like a star shining in the heavenly darkness, the 'Child' of the Mother-Father.[126] The words 'Christ' and 'Christhood' refer to the two aspects of Light — Divine Wisdom and Human Knowledge respectively — whilst the 'Golden Age' means an era of enlightenment or knowledge of God, which is also synonymous with peace, mercy and justice (or righteousness)[127], of which the Christly Melchizedek is King.[128] In classical Greek mythology the two 'parental' aspects were represented by Apollo and Pallas Athena, the principles of Wisdom and Intelligence; their hermaphrodite son (sun), Esclepius, being the principle of Knowledge or Revealed Truth. The Roman terminology for these three was Apollo, Minerva and Aesculapius; whilst the British terminology was Hu ('the Light'), Britannia ('the Chosen of Anna' — *i.e.* the Virgin Mary) and Ies-Hu (or Jesu, 'the embodied Light' or 'revealed Light'). The 'Mother' aspect in Hebrew tradition was called *Sabbaoth*, which is derived from the word for the day of rest or peace — the 'Sabbath' — which is Saturn's (or Satan's) day. The whole truth concerning Satan, 'the Adversary', (who was also created as the Crown of God's Creation), and the redemption of Satan from darkness to light, is the secret of divine Intelligence, pierced by the rays of Wisdom shining from the divine Heart and being transformed by the processes of life from the darkness of ignorance to the light of knowledge — a mystery that the symbol of St. Michael and the Dragon, or St. George and the Dragon, portrays.

Astraea, the Virgin Queen, is the *Sabbaoth* of the Jewish Mysteries, the Divine Mother (or Virgin Mary in Christian terminology) — the pure state of peace, harmony and intelligence in which (or from which) light can be born or manifested. As Sovereign of England, the focus of the nation, Elizabeth was thus seen to represent and summarise the potential and ideal state of the realm — peaceful, intelligent and understanding — in which the true light of Christ could be born and made manifest. The additional factors that (a) the land of Britain has from the most ancient times been especially revered as the land of the Virgin Mother or Queen, and (b) that the particular time-span of the Renaissance, culminating in the Reformation in England, corresponded to the 'Virgo' period of the whole Age of Pisces (see F.B.R.T. Journal I/2, *The Virgin Ideal*), and (c) that the supernova had appeared in the Sign of Cassiopiea, made the reign of Queen Elizabeth particularly significant.

This whole symbol of Astraea-Virgo- Minerva- Athena- Mary, the Virgin or 'Faerie' Queen, was made the focal point of the nation and built up to immense proportions by the poet-knights of the realm, not least of all by Francis Bacon, the poet-prince of all learning (or knowledge) and son of the Queen.

Elizabeth seems to have been acknowledged as a personification of Astraea, the Virgin Queen, from the moment of her accession, and it was an image acceptable to both Protestant and Catholic, for varying reasons. This regal image trod the path of the middle way, passing between and harmonising two extremes: the *"Mediocria Firma"* of Sir Nicholas Bacon. From the start of the reign it became the popular custom to enthusiastically ring the town and church bells, have special services and sermons, light bonfires and make merry with feasting on the Queen's Accession Day Anniversary. The natural need for regular celebration that had once found an outlet in the Roman Catholic feast-days, now suppressed, was channelled into a feast-day celebrating the accession of the Faerie Queen to the Throne of Merrie England and to other events honouring the Queen. The Queen's Accession Day (November 17th) became, by popular demand and custom, a national festival. It is said that Oxford University and City precipitated the initially modest celebrations into something grander, in the year 1570, the celebrations having gained momentum in the wake of the Duke of Alva's Spanish invasion of the Netherlands in 1567, the attempted Northern rebellion of 1569 and the papal excommunication of Elizabeth in 1570. These and subsequent crises, together with events like the supernova, helped to generate a popular and fervent cult of Elizabeth and ensured that the Accession Day celebrations, as Holland wrote, "flowed by a voluntary current over all this realm."[129] No government legislation was needed, no decree was issued to make Elizabeth's Accession Day a national, if not *the* national festival of the year. Throughout the realm civic and guild authorities organised spectacular processions, ceremonies and pageants in honour of the Queen and the Virgin Ideal; whilst poems, prayers, hymns, ballads and various literary tributes and tracts were written on the theme.

November 17th, once celebrating St. Hugh of Lincoln, amongst others, and being reasonably close to All Saints and All Souls Day (1st and 2nd November) which were substituted for the original Festival of Peace, or Celtic *Samhain* (the major festival celebrating the end of one year or cycle and the beginning of a new one), was easily assimilated by the new Reformed or Anglican Church as an official feast day. Special service books, containing psalms, prayers and readings giving thanks to God for the reign of the Queen who had delivered the English people "from danger of war and oppression, restoring peace and true religion, with liberty both of bodies and minds", were soon to be issued each year (the first such book was published in 1576), and in 1571 the Church Convocation ordered that a copy of John Foxe's *Acts and Monuments* should be installed alongside the Bible in every cathedral and parish church, and in the chapels of noblemen's houses, so that any person might come and read either book freely. The *Acts and Monuments*, designed for easy reading with many woodblock illustrations, was intended for both the learned and the ignorant, and provided "the light of history" alongside the 'light' of the Sacred Word.[130] These official publications and the inauguration of the special Accession Day Tournament were the first signs of any official government action in connection with the festivities — and what a gift to the government the festival was!

It is not definitely known when the Accession Day Tournaments were begun, although Sir William Segar, painter and herald, tells us that they were inaugurated at the beginning of Elizabeth's reign by Sir Henry Lee. Segar probably means "in the early years of Elizabeth's reign," for if Sir Henry inaugurated them it could not have been until the beginning of the 1570's after his return from the Continent. The first recorded account of an Accession Day

(12) QUEEN ELIZABETH I — woodcut, from Gylles Godet's *Genealogie of the Kinges of England* (1560).

(13) SIR HENRY LEE, aged 35 years — oil painting painted (1568) by Antonio Mor.

Tournament is of one that took place in Whitehall in 1581, to honour the Queen and impress the visiting Duke of Anjou, the supposed prospective bridegroom of the Queen. This was a particularly elaborate and public spectacle and, once begun, the strong and popular impetus carried on such spectacles year after year.

The Accession Day of Queen Elizabeth was celebrated on an increasingly lavish scale, the main features of the celebration being the triumphant return of the Queen in royal procession to London for the winter months, followed by the Accession Day Tournament in which the Queen's knights jousted in tilts set up in Whitehall[131] in the midst of elaborate and symbolic pageantry, as public stage-shows for the Queen, her Court and her people. They developed into great works of art, enormously expensive but utilising the very best that any man could give, be he (or she) a knight, poet, scholar, writer, artist, craftsman, actor, entertainer, wise man or fool. All talents were employed, for the most part anonymously, the chivalrous knight being attired and endowed with the ideas, words, music, apparel, scenery, retinue and play-action of those who assisted him. The world of creative imagination and allegory was made visibly real, and used as a potent and romantic force to unify and tame a world of violent divisions and cruelty.

Sir Henry Lee appeared at the 1581 Accession Day Tournament as the Queen's Champion, and was present at that and every succeeding one (until his retirement from this office in 1590) not only as Champion but organiser as well. Even after 1590 he attended on the Queen's orders, "there to see, survey, and as one most skillful to direct them." The first record of Sir Henry in the tiltyard (although not on Accession Day) is in 1571, the year in which the Queen appointed him Lieutenant of the Royal Manor of Woodstock in Oxfordshire. Evidence suggests that tilts had begun in a casual way as informal jousts staged by the gentlemen knights of the Court in the Queen's honour (not necessarily on her Accession Day), but at some point in the 1570's had been deliberately organised and developed into a complete pageant and public spectacle glorifying the Queen and proclaiming the Virgin Ideal in the most powerful way then possible. The diplomatic use of chivalry and magnificent pageantry to bring about rapprochement, used by Henry VIII and François I in 1515 at the Field of the Cloth of Gold, and by Catherine de Medici in a series of *Magnificences* at Fontainbleau in 1564, Bayonne in 1565 and the Tuileries in 1573, was deliberately revived and used in England in an attempt to unite the different factions of the country: to unite Church and State in one body under one head, to unite all idealists and political parties under one all-embracing ideal and overall policy, and to unite countries and nations in common friendship through entertainment, idealism and high art. It was a magnificent concept and project, an old idea revived, and it may have had its concrete beginnings or prototype in the Kenilworth and Woodstock Entertainments of 1575.

The Kenilworth and Woodstock Entertainments were lavish spectacles provided for the Queen and her Court during the Queen's Summer Progress. The former was staged by the Earl of Leicester during July-August 1575[132] at his 'royal' home, Kenilworth Castle, the site of the earlier revival of the Arthurian tournaments in Edward I's time. The latter was staged by Sir Henry Lee during September 1575[133] at the Queen's royal manor of Woodstock, in Oxfordshire, continuing the theme that was introduced in the Kenilworth Entertainment. It was at this Woodstock Entertainment that the Faerie Queen made her first known and publicly recorded appearance in English literature and drama.

Sir Henry Lee (1533-1611) is an extremely important figure in this whole story who, without doubt, like Sir Nicholas Bacon, had a potent influence on the mind of young Francis Bacon, not to mention his influence through the Tournament pageantry on the minds and ideals of others. Sir Henry was born the eldest son of Sir Anthony Lee and Lady Margaret, daughter of Sir Henry Wyatt of Allington Castle and sister of Sir Thomas Wyatt, the poet. As his father died in 1549 when he (Henry) was only sixteen, Henry was brought up in the household of his poetic and literary uncle. Henry's grandfather, Sir Robert Lee, had married twice, his second wife being Lettice Peniston, widow of Sir Robert Knollys, Gentleman Usher to both Henry VII and Henry VIII. Sir Francis Knollys, Lettice's son by her first marriage, married Catherine Carey[134], daughter of Mary Boleyn and first cousin to the Queen. Sir Francis was made Privy Councillor and Vice-Chamberlain of the Royal Household on Elizabeth's accession. From 1572 until his death he was Treasurer of the Royal Household. Sir Francis's daughter, Lettice Knollys, first married Walter Devereux, Earl of Essex (died 1576), and then later (in 1578) she secretly married Robert Dudley, the Earl of Leicester (the Earl thus marrying bigamously).

Sir Henry Lee married (in 1554) Anne, daughter of William, Lord Paget, one of Queen Mary's most trusted advisers. With his marriage and family connections (Lee was also connected with the Coke, Cecil and Hunsdon families) he managed to serve "five succeeding princes"[135], and kept himself right and steady in many dangerous shocks and three utter turns of state".[136] He was knighted in 1553, served in the Scottish border fighting of 1558 and 1572. In the 1560's he was employed by Sir William Cecil on various diplomatic missions on the Continent, travelling widely in Germany, Italy and the Low Countries, and returning "a well-formed traveller, and adorned with those flowers of knighthood, courtesy, bounty, valour." It is these "flowers of knighthood" that helped to make him the Queen's Champion, as well as his sheer skill in arms, bravery and charisma. Added to which he seems to have been the acknowledged instigator and organiser of the Accession Day Tournaments.

> ". . . these annual exercises in Arms, solemnised the 17 day of November, were first begun and occasioned by the right vertuous and honourable Sir Henry Lee . . . who . . . in the beginning of her happy reigne, voluntarily vowed . . . during her life, to present himselfe at the Tilt armed, the day aforesayed, yeerly . . ."[137]

Whilst abroad, Sir Henry may have seen something of the French *'Magnificences'* and/or met some of the inspirers and artists of these grand works of philosophy and art. Certainly the movement of refugees from the Continent to England brought with it the Burgundian influences of the French high art, developed under the patronage of Catherine de Medici, which were closely interwoven with the 'Family of Love'[138] influences. Familist and Humanist influences were particularly strong in early Elizabethan England among the refugee artists, scholars and craftsmen who fled to England to escape the horrors of continental wars that were waged in the name of religion. They played a not inconsiderable part in helping to cultivate the arts and sciences, peace, charity and toleration in England, as well as in France where they were a major factor in Catherine de Medici's efforts at peace and culture. There was hardly an English nobleman's house that remained unaffected by the style and art, or by the philosophy that came with it, of these continental refugees and visitors; which, added to the natural style, habits and philosophies of the English themselves and the extensive continental travelling done by so many young English noblemen of the day, makes it hardly surprising to find that in England a rich cornucopia was quickly gathered — the cornucopia of goodly things also being one of the symbols of the Virgin, the significance of which was not missed by the Elizabethans.

In 1571 Sir Henry was appointed Lieutenant of the Royal Manor of Woodstock, Oxfordshire, by the Queen. She visited him there in 1572 and again in 1574, in which year he was made Master of the Leash. From 1574 Sir Henry appears at all Court fêtes, including the fête on Accession Day which seems to have been the forerunner of the Accession Day Tournament or *'Magnificence'*. In 1575 he entertained the Queen at Woodstock with the famous Woodstock Entertainment, and became her Champion at Arms, challenging all comers at each royal tournament until his retirement in 1590. From 1578 until his death he was Master of the Armoury at the Tower of London, in charge of all army equipment as well as armour and trappings for ceremonies of State. At some time between 1581 and 1592 he purchased Ditchley Manor, near Woodstock, and here he staged his last great 'Entertainment' for the Queen in 1592, in which he rounded off the story in pageant that he had begun at Woodstock in 1575. He wife died in 1590, their three children having predeceased her, and Sir Henry lived thereafter with his mistress, Anne Vavasour, a former Gentlewoman of the Bedchamber to the Queen and one-time mistress (in 1581) to Edward de Vere, Earl of Oxford. When James I came to the Throne of England, Sir Henry quickly gained the favour of the new sovereign who loved to hunt in the grounds of Woodstock. He eventually died in 1611, having been without doubt one of the 'stars' of the Elizabethan era.

The Kenilworth and Woodstock Entertainments

Although the first recorded Accession Day Tournament was that of 1581 (held at Whitehall), the origin and prototype of this special festival entertainment, in honour of Elizabeth as England's Faerie Queen and 'Minerva', was almost certainly the Woodstock Entertainment of 1575. Further, it is possible to see that all the Accession Day Tournaments were a continuation of the theme that was introduced at the Woodstock Entertainment, and that they not only developed and elaborated the initial theme but unfolded the story (or romance) step by step in true 'Arthurian' or Bardic fashion. From the start Sir Henry Lee was the Queen's Champion and knight-hero of the romance, culminating in his retirement as Champion (in 1590) and his Entertainment provided for the Queen at Ditchley Manor (in 1592) in which he 'rounded off' the chivalric romance, as far as he was concerned, that he had begun seventeen years earlier by harking back to its beginnings at Woodstock and recalling all that had taken place since.

The essential story of this romance is a cyclic one — it is always continuing, and many different persons may assume the chief characters or roles in the story, each in their turn. The structure of the story is based upon a profound knowledge of the ancient 'caste' system, which is in effect a structure of initiation, this being the true basis of the Arthurian mythology. The story concerns characters who personify each of the castes or degrees of initiation: the rustic (or servant), the squire (or yeoman's son), the knight, the sage and the sovereign. The story is about how a young squire, who aspires to knighthood, arrives at a hermitage on a hill and meets there the hermit or sage. The sage, who had himself once been a knight, is able to instruct the squire in the rules and order of chivalry, and to teach the young man all that he (the sage) has learnt from his own life's experiences. Once instructed, the squire then sets out on his adventures and proves himself as a knight, dedicating his chivalrous exploits to the Faerie Queen (*i.e.* Minerva or Pallas Athena, Goddess of Virtue and Wise Intelligence, the Muse of all knight-heroes and Patronesse of all learning and arts). In course of time the knight passes heroically through many varied adventures, both painful and joyful, until he comes to the age when he wishes to retire from the active life and rededicate himself to a more private and contemplative life. The wise and experienced knight thus becomes a hermit, a sage to whom other squires and knights come in their turn for instruction and guidance. Over all this drama or romance presides the Faerie Queen watching the events and encouraging her subjects, protecting, supporting and ruling with virtue and justice.

Besides the 'handed-down' Arthurian initiatory tradition, another prominent source of inspiration for the renewal of this ancient saga in the context of English chivalry is *The Book of the Ordre of Chyvalry,* being one of the five books printed by William Caxton between

1483-85. This English version was a translation into English by Caxton of the book written by the medieval Catalan philosopher, Ramon Lull, which is based upon the traditional 'Arthurian' story of the hermit-knight. Caxton, who was keenly interested in chivalry as an initiatory path and who also printed Malory's *Morte d'Arthur,* makes an urgent appeal at the end of Lull's book for the revival of chivalry in England, just as he did in Malory's book. Caxton advises the reinstitution of regular public jousts or tourneys as the best means of reviving the spirit of chivalry with its ideals of virtue, heroism and service. Lull, in his book, places chivalry in its cosmological context, introducing some of the cabbala that underlies the grand concept. Thus, he states, God rules over the seven planets, whilst the seven planets dominate all terrestrial things. By analogy (or imitation) the Prince, like God, rules (or should rule) over the knights (the planets), who in turn rule over all lower orders (or terrestrial bodies). The Prince stands for Virtue, and thus the principal ethic in which the hermit instructs his knights is the Classical[139] ethic of the mean (*i.e.* the balance or harmony point): "Virtue and measure abide in the middle between two extremes", and knights must be virtuous by "right measure".[140] This is the same ethic as stated by the words, *"Mediocria Firma"* — "Mediocrity (the middle path) is firm (strong, safe)" — the motto of the Bacons, paralleled by Lord Burghley's motto, *"Cor unum, unum via"* ("One heart, one way"), which became a favourite expression of Francis Bacon.

The Woodstock Entertainment began with a combat between two knights, Contarenus and Loricus, the latter being played by Sir Henry Lee.[141] This combat formed the opening scene of a romance entitled *The Tale of Hemetes the Heremyte* (*i.e.* Hermit). The Hermit Hemetes intervened to stop the fight and then began his Tale, addressing himself to the Queen. The Tale is a mythological love story, in which the only daughter and heir of the mighty Duke Occanon falls in love with a knight, Contarenus. The Duke Occanan tries to prevent his daughter, Gaudina, from marrying the knight, but she goes off in search of her lover. Arriving at the grotto of the Sibyl, she meets Loricus, who is in love with a lady of high degree. The Sibyl advises Gaudina and Loricus to travel together, and that when they came to the best country in the world, with the most just ruler, all would then be well. As for the Hermit, he too had once been a famous knight, loved of ladies, but was now old, wrinkled and "cast into a corner", living in a hermitage on the hill nearby. Having arrived in this best of all countries, met the Hermit and being in the presence of the most just Sovereign, the two constant lovers, Contarenus and Gaudina, come together again; and Hemetes the Hermit, who had been blind, now receives his sight.

At the end of the Tale, the Hermit led the Queen and her suite to a delightfully adorned banqueting house specially built for the occasion on a little hill in the wood, nestling beneath an oak tree hung with emblems and posies. Here the royal party feasted. To the sound of music the Faerie Queen made her appearance and what seems to have been her first speech in English literature. On the way home after the feast the 'Song in the Oak'[142] was heard, echoing the 'Song of Deep Desire' heard a few weeks earlier in the holly bush at Kenilworth. The next day (the second day of the entertainment) a play was performed which brought the story to a rather unexpected conclusion, wherein Gaudina gave up Contarenus for reasons of state.

This little play apparently moved the audience profoundly, "in such sort that her Grace's passions, and other the ladies could not but show itself in open place more than ever hath been seen."[143] When one considers the probable allusion in the story to Queen Elizabeth's own predicament, as seen by the general public, in which she had apparently sacrificed the possibility of a marriage with the Earl of Leicester for reasons of state, and add this pointed allusion to the actual public presentation of the Queen as the just and virtuous Faerie Queen surrounded by her loyal knights, then the reaction is truly understandable.

In addition, it seems most likely that it was during the Woodstock Entertainment that Sir Henry Lee, as Loricus, made his vow to present himself annually, on the Queen's feast-day, "at the Tilt armed" to meet all comers as the Queen's Champion. Thus was born the very special series of annual Tournaments that were to take place on the Queen's Accession Day — great pageants that helped to mould the minds and habits of the nation, and to influence the direction and policies of state. It was a revival on a gigantic scale of the work performed by the Cathars through the Troubadour movement, and of the even earlier grand Mysteries of Classical and Arthurian philosophy.

Concerning the *Tale of Hemetes the Heremyte,* Frances Yates wrote in her fine book, *Astraea:*

> "The Tale of Hemetes may seem to us somewhat prosy, but it was received as a striking novelty and made a great impression. It was rumoured to contain important secret meanings. The Queen desired to have a copy of it, and of the whole entertainment. George Gascoigne — eagerly following up a possible line into royal favour — presented the Queen the next Christmas[144] with a copy of the Tale as given in English at the Entertainment, to which he pedantically added Latin, Italian and French translations of it. This shows that it was admired as an example of style. Gascoigne, however, says definitely that he was not the author of the Tale. Who then was its author?"[145]

In fact, Gascoigne states in his dedication to the Tale that he was only responsible for translating the original English version into Latin, Italian and French, which he presented together with the English original to the Queen on 1st January 1576:

> "I will say then that I fynd in my self some suffycyency to serve your highness, which causeth me thus presumpteowsly to present you with theis rude lynes, having turned the eloquent tale of Hemetes the Heremyte (wherwith I saw your lerned judgment greatly pleased at Woodstock) into latyne, Italyan and frenche, nott that I thinke any of the same translations any waie comparable with the first invencion, for if your highness compare myne ignorance with thauctors skyll, or have regard to my rude phrases compared with his well polished style, you shall fynde my sentences as much disordered as arrowes shott owt of ploughes, and my theames as inaptly prosecuted as hares hunted with oxen, for my latyne is rustye, myne Itallyan mustye, and my frenche forgrowne."[146]

The Tale was not actually printed until after Gascoigne's death (7th October 1577), when it appeared, in 1579, in conjunction with a pamphlet stated to be by Abraham Fleming. The whole Entertainment, incorporating *The Tale of Hemetes the Heremyte,* was eventually published in 1585 under the title, *The Queenes Maiesties entertainment at Woodstock,* printed anonymously but attributed to 'George Gascoigne'.[147]

A manuscript copy of the Tale still exists amongst a manuscript collection of Sir Henry Lee's devices for tilts and other entertainments, containing speeches that were almost certainly composed for Accession Day Tournaments. Frances Yates points out that their style is very like that of the Hermit's Tale of Woodstock, and that the Hermit's Tale "takes a not unimportant place in Elizabethan literary history, for in its mixture of Greek and chivalrous romance, its ramblingly attractive prose style, the Tale of Hemetes foreshadows the *Arcadia.*" In considering the whole Entertainment, Frances Yates (with others) goes on to suggest perceptively that "we are here probably close to the living springs in living pageantry whence both Sidney's *Arcadia* and Spenser's *Faerie Queene* drew their emotional nourishment. She suggests that Sir Henry Lee was the author of the Tale and other speeches found in his manuscript volume, as well as the overall inspirer of the pageant.

But (and a big "**but**") this is not necessarily so. Sir Henry Lee was the organiser, promoter and principal instigator of the Accession Day Tournaments, that seems clear, but we should not lose sight of the importance of symbolic meanings to the Elizabethans and the symbolic accuracy with which they were displayed and adhered to. Lee portrayed himself as an Arthurian knight, a Champion of the Queen, but who was himself instructed by a Hermit who gave to him the 'book of rules' and masterminded the whole romance from behind the scenes, in his "corner". This Hermit is found again in a later Accession Day Tournament[148] petitioning the Queen on behalf of a homely, rude company, no better than shepherds, who are led by a knight who is "clownishly clad" (*i.e.* a shepherd-knight). The Hermit states again that he was formerly also a knight but is now old and "cast into a corner". On the Hermit's advice the shepherd-knight had withdrawn from the Court to live in the country, the crowd of country people (shepherds) attaching themselves to him. This is the imaginative world of Arcadia; and if we take note that the term 'shepherd' was employed by the 'English Areopagus' of poets to portray the contemplative state[149] and, more particularly, a poet who is also a teacher and 'shepherd' of the people, and that Sir Philip Sydney was not only the acknowledged leader of the Areopogus but also the ideal personification of the 'Shepherd-knight' for the public, then we have a definite clue as to the identity of the real inspirers and motivators of these vitally important 'Entertainments'.[150]

The English Areopagus came into public view in 1579 with the first publication of *The Shepherds Calender* (5th Dec. 1579) by 'Immerito', who was one of the little group of 'shepherds' and the 'prince' of the poets. 'Immerito' (meaning 'the blameless one' or 'spotless', 'pure', 'undeserving of blame or dishonour') was, as clearly shown by cipher and hinted at by historical and literary evidence, a pseudonym adopted by Francis Bacon after his discovery that he was in fact a concealed son of the Queen, who undeservedly blamed Francis and his untimely birth for all her secret trials and tribulations. Later, when the first part of *The Faerie Queen* was published in 1590, Francis used the mask of 'Ed. Spenser' for the first time for his 'Immerito' writings. That same year, 1579, also saw the publication of *The Tale of Hemetes the Heremyte* and *Eupheus, the Anatomy of Wit,* the latter eventually being attributed to and masked by John Lyly. All three publications (*The Shepherds Calender, The Tale of Hemetes the Heremyte* and *the Anatomy of Wit*) would seem to have two fundamental things in common, besides the same year of publication: namely, common authorship concealed by similar symbolic pseudonyms, and common purpose — **to instruct and lead men to discover and practice virtue and truth in an enjoyable and appealing way.** In other words, they are perfect examples of the Baconian grand project in action.

The timing should be noted, for it was during 1573-4 that Francis clearly perceived the problem afflicting mankind and conceived the great vision or idea about how to remedy the situation; then during 1574 and the first months of 1575 he was busy working out his ideas as a definite project and plan of action with his brother, Anthony Bacon, eventually sounding out and drawing on the aid of others to help in the initial stage of the project — this initial stage being the creation of a grand drama, creating and building up step by step a living mythology or 'Mystery' centred upon Elizabeth I as the personification of the Faerie Queen, grand Patronesse of scholar-knights whose pursuit is virtuous fame and truth. The Bacon brothers did not have far to seek for help in the project, nor was it necessary to explain all that it was about. It was sufficient simply to find the ear of, for instance, the Earl of Leicester (Francis' real father and a notable patron of scholars and poets) and Sir Henry Lee, and for varied reasons urge them to the heroic enterprise, supplying them with the

(14) ROBERT DUDLEY, Earl of Leicester, aged 44 years — miniature painted (1576) by Nicholas Hilliard.

poetic means whilst the knights played the visible and glamorous roles. It is practically certain that both Leicester and Lee (as well as many others) would have jumped at the opportunity to champion themselves before the Queen — Leicester to obtain favours, fame and public recognition as the Queen's Consort, and Lee to be the Queen's Champion perhaps for more philanthropic and heroic reasons. Others, such as Sir Nicholas Bacon and Lord Burghley, would have seen the opportunity to mould public opinion, forge diplomatic links and consolidate the Reformation in England. At the same time, through their university and family contacts, the Bacons knew many good or aspiring poets and writers who would welcome any such opportunity with open arms and great excitement. All it needed was for the seed-thought to be sown and a little push given: the ground was peculiarly fertile, and there were well-known precedents to be encouraged and inspired by in the *'Magnificences'* of the French Court and, above all, in the British Arthurian tradition.

* * * * *

The Woodstock Entertainment should not be seen in isolation from the Kenilworth Entertainment. The grand and sumptuous Kenilworth Entertainment — the first to be put on for the Queen in England on such a grand scale — was provided specially for the Queen by the Earl of Leicester with the object of persuading the Queen to publicly acknowledge her love for Leicester and to publicly marry him. At the same time it was a carefully planned and deliberate revival of the Order of the Round Table, set at the very place where Edward I had renewed the Arthurian tournaments. Kenilworth Castle, one of the grandest of the royal castles and former seat of John of Gaunt, was given to Robert Dudley by the Queen in 1563, together with a gift of £50,000: this lavish and much sought after present being followed the next year with the bestowal of the title of Earl of Leicester (a royal peerage). By the time of the Kenilworth Entertainment, Leicester had made many notable alterations and additions to the medieval castle, turning it into a magnificent and comfortable palace with a block of grand apartments for distinguished visitors, a grand new gatehouse on the town side of the castle, a pleasure garden with a terrace and an aviary, and a tilt-yard. The Entertainment itself lasted for nine consecutive days. It consisted of various 'devices', each invented and written by different people but woven together according to an underlying master-plan. Besides the Queen, thirty other noble guests had been invited to the Entertainment by Leicester, including Sir Henry and Lady Mary Sydney, their son, Philip Sydney, and probably his sister, Mary Sydney (who was certainly present at the Woodstock Entertainment), and the Countess of Essex.

The Queen arrived at Kenilworth on the evening of Saturday 8th July, 1575, and was "met on her way, somewhat near the Castle" by Sibylla, who proceeded to "fore-shew what shall be". Before the castle gate itself she was met by Hercules, as the Porter. Then, inside the castle, when the Queen had passed the gate and entered the "base court", she was met by the Lady of the Lake who was conveyed across "the pool" to greet the Queen with a poem in which she stated that she and her royal court had lived at the castle, since the time of King Arthur, in concealment and "restless pain", afraid to come forth from the lake; but that now at last she could appear again, on this occasion of the Queen's third visit to the castle. As the Queen proceeded to the inner court she was greeted with more poetry, sweet music and many delightful and symbolic sights, till at length she retired to her lodging in the new block of grand apartments.

The next day (Sunday) nothing was done till the evening, when "there were fireworks shewed upon the water which were strange and well executed". On Monday 10th July, her Majesty went hunting in the forests of the deer park, and as she returned from hunting she was met by Jupiter, Echo and a Wild-man, who advocated Leicester's goodness. This was followed by further devices culminating in "the device of the Lady of the Lake", taking place on the water at night, in which the Lady of the Lake was supposed to be rescued by the Queen herself (who alone could perform this deed) — but apparently the original idea was not wholly carried out and seems to have miscarried somewhat. Over several days many other elaborate and imaginative entertainments were provided, with some tilting in the castle tilt-yard and many Arthurian devices. One very important character was a fourteen-year-old minstrel — a 'bard' who related the key stories about King Arthur and his knights of the Round Table.

Sadly for Leicester, there was one "shew" that was prepared but "never came to execution" because of "lack of opportunity and seasonable weather", in which it was to be boldly suggested to the Queen that she leave her "virgin" condition and marry the Earl of Leicester. Furthermore, the Queen hastened her departure from the castle, leaving on Wednesday 27th July. The ommission was in part corrected by the "Farewell", in which "Master Gascoigne", at the command of Leicester, clad himself as Sylvanus, god of the woods, and spoke *ex tempore* to the Queen whilst she rode for hunting, he running beside the Queen's horse. The oration, which was a recommendation to the Queen to marry Leicester, was extremely lengthy and allegorical, and it is certain that it was neither the style of Gascoigne nor likely for him (aged 50) to have delivered the discourse in such an arduous fashion. It could only have been the work of an exuberant and 'euphuistic' young man, the one who was masked by Gascoigne, who was responsible for the idea of the Entertainment and therefore wholly to be expected to be the one commanded by Leicester "to devise some farewell worth the presenting" in an attempt to save the day: for clearly the purpose that Leicester had in mind for his Entertainment was to persuade the Queen to a public marriage. For Leicester's purposes, the Entertainment fell flat, but it prepared the way for the full rebirth of chivalry and appearance of the Faerie Queen at Woodstock.

The Kenilworth Entertainment was recorded in *The Princelye pleasures at the Courte of Kenelwoorth,* published 26th March, 1576, anonymously, but bearing the same motto, *"Tam Marti quam Mercurio"* ("As much by Mars as by Mercury"), that was used as Gascoigne's literary signature. The work was ascribed to Gascoigne and was later included in a complete edition of his works published after his death. The author of the publication is in fact the "Printer" and editor, who had collected together copies of the "sundrie Pleasant and Poetical Inventions" presented at the Kenilworth Entertainment (all except for one) so as to publish them. Previous to this publication another account of the Entertainment had appeared, in the form of a letter signed by one 'Robert Laneham', an obscure personality who had been made Keeper of the Council Chamber Door by request of his patron, the Earl of Leicester. This letter is a highly elaborate and entertaining account of some of the Kenilworth festivities, written in a jocular, tongue-in-cheek vein as impressions and speculations concerning the Entertainment. It would appear that 'Laneham', like 'Gascoigne', was a mask for the youthful Francis Bacon, the former name and personality being invented, the latter being a real person who assisted Francis in return for Francis' help. In the letter, 'Laneham' reported that, besides all the Arthurian and other entertainments, there was much "Philosophy, both morall and naturall", poetry, astronomy (which included astrology) "and oother hid sciencez"; also, "the Sheperdz kalender,[151] The Ship of Foolz, Danielz dreamz, the booke of Fortune. . ." and many others. Twenty years later Francis was to incorporate much of what he saw (and perhaps wrote or inspired) at the Kenilworth Entertainment in his play, *A Midsummer Night's Dream,* which he specially wrote for the wedding of his 'niece', Elizabeth Vere, granddaughter of Lord Burghley, when she married William Stanley, 6th Earl of Derby, on 26th January, 1595, the wedding festivities taking place at Greenwich Palace.

It might be of interest to note in passing that the Queen, on her way to Woodstock, stayed first at Lichfield for eight days and then went on to Chartley, the Earl of Essex's Staffordshire estate and principal family home. The Earl (Sir Walter Devereaux) was absent in Ireland, but the Queen was received by her cousin Lettice, the Lady Essex (the daughter of Sir Francis Knollys and niece of Anne Boleyn), who had also been at the Kenilworth festivities. It was into Lady Essex's care that the Queen had placed her second son by Leicester — Robert — when he was born in 1567.

Like Lady Anne Bacon, Lettice (then Lady Hereford) had been the Queen's Principal Lady-in-Waiting at the time of the child's birth, and Robert was adopted by Lettice and her husband just as Francis had been adopted by Sir Nicholas and Lady Anne Bacon six years earlier. Robert was just eight years old at the time of the Kenilworth and Woodstock Entertainments, and had recently been placed under the guardianship of Lord Burghley, at the request of the Earl of Essex (and most likely at the command of the Queen), in respect of his "direction, education and marriage."[152] The Queen stayed at Chartley for a fortnight, then proceeded to Stafford Castle, Dudley Castle, Hartlebury Castle and the city of Worcester. Then she went to Woodstock whence, after visiting Reading, she returned to Windsor.

Hemetes and the
New Rosicrucian Birth

Certain important symbolic allusions should be noted in the Kenilworth and Woodstock Entertainments, from which was born (or rather, which announced the birth of) the great Elizabethan Renaissance of culture and learning, and the new (or renewed) Christian 'Order' of the Rose Cross.

Most important of all, there was the hermit-sage who initiated the whole project and who taught the knights, giving them the rule-book of chivalry. He was called 'Hemetes the Heremyte'. Hemetes is quite clearly the same as Hermes, the Greek name for the Egyptian Thoth or *Tot*. Thoth-Hermes refers, quite simply, to the Divine Wisdom or Word of God, which the later Greeks called *Logos* (Word of Wisdom) and *Christos* (Light of Wisdom); and, as with the title of *Christos* or Christ, it was also applied to any individual soul or group of souls who manifested this divine Idea or Truth and who were thus 'illumined' or 'anointed' by its Light. In Ancient Egypt the greatest of the high initiates — the pharoahs and their closest councillors, the principal high-priests — were known collectively as *Tat*, *Tot* or *Thoth,* as also were their teachings and decrees. The Greeks referred to this same line of teachers and teachings as *Hermes;* whilst in esoteric Christianity (which was launched into a Greek cultural world) the same rule applied, using the name *Christos* or Christ. The Ancient British used the name *Myrddin* or Merlin: the Romans used *Mercurius* or Mercury. *Hemetes* is of this same line, uniting the sacred *Tau* (symbol of Light) of the word *Tot* with the Greek word *Hermes* so as to emphasise the intended meaning and the pointed reference to the Hermetic teachings handed down from Ancient Egypt — periodically reinterpreted and re-enacted — from which the Western world derives most of its 'light'.

In Ancient British mythology Hermes is *Hu* the Mighty, 'the Glorious Light', (whence the name *Ies-Hu* or *Jesu,* 'the embodied/revealed Light'), Merlin (or *Myrddin*), 'the Star of the Sea', with his little black dog, and also Herne the Hunter, the Oak God — the Oak being the 'Tree of Light' from which the Druids constructed their symbol of the sacred *Tau* or Cross of Light, up which entwined the Mistletoe as the Gnostic Serpent or 'Son of Man'. The Oak represented the Light, Wisdom, the Law, Truth, Strength, Protection, Courage and Steadfastness in Faith and Virtue.

In the Woodstock Entertainment not only did Hemetes appear, but also the Oak tree stood out as the central feature of the stage-set, being the great tree set on a man-made hill under which the Queen sat, ate the banquet and watched the proceedings, and where the Faerie Queen appeared in order to give her first known speech in English drama and literature. From this Oak the important 'Song in the Oak' was heard. Neither would anyone have missed the allusion to the great Druidic and Arthurian centre of Windsor, surrounded by its sacred oak forests in which Herne hunted with his hounds. The hounds of Herne are the capable and trained seekers after truth, the companions of Herne and the conductors of the other 'ordinary' huntsmen. These hounds are the initiates whom the Egyptians represented under the archetypal form of Anubis, the jackal-headed god, Conductor of the souls of the dead to the Place of Truth, the Guardian and Interpreter of the Mysteries. In the landscape, near Windsor Castle, is laid out the gigantic form of Anubis, guarding the secret entrance to the heart of what the Druids once called *Caer Troia,* the royal seat and landscape temple of their High King (who was also Supreme Arch Druid), of which Westminster and London City are small but vitally important parts.

Whilst the Oak, the Faerie Queen and Hemetes the Hermit made their appearance at Woodstock, at Kenilworth they were forshadowed by the Holly Bush (from which was heard the 'Song of Deepe Desire'), the Lady of the Lake and the fourteen-year-old minstrel boy who recited the key story of King Arthur and his knights of the Round Table. The Holly was another tree used in the symbology of the Cross by the Druids, but in this case it represented the Cross or Tree of Goodwill, Joy, Health and Happiness — the fulfillment of "deep desire".[153] From the initial (and initiating) desire springs thought; and when the desire is joy and goodwill (which is pure love), then the thought will be light — it will be the thought of truth, the light of understanding. First, sorrow is turned to joy. From joy springs forth light: from the Kenilworth Entertainment springs forth the Woodstock Entertainment. This was the message conveyed. The message was further emphasised by the minstrel or Bard of Kenilworth, signifying the world of joys and initiating desires, in contradistinction to the Hermetic sage of Woodstock who corresponds to the Vate or Seer of the Druids and who is representative of wise thought and true perception. Similarly, the Lady of the Lake who made her appeal at Kenilworth became transformed to the Faerie Queen at Woodstock: the former being the Queen of the Water Element that represents the Emotions, and the latter being the Queen of the Air Element that represents the Thoughts. The Lady of the Lake provides the sword of goodwill, the initiating power of joyous love or "deep desire", whilst the Faerie Queen provides the shining helmet and crown of illumination to her knight-heroes who search for and find truth.

These two interlinked Entertainments represented the first organised effort on a grand scale in England to develop and channel the powers of imagination, the fires of emotion and the aspirations of men's thought towards a communal and high ideal, using all the God-given factors that were then uniquely available. The roots and purpose of the project were firmly based in esoteric Christianity, as represented by the ancient symbols of the mystical Rose Cross which gave rise to the Rosy or Red Cross emblem of St. Joseph of Arimathea and the early British Church, priestly guardians of the Holy Grail of Truth; of St. George, the ideal Rose Cross Knight; and of King Arthur and his knights of the Round Table, chivalrous seekers after the Grail of Truth. The Entertainments were the beginning of a deliberate exercise to revive the ancient and true Rosicrucianism, not only in its knightly or chivalrous aspect but also in its priestly or illumined aspect, and moreover in the sacred country of the Virgin,[154] bearer of the Christ Child — the Land of the Rose, the Land of Merlin[155] and Arthur, the Sun King. Fate[156] had provided England with a Queen who was compelled by circumstances to play the role of a Virgin Queen in the orthodox sense, and who willingly allowed herself to take on the role of Virgin Queen in the mythological and poetic sense. Reigning in England at the height of the Renaissance in Europe, Elizabeth's sovereignty made it possible to unite all the hitherto seemingly disparate sacred traditions in her single symbolic role, demonstrating the real oneness underlining all sacred traditions. She was at once Britannia, Mary, Minerva, Pallas Athena, Isis, Demeter, Astraea, Gloriana, Ceridwen — the Virgin Queen of all traditions who brings forth the Christ Child, the 'illumined one'. All it needed was a new Hermes or Merlin — a new Initiator and

Teacher — to seize the opportunity and make it so. This is precisely what the fourteen-year-old Francis did — and, moreover, he was (without yet knowing it) the Virgin Queen's son!

The atmosphere and images of the Kenilworth-Woodstock Entertainments set the scene for the development of the Arthurian and Arcadian imagery of the English Areopagus, notably in the principal literary works of Philip Sydney's *Arcadia*[157] and 'Immerito's'[158] *Shepheards Calendar* and *Faerie Queen,* which permeated and influenced the whole development of Elizabethan pageantry, culture, learning, politics and national belief. An initiatory programme and 'school' was established, leading the crude and uncultured 'swain' to the culture and chivalry of the courtier's 'schoolroom', and the valorous and courteous 'knight' to the philanthropy and illumination of 'priesthood'; all under the aegis of the Great Virgin Goddess, the Muse of Wise Intelligence who, with the help of her knight-heroes, shakes her spear of illumination[159] at the formidable Dragon of Ignorance, Vice and Selfishness.

In the Arcadian imagery the shepherd-knight is the Christian Rose Cross (or Red Cross) knight — the gentle, holy man who is introduced in English literature in the first book of *The Faerie Queen* and whose sacred Muse is the "holy Virgin, chiefe of nine",[160] "*Gloriana, . . .* Queene of *Faerie* land."[161] He, together with eleven other aspects or embodied virtues, comprise the great King, Arthur, the highest of the initiates and who is also become, by initiation, a Merlin, embodying and proclaiming the Word of God. He is St. George, the virtuous knight and dragon-slayer, who 'shakes' his spear of light at the dragon of darkness in imitation of his divine Muse.

On the frontispiece of the manuscript of *Hemetes the Heremyte* is a pen-and-ink drawing of a knight kneeling before the Queen and presenting her with a book. In his left hand he holds a spear, whilst his right hand is offering the book. This is a symbolic picture of the poet-knight — the knight who has become the bard or sage, Hemetes himself. Cabbalistically the book of wisdom, synonymous with a ray of light or word of truth given forth by the prophetic teacher of truth, is meaningfully placed in the right hand of Wisdom and Mercy; whilst the spear of illumination, wielded by the chivalrous seeker and defender of truth, is placed in the left hand of Intelligence and Judgment. In this cabbalistic drawing Mercury, the Seer-Teacher, and Mars, the chivalrous Warrior, are thus portrayed in equal balance in the one person and thereby express pictorially the motto hung over the head of the poet-knight and which became associated with other works of 'Gascoigne': namely, *"Tam Marti quam Mercurio"* — "As much by Mars as by Mercury". This idea formed one of the ruling principles of the Baconian (and hence the Rosicrucian) work: that the Mercurial and philosophical aspect of man should be carefully balanced with the Martial and practical aspect — the initiatory revelation with the evolutionary adventure — the two in balance producing the harmony of the 'middle path'. This idea was perceived and expressed in various ways by Francis Bacon, particularly in terms of contemplation and action, speculation and operation, truth and usefulness, truth and goodness, wisdom and power, philosophy and charity, the study and the practice of both the 'Book of God's Word' and the 'Book of God's Works' — the realms of Ideation and Manifestation:

> "Let no man upon a weak conceit of sobriety or an ill-applied moderation think or maintain that a man can search too far, or be too well studied in the **Book of God's Word,** or in the **Book of God's Works**— Divinity or Philosophy. But rather, let men endeavour an endless progress or proficience in both; only let men beware that they apply both to CHARITY, and not to swelling; to USE, and not to ostentation; and again that they do not unwisely mingle or confound those learnings together."[162]

(15) FRONTISPIECE TO *HEMETES THE HEREMYTE* — mss presented to Elizabeth I by George Gascoigne in 1575 (published in 1579).

"It is so then, that in the work of the creation we see a double emanation of Virtue from God; the one referring more properly to Power, the other to Wisdom; the one expressed in making the subsistence of the matter, and the other in disposing the beauty of the form. This being supposed, it is to be observed that for anything which appeareth in the history of the creation, the confused mass and matter of Heaven and Earth was made in a moment; and the order and disposition of that chaos or mass was the work of six days; such a note of difference it pleased God to put upon the works of Power, and the works of Wisdom . . ."[163]

"The Understanding of Man and his Will are twins by birth as it were, for the purity of illumination and the liberty of will began together. Nor is there in the Universal Nature of things so intimate a sympathy as that of Truth and Goodness."[164]

"But this is that which will dignify and exalt knowledge: if contemplation and action be more nearly and straightly conjoined and united together than they have been: a conjunction like unto that of the highest planets, Saturn, the planet of rest and contemplation, and Jupiter, the planet of civil society and action."

"And these two rules, the practical and the contemplative, are the same thing; and what is most useful for practice is also most true for knowledge."[165]

"The unlearned man knows not what it is to descend into himself, or to call himself to account . . . The good parts he hath he will learn to show to the full, and use them dexteriously, but not much to increase them: the faults he hath he will learn how to hide and colour them, but not much to amend them: like an ill mower, that mows on still and never whets his scythe. Whereas with the learned man it fares otherwise, that he doth ever intermix the correction and amendment of his mind with the use and employment thereof. Nay, further and in sum, certain it is that *Veritas* and *Bonitas* differ but as the seal and print: for Truth prints Goodness . . ."[166]

"Let men know . . . what a difference there is between the phantoms of the human mind, and the Ideas of the Divine Mind. The former are nothing but fanciful abstractions: the latter are the true signatures of the Creator upon creatures, as they are impressed and limited in matter by real and exquisite lines. And so the chief things of all are, in this kind, TRUTH and USEFULNESS; and effects themselves are to be accounted of more worth in so far as they are pledges of truth, than as they give the comforts of life."[167]

"This Janus is bi-fronted . . . The face towards Action hath the print of Goodness, the face towards Reason hath the print of Truth."[168]

In this last quotation *Janus* is used by Bacon to signify Imagination, which is the mediatating principle between the divinity (or inspiration) of the heart and the philosophy (or reason) of the intellect, and also between the intellect and the body which performs the actions. *Janus* is a Roman name for the Divine Imagination and its Vision of Truth, which Vision is the concept, idea or thought-form of Light. He is the God of all gods, *Janus Pater,* the Creator (*i.e.* creative Idea) of all things. Janus took precedence over all other gods of the Roman pantheon, and the first month of the year is dedicated to and named after him. "Ovid relates that *Janus* was called *Chaos* at the time when air, fire, water and earth were all a formless mass. When the elements separated, *Chaos* took on the form of *Janus:* his two faces represented the confusion of his original state."[169] More than this, the two faces symbolise the opposing polarities harmonised and united in one 'head' or supreme state of consciousness: the balance between non-Being and Being, between Chaos and Cosmos, between the Unmanifest and the Manifest, between the pure impulse of heart desire (or love) and its manifestation in the form of the living soul. *Janus* is thus known as and means the Mediator, Advocate and Gatekeeper who stands on the horizon between two states of Being (or non-Being). He holds the keys to (*i.e.* the control and interpretation of) Heaven and Earth — the two polarities of Being. The Egyptians used the symbol of the falcon, *Herakhty*[170], sign of the rising sun on the eastern horizon and thus of the dawn of Light, the First 'Day' or *Alpha* of Creation, the creative Vision or Idea (*Ra*) of God. *Herakhty* also represents the setting sun on the western horizon and thus the fulfillment of Light's purpose, the fruit and crown of its creativity, the *Omega*. *Herakhty* is also especially identified with *Horus,* which is the name of Man himself, and this leads us to the very secret and purpose of Man which is at the heart of Rosicrucianism.

> "And God said, 'Let us make Man in Our image, after Our likeness, and let them have dominion . . .
> "So God created Man in his own image, in the image of God created He him; male and female created He them.
> "And God blessed them, and God said unto them, 'Be fruitful, and multiply, and replenish the earth[171], and subdue it: and have dominion over the fish of the sea, and over the fowl of the air, and over every living thing that moveth upon the earth.'"[172]
> "And God said, **'Behold,** I have given you every herb bearing seed, which is upon the face[173] of all the earth, and every tree, in the which is the fruit of a tree yielding seed; to you it shall be for meat . . .' "[174]

'Man' means 'the Thinker', and the purpose of all thought is to see, understand and know. 'Seeing' is synonymous with 'comprehending'. The *manna* or meat (bread) of heaven is the very Thought or Idea of God, the Light of heaven, the Holy Wisdom of God's Mind[175], the Word[176] or Spirit of God. The 'herb' and the 'tree' are symbols of the divine Ideas, all of which contain the procreative forces or 'seed' of life. The 'tree' in particular refers to the Tree of Life — the whole Archetype of Creation which only Man can behold and feed from. All other creatures feed only from the lesser living Thoughts or 'green herbs'[177] and even then they do not 'see' those creative ideas. Only Man has this privilege, as the crowning point of Creation.

Man can think, he can see and, by putting his understanding into action, he can know. Man's spirit is the actual Idea that God has of Man, created in the Light of the divine Mind. It is Man's ruling Archetype or Blueprint. Man's soul is his mind — that which thinks in order to see and know the spiritual Archetype or Truth; and this soul evolves in consciousness from its primal ignorance to a complete revelation and embodiment (in action) of the truth. By means of his body of action, Man may operate the laws of whatever truth he has perceived, and thus by experience come to *know* the truth. Man begins as Adam, a living soul, and culminates as Christ, a life-giving spirit, having spiritualised all matter and thereby knowing all things.

The knowledge of Truth (*i.e.* the knowledge of God) and the mind which contains this knowledge is called the Holy Grail. As the mind or soul of man it is *san greal,* the 'holy vessel', and as the spiritual knowledge it is *sang réal,* the 'royal blood' or wine. Turning the water into wine is the act of turning ignorance into knowledge. The pure 'dew' or 'waters' of the soul are transformed into the 'rose dew' or 'wine' of the Grail. The transforming power is divine love, well understood and put into action.[178] In Grail symbology the 'water' and 'wine' states of being are also represented by the while lily and the red rose respectively, or by the white (or crystal) cross and the red (blood-stained) cross. Only by loving, understanding and serving in a self-sacrificial way is the Grail achieved. The very word 'rose' has an intimate symbolic association with the Latin word *ros*, meaning 'dew'. The dew of heaven is the spiritual knowledge of a spiritualised mind or illumined soul. The dew of heaven is the 'rose-dew'. The rose is also associated with the resurrection that occurs after the self-sacrifice has taken place. What arises as 'a rose' is the spiritualised soul in its transmuted body of light. A Rosicrucian is one who has arisen from the sleep of ignorance and deadness of inertia to the awakeness of God-consciousness and livingness of God-activity. For such a one, his rose has bloomed (and continues to bloom) upon the cross of sacrifice — the cross of charity or loving service. From being a disciple who was learning how to love, understand and serve truth, he has now become the master who loves, understands and serves perfectly, knowing and living in truth. Jesus of Nazareth was known esoterically as the Rose of Sharon: Francis Tudor became known as the Rose of England — the Tudor Rose — a symbol which he adopted as his personal emblem as well as it representing his work.

Jesus (Hebraic: *Jeheshua*), the name given to a fully illumined soul and identical to the Ancient Egyptian name, Horus (or *Heru*), and the Ancient British name, Jesu (or *Ieshu*), is the Holy Grail or Rose. The meaning of the name is 'the embodied Light' or 'the ensouled Light' — the Soul of Light, or 'Illumined Soul'. More precisely it means 'the Jewel of Light', 'the Face of Light', 'the Countenance of the Lord', or 'the Visible Appearance of Light, the Son of God'. To one who is illumined in the fullest sense possible the title of Christ (in the West) or Buddha (in the East) is applied, for such a soul is fully anointed or illumined by and become as one with the spiritual Light of God.

The significance of the 'face' or 'countenance' when referring to the soul of man is important. The Archangel *Michael* is also known as 'the Countenance of God' and 'He who is like unto God'. In *Michael* we have the actual spiritual Archetype or divine Idea of Man, which humanity (individually and collectively) is destined eventually to express and manifest fully. *Michael* is supremely the 'Shear-shaker' and 'Dragon-slayer' — the Archetypal image which St. George imitates and embodies as a human soul. *Michael* is the Face of Light that veils and at the same time reveals the Heart of God. *Michael* is the 'head', 'prince' and 'commander-in-chief' of the Host of Light. The command faculty is related to the functions of the head, as is thinking, seeing and knowing. Through over-seeing and knowing his kingdom, man may have dominion or command over it. His very illumination is expressed by the opening of the 'Third Eye' in the head and the light-radiation through and from the head which is called the halo or corona — the glory or crown of light. This 'Third Eye' is man's Eye of Imagination, the Eye of the Mind. It is called the 'Third Eye' because it is the *Janus* or Mediator between two polarities or points of view — inwards and outwards, heavenwards and earthwards, individual and universal — thus symbolically it is set in the centre of the forehead between the right and the left eyes, and is the centre point from whence radiates the halo of light.

> "As for me, I will behold Thy face in righteousness:
> I shall be satisfied, when I awake, with Thy likeness."[180]

The mystery of Man and the Holy Grail is also wound up in the symbology of the 'Stone' or 'Jewel' and with *Lucifer*. Initially the Archangel *Lucifer* was the chief of the Angelic Host. His name means "the Bearer of the Fire of God' or, simply, 'Light-Bearer'; for the divine Fire is the hidden Light or Wisdom of God. That which bears this Fire is divine Intelligence — the *Sophia*. However *Lucifer* was cast out of Heaven and fell "like lightning" into the dark abyss of primal matter. Why did this happen? Certainly not because something went wrong with God's Plan! *Lucifer* said in his heart, "I will ascend into Heaven, I will exalt my throne above the stars of God: I will sit also upon the Mount of the Congregation, in the sides of the North: I will ascend above the heights of the clouds; I will be like the Most High."[181] But *Lucifer* was already in Heaven when he uttered that thought, and not just as any one of the 'stars' (*i.e.* angels) of the divine Light but as the co-equal of *Michael*. For *Lucifer* to be able to **ascend** into Heaven he must first **descend** into the Earth of universal matter. For *Lucifer* to be able to **exalt** his throne (and 'throne' is a synonym for 'soul') he must first **form** his throne out of matter. To be all that he says he will be, *Lucifer*, a pure spirit, must incarnate into matter and generate a living soul — and that soul, in order to achieve all that he says and to become "like the Most High" must be the soul of man. In other words *Lucifer* is no more nor less than the incarnating spirit of man — man's own ego or spiritual identity.[182] *Michael* is *Lucifer's* counterpart — the Archetypal Idea that resides in the divine Mind, whilst *Lucifer* is that generative part of the Idea which incarnates into matter, uniting with matter as its 'breath of life' and thereby forming the soul of man, called Adam.

The casting of *Lucifer* out of Heaven and into the Earth by *Michael* and his angels is, as it must be, a complete act of God or Good-will. The 'fall' of *Lucifer* is a voluntary act of self-sacrifice — God's sacrifice — in order to fulfil the plan of Creation, whilst *Michael* is the divine Consciousness or Plan that initiated and continues to direct the process from "on high". By falling into matter, *Lucifer* causes that matter to move and tremble, vibrating and ordering it into forms that will manifest the Idea. Because of the effect on matter of this incarnating life force, like a wind which moves water and makes it swirl and ripple, or like a sound which causes water to vibrate into moving, oscillating patterns, the ancient sages used the appropriate image of a snake or serpent to represent this natural state of being. The result of the vibration is to order the universal matter according to the divine Archetype, and so the four 'Elements' are formed as four states of manifestation within the Cosmos that parallel the four overall states of divine Existence. *'Fire'*, paralleling the World of Emanation in which Love as the primal Desire-to-Be enamates from the Heart of Divine Being, bringing all else into existence, is the essential 'desire' state of the Luciferian life force in the Cosmos — the incarnate ego of man. *'Air'*, paralleling the World of Creation in which Light is created in the Mind of God, is formed as the heavens or spiritual realm of the Cosmos, corresponding to the intellect of man with which he may acquire vision, understanding and eventual knowledge of truth. *'Water'*, paralleling the World of Formation in which the living soul is formed by the spirit working in universal matter, is generated as the lower heavens or psychic (astral) realm of the Cosmos, corresponding to the psyche of man in which are engendered feelings and emotions. *'Earth'*, paralleling the World of Action in which the living soul acts and works out its destiny as an increasingly useful and god-like being, is precipitated as the dense etheric and physical realms of the Cosmos, corresponding to the earthly or mortal body of man in which sensations are received and with which man acts and experiences the results of his actions. For this reason the 'Serpent' is also called a 'Dragon' — a mythological creature that symbolises all four of these Elements or states of manifestation — the 'fire' being the essential life principle and 'breath' of the Dragon. In the East the Dragon is called *Kundalini*. One of the Western names is *Leviathan*.[183]

By involving itself in matter the incarnate spirit or life force becomes sluggish and sleepy, but it responds to the commands or stimulation of the spiritual Light. In its sluggish, sleepy state the Dragon is the antipathy of the vitality and awakeness of the spiritual Light and, when it is first stirred by the impulses of Light sent by *Michael* to awaken it, its response is normally surly, bad-tempered, aggressive and purely selfish. In this condition the Dragon is known as *Satan,* the Adversary to the Spirit. Eventually these and other negative responses to the spiritual decrees (which are spoken in the still, small voice of the heart and heard as intuition) are overcome by continual effort and replaced by positive, helpful responses. The ego-centred Dragon is slain and the Virgin is released: that is to say, the Dragon becomes the Virgin, a state of soul purity and genuine receptivity to the Spirit, enabling the beautiful soul to manifest or give birth to the Christ Light — its 'whiteness' of purity becoming stained with the 'redness' of spiritual Fire. The lily becomes the rose. The pure water becomes rich wine. The ignorant Dragon is now become the wise and royal Dragon, raised upon the Cross of self-sacrifice, harmless as a dove.[184]

> "In that day the Lord with His sore and great and strong sword shall punish Leviathan the piercing serpent, even Leviathan that crooked[185] serpent; and He shall slay the dragon that is in the sea. In that day sing ye unto her, 'A vineyard of red wine!'"[186]

When the Grail is likened to the jewel that is struck off the forehead of *Lucifer,* which falls into the Abyss and which is there ministered unto by angels of light, who gradually (*i.e.* degree by degree) raise it up again to heaven to become the crown jewel of the Christ, then it is but another way of telling the same story. This stone or jewel is described as an 'Emerald' — not an ordinary emerald but a pure white (or transparent) crystal having the livingness (or 'greenness') of the spirit. In the Ancient British language, for instance, the word *glas* meant both 'green' and 'clear', *i.e.* crystal-clear or pure. The Stone of Merlin — the Philosopher's Stone — was the Green Diamond. This Green Stone, recovered from the Abyss of materialism, was carved by the angels of light into a many faceted and beautiful jewel — the Green Diamond or Holy Grail — the spiritual Soul of Merlin. Likewise the Emerald Tablet of Hermes Trismigistus ('the thrice greatest'), upon which are engraved the Hermetic teachings, is the Holy Grail. For each man the Grail is his own individual, clear perception and knowledge of truth. For mankind as a whole the Grail is the total tradition of knowledge, added to and shared by all in the holy communion of the real 'Last Supper' or 'Brotherhood Feast', in which the greatest truths are lovingly shared and assimilated.

The secret of man lies, therefore, in his ability to see, or to **behold** truth, followed by his action in carrying out the precepts of truth. Man is made to both contemplate truth and to practice truth.[187] He is destined to be both a poet and a knight, mercurial and martial, philosophical and practical. Thus when we see, on the very frontispiece of the manuscript *Hemetes the Heremyte,* the words, "**Beholde good**", we should recognise that here is the fundamental maxim of Rosicrucianism (or esoteric Christianity) — the unique command of God to man that makes man what he uniquely is and what he has the responsibility to become. When we turn the page of the manuscript it becomes blatantly obvious that this command or request is intended to be set apart, duly noted and thought upon; for the next page is the first page of the text, beginning **"Beholde (good Quene) a poett with a speare"**. The "Beholde good" thus forms the usual catchword or catchphrase that was placed at the end of one page in order to introduce the next page of text. Practically this simple device ensured that the pages of the volume would be assembled in their correct order before they were numbered. But the bracket is missed out in the catchphrase, thus making it into a definite and complete sentence — a command — in its own right, and the first or opening sentence of the Tale.

Furthermore, in the second sentence (the first of the page of text), the author is asking the good Queen to "Beholde a poett with a speare," thus drawing attention in a purposeful way to what the poet-knight, Hemetes, represents. He is the Christian Rose Cross philosopher and knight, the *Spear-shaker,* the teacher and defender of truth who unites Mercury and Mars in himself. The sentence links in meaning with an old adage carved above the ancient

Temples of Initiation: "Man, know thyself." The presence of a sphinx (with the golden 'ring of truth' held in and issuing from its lion's mouth) rather than a lion supporting the throne further confirms this meaning — and another meaning also.[188] In the drawing Hemetes is shown offering the book of poetry — the Hermetic or Rosicrucian teachings, which constitute a revelation of spiritual Truth or Law — to the Faerie Queen. Tucked above his ear is his quill pen, emphasising that he is indeed the poet who wrote down the inspired teachings and who thus is, like Hermes or Moses, the law-giver and prophet of God. He kneels before the Queen in service, offering the sacred book of the Law to her, whilst the Latin sentence woven into the tapestry of the Queen's canopy over her head proclaims ambiguously, "*Decet Regem Regere Legem*", which can mean either "It is fitting for the Sovereign to direct the Law" or "It is fitting for the Law to direct the Sovereign". Even at that early age Francis was a master of specialised ambiguity carefully employed to teach truth and yet to disguise it at the same time, so as to make people think yet to leave them to think what they pleased. Truth was (and is) thus offered to all, either hidden or revealed according to whether the viewer-listener has the eyes to see and the ears to hear.[189]

To make this plainer Hemetes wears a cloak (or "gowne"), symbol of the mystical veil, which is awkwardly draped over his right shoulder yet thrown off his left shoulder, thereby revealing his left breast (and natural heart) but concealing his right breast (and spiritual heart). This otherwise extraordinary feature is pointedly mentioned in the introductory poem ("his gowne haulffe off") but with no textual explanation as to why it should be important enough to both portray and mention it. What the note does is to draw our attention to the intended symbology, for us to discover and read if we will. Thus the inspired nature of the poet-teacher (represented by the right-hand side) is shown lying veiled, whilst the chivalrous nature of the courtier-knight (represented by the left-hand side) stands revealed. Of the two natures the former has precedence, for action depends first on thought. Action follows thought, and good action follows good thought.

> "But to speak truly of things as they are in worth, rational knowledges are the keys of all other arts, for, as Aristotle saith aptly and elegantly, 'that the hand is the instrument of instruments, and the mind is the form of forms'."[190]

> ". . . We see an image of the two estates, the contemplative state and the active state, figured in the two persons of Abel and Cain, and in the two simplest and most primitive trades of life; that of the shepherd (who, by reason of his leisure, rest in a place, and living in view of heaven, is a lively image of a contemplative life,) and that of the husbandman: where we see again the favour and election of God went to the shepherd, and not to the tiller of the ground."[191]

The favour and election of God is given to Abel, the shepherd, rather than to Cain, the husbandman, because it is through thought that the action is guided and governed.[192] Action follows thought, not *visa versa* (although action can give rise to new thought), and hence the contemplative state has precedence over the active state: not for any other reason. It was this biblical image of the shepherd, Abel, and the husbandman, Cain, and of their royal descendant, David, who united both the poetic and the martial qualities in one person, that clearly gave rise to the Arcadian image of the shepherd-knight. Moreover, in Christian symbology the good Shepherd is inextricably linked with his sheep, and the spotless or unblemished 'lambs of God' that are sacrificed for the expiation of sin are also synonymous with, or an aspect of, the Christ. As we will see, Francis' choice of '*Immerito*' for a pseudonym embodies this image of Christian truth. Christ is, above all, the Shepherd of all shepherds and the *Imperator* of all knights, and the Christly shepherd-knight is Hemetes, the Christian Red (Rose) Cross knight who is wed to Una, the Truth.[192]

(16) WILLIAM CECIL, Lord Burghley, and his son ROBERT CECIL, Earl of Salisbury — oil painting painted after Robert was created Knight of the Garter, and showing both father and son with Garter Georges. Artist unknown.

Royal Revelation and Banishment

Sometime between the Kenilworth and the Woodstock Entertainments — either at Stafford Castle, Dudley Castle or Hartlebury Castle — Francis Bacon received a devastating blow which affected his whole life dramatically. Francis discovered who his natural parents were and who he rightly was. The knowledge did not change his philanthropic views and purposes, but the manner of the discovery and what it signified made an immense difference to his thoughts, emotions and the manner in which he could afterwards conduct his life and work. The sudden and unintended revelation burst explosively into his (and others) consciousness with an uncontrolled force that caused dread lips to be sealed for evermore, inaugurating a chain of crucifying events that were to have the profoundest effect on mankind.

The story of this royal revelation is, naturally, only to be found told by Francis himself in his ciphers, but he recounts the story in full and in at least two different ciphers[193], each checking against and enhancing the other. In his 'Word Cipher' account the dire emotional impact is experienced to the full, as he pours out his heart to his future readers in a faithful and detailed report of the events as they occurred and as they were actually experienced by him at the time, hiding very little. In his 'Biliteral Cipher' account he provides summaries and additional details, but in this latter cipher the emotion is on the whole reserved and kept out of the way.

Not all that Francis recorded in cipher has been deciphered yet, and there are various details and missing sections of the story here and there which, if we cannot supply them from known history, we have to form an educated opinion about. One such detail is the date and place of the royal birth revelation. Checking what Francis tells us with our present knowledge of history would appear to place the date of the royal revelation a little before the escape of Françoise de Valois, Duc d'Alençon, from the Court of his brother, Henri III, King of France, on the 15th September 1575, but not before the end of the Kenilworth Entertainment (8th July — 27th July 1575) which we feel certain from other evidence that Francis witnessed. We also know that the revelation took place in a castle, with postern gates, and with a "gothic mansion" within a short walking distance at which Sir Nicholas and Lady Anne Bacon, with Francis, were lodged whilst the Queen resided at the castle. It also appears certain that this mansion was in a town or walled "city" attached to the castle, with a range of shops that included an apothecary's.

Into every generation is born a soul filled with malice, deformed spiritually and often physically as well, as a kind of countermand and adversary to any generous and light-filled impulse. For every movement we make, there needs to be an opposite movement of resistance to make this step possible. Every Jesus has his Caiaphas, every knight his dragon. Francis Bacon had Robert Cecil.

From childhood Francis and Robert had detested each other, Francis loathing the malice in Robert and Robert loathing Francis through envy and jealousy. Both boys were born with highly developed intellects, but there the likeness stopped. Francis had also a kind heart, filled with love and good-will, a generous disposition, and was endowed with a noble spirit, a handsome form and bearing, and was well liked by people of all walks of life. Robert Cecil, on the other hand, was born deformed, an unsightly hunchback who was feared and disliked by his contemporaries, and whose heart was filled with envy of the fortune of others and malice towards them in revenge for his own mishap. As the two boys were frequently in each other's company — Francis being the adopted son of Lady Anne Bacon and Robert being the elder child and only son of Lady Anne's sister, Lady Mildred Cecil, wife of Sir William Cecil (Lord Burghley) — it was Francis with all his gifts who became the principal target of Robert Cecil's poisoned mind. Both Francis and Anthony Bacon were hated by Robert Cecil, and Robert early on made a vow, publicly expressed, to see the ruin of those two devoted brothers, his cousins. To this 'short-list' he added that of Robert Devereux, the Queen's second son, when he discovered that Francis and Robert Devereux were in fact royal brothers, rightful sons of the Queen.

The whole cause of the royal revelation lay in the maliciousness of the twelve-year-old Robert Cecil who had wanted to avenge himself upon the Queen and one of her ladies-in-waiting, due to an incident in the hall of the castle where the Court was assembled one fateful day in August, 1575. Even Robert Cecil had not entirely foreseen what his spite of the moment would lead to. But let us allow Francis to tell the story in his own words — first in the Biliteral Cipher[194]:

> ...... Queen Elizabeth, who as hath been said previously, publicly termed herself a maiden-queen, whilst wife to the Earl of Leicester. By the union, myself and one brother were the early fruits, princes by no means basely begot, but so far were we from being properly acknowledged, in our youth we did not surmise ourselves other than the son of the Lord Keeper of the Seal, Nicholas Bacon, in the one case, and of the Earl of Essex, Walter Devereux, in the other.
>
> Several years had gone by ere our true name or any of the conditions herein mentioned came to our knowledge. In truth, even then the revelation was in a measure accidental — albeit 'twas made by my mother — her wrath over one of my boy-like impulses driving her to admissions quite unthought, wholly unpremeditated, but, when thus spoken in our hearing, not to be retracted or denied .....[195]
>
> I am indeed, by virtue of my birth, that royal though grossly wronged son to our most glorious yet most faulty — I can find no stronger terms — Queen Elizabeth, of the stock that doughty Edward truly renowned. Of such stock Henries Fifth, Seventh and Eighth, historic battle kings came, like branches sent from the oaks. My true name is not as in some back pages it was given, but Tudor. Bacon was only foster parent to my early youth, yet was as loving and kind to me as his own son, careful of my education, and even aspiring to my high advancement. But to Mistress Anne Bacon, ever quick with her sympathy and wise to advise, I do owe a greater or warmer gratitude, since she did much more truly and constantly guard, guide, protect and counsel me.

Moreover, to her I owe my life, for though she did not rear me, not being, *de facto*, my mother, it was by her intervention that the hour of nativity did not witness my death.[196] Her majesty would truly have put me away privily, but Mistress Bacon, yearning over helpless baby-hood, saved me, having held over me a hand of protection. My attempts in after years to obtain my true, just and indisputable title of Prince of Wales, heir-apparent to the throne, must not however be thought or supposed to indicate that I held myself disinterested of these obligations, offered affront to these kind benefactors, or in any way conducted myself in such manner as would either cast reflections upon my breeding or do discredit to my birth.

It may clearly be seen that it was but the most commonplace of ideas — an action barely ambitious because 'twas simply natural. But it failed most sadly, for the would-be Virgin Queen, with promptness (**not liking our people's hearts to be set upon a king**), before my *ABC*'s even were taught to me, or the elements of all learning, instructed my tutors to instill into my young mind a desire to do as my foster father had done, aspiring to high political advancement, look for enduring renown there; not dreaming, even, of lack wherein I should look for many honours, since I was led to think I was born to nothing higher.

Of a truth, in her gracious moods my royal mother showed a certain pride in me when she named me her little Lord Keeper, but not the Prince — never owned that that be truly the rightful title I should bear, till Cecil did sorely anger her and bring on one of those outbreaks of temper against one of the ladies of her train who, foolish to rashness, did babble such gossip to him as she heard at the Court. In her look much malicious hatred burned towards me for ill-advised interference, and in hasty indignation said:

"You are my own born son, but you, though truly royal, of a fresh and masterly spirit, shall rule nor England nor your mother, nor reign over subjects yet to be. **I bar from succession forevermore my best beloved first-born that blessed my union with** — no, I'll not name him, nor need I yet disclose the sweet story concealed thus far so well men only guess at it, nor know of a truth of the secret marriages, as rightful to guard the name of a Queen as of a maid of this realm. It would well beseem you to make such tales skulk out of sight, but this suiteth not your kingly spirit. A son like mine lifteth hand nere in aid to her who brought him forth; he'd rather uplift craven maids who tattle whenever my face (aigre enow ever, they say) turneth from them. What will this brave boy do? Tell *a,b,c*'s ?"

Ending her tirade thus she bade me rise. Tremblingly I obeyed her charge, summoned a serving-man to lead me to my home and sent to Mistress Bacon.[197]

But now let us read the story in Francis' words in the Word Cipher, and thus re-experience the great drama and emotional impact that this most powerful event had on Francis' life, placing as it were a "causeless curse" upon his head from which his entire life from then on "felt the blight":

When I consider the riddle of this monster's [*i.e.* Robert Cecil] dominion over her [*i.e.* the Queen] I must indeed admit it to be one of the things for which there is no solution, being, as he was, rude exteriorly; his head, by its own weight and heaviness, turning his neck over on one side, and upon it he had a mole, a sanguine star, that was a mark of wonder. His limbs were so abortive, defective and loose jointed that he staggers in his feeble step.

Taking note of his abhorred and beastly, prodigious face, women were as afraid of him as of the devil. And when they talk of him they shake their heads and whisper one another in the ear; and she that speaks doth grip the hearer's wrist; whilst she that hears makes fearful action, with wrinkled brows, with nods, with rolling eyes; and if they be by chance left alone with him, they shortly weep and howl.

I saw him once break into a mad passionate speech and entreat the Queen to dismiss them [*i.e.* the Queen's ladies-in-waiting] from the Court. But she condoles his mishap and smiling said they were a company of merry fools: "Let them laugh and be merry; they had rather lose a friend than a jest, and what company soever they come in they will be scoffing, insulting over their inferiors. God in heaven, man, you have no cause to complain. They would make me the subject of a calumny, a scurrilous and bitter jest, a libel, a pasquil, satire, apologue, epigram, stage-play or the like, for want of change. They live here solitary, alone, sequestered from all company but heart-eating melancholy, and they must crucify someone.

"Every one of these creatures pities you, and if thou didst but hear them play and dance I know thou wouldst be so well pleased with the object that thou wouldst dance thyself for company. Thou wilt without doubt be taken with such companions, and they will be especially delighted to let thee be in company with them".

And with her ivory hand she wafts to her a fair maid, the worst one of her merry company of women and the one most adverse to the part, and addressing her said: "This good gentleman is not ashamed to confess that he takes an infinite delight in singing, dancing, music, women's company and such-like pleasures, therefore he would have thee dance. And, fair goddess, fall not deep in love with him."

Saith the lady: "Does the lamb love the wolf? Give me good excuse, Madam, for I am sick and capable of fears, a woman naturally born to fears, and therefore full of fears: and though thou now confess that thy Highness didst but jest with my high spirits, I cannot but quake and tremble all this day. If he were but grim I would not care. I then would be content, for then I should love him. But, as all may witness, he is fair".

Tickled with such good answer, the Queen said: "He is a happy man; take his arm and go along with him; enjoy the brightness of this clear light and those nimble feet."

It is an old saying: a blow with a word strikes deeper than a blow with a sword, and he was more galled with his own royal mistress' wit than he was with his merry companion. He went with the poor maid and, buried in silence, stood like a blasted tree amongst them; and they in consternation like bashful, solitary, timorous birds, avoided him. They broke away as if he were a mad dog which must by all means be avoided .....

Cecil, who from the hour of his birth was weak, sickly and deformed, stood like a hapless, wretched, mis-shaped and sullen knave, plunged in melancholy, while his companions were busily discoursing about him behind his back. He distasted this kind of company out of a sinister suspicion that such an infinite company of pleasing beauties obscured his sickly and unnatural body. For this reason such a saucy companion oppressed him with fear that, by her wanton image, she might provoke and tempt the spectators to laugh.

Every base knave has a wolf's nature, and this foul devil, I promise you, was as hard-hearted, unnatural a monster as the devil and his ministers need to have. Cunning hath made the devil so sly that he devises a way to be revenged upon the soft, silly maid, and withal at the same time to be honoured, admired and highly magnified. To do this the monster of a man cheats his fair companion into covert rubs of the worth and honour of the Queen.

The complexion of the maid changed from pale to red, and from scarlet to pale, when he with big, thundering voice cried twice: "All this condemns you to the death to so much dishonour the fair Queen."

As falcon to the lure flies the Queen to him to ask what he had heard.

"Madam, this innocent and pure model, moved by love for thee, told me that thou art an arrant whore, and that thou bore a son to the noble Leicester. I pray that thou give her chastisement. Either thou must, or have thy honour soiled with the attainder of her slanderous lips."

Holy St Michael, what a change was here! As a painted tyrant the Queen stood, and like a neutral to her will and matter did nothing. But as you often see against some storm a silence in the heavens, the wrack stand still, the bold winds speechless and the orb below as hushed as death, anon the dreadful thunder doth rend the region. So, after a pause, upon mine honour you should have heard the great Queen roar against the fair daughter of Lord Scales.

"By Holy God!" in uncontrollable rage said she, "Thou liest, dishonourable, vicious wench! We were married to him by a friar — a tried holy man — and if our dear love were but the child of state, it should be told. The world should know our love, our master and our king of men. Small glory dost thou win to frame this public, foul reproach. Behold the open shame which unto us day by day is wrought by such as hate the honour of our name. And shalt thou do him shame? By God, we will cut and mince the throat that doth call us a common whore! Like to a Turkish mute thou shalt have a tongueless mouth."

With shrilling shrieks the wretched lady turned and in a twinkling, like the current, flies in violent swift flight from her fair foe. After her, in rage and malice, the great Queen chases. As she doth bound away her sunny locks hang o'er her temples like a golden fleece, and as she flies, inflamed with rage, her gown slipped from her, and in her shift she springs along. In a circle they take their flight, and after long pursuit and vain assay, whether fear, wicked fortune or cruel fate the girl mislead, by some unfortunate hap or accident down she did tumble, and, being a woman, there did lie.

The angered Princess, as she lies, above her lily arms turned her smock, and in her hair her hands she dived and hales her up and down in cruel wrath. She said: "I'll unhair thy head; thou shalt be whipt with wire and stewed in brine, smarting in lingering pickle. I'll spurn thy eyes like balls before me. I will teach thee to slander me — thou hast lived too long."

And then, from one that before her bends, she draws a knife. The lady had taken advantage of the time, and with arms outstretched essays to fly, but eclipses crooked against her fight, and the Queen, who in her hand the foul knife grasps, did jump upon her, and they both together fall upon the slippery floor. Unmoved with her plenteous tears and prayers, the despightful Queen at the maiden's heart and snow white breasts did strike and tilt. O she did plead and her intreat for mercy, crying, "O, thou wilt kill me; forgive me; kill me not," and conjured her to spare her life. But the cruelty of womanhood is such the Queen her heeded not, and because of the slippery floor that would not let her stand, she presently did sheath her dagger and stamped upon her breast, for those milk paps that through her windows barme peep at men's eyes are not within the leaf of her pity writ.

At last, when all her speeches she had spent, nature, in sad despair, her senses swooned; and like a wearied lamb she lies panting there. My boding heart pants, beats, takes no rest, as with the rest of the royal Court I in painful silence stood, tears in mine eyes, being grieved that I, a youth, must mine eyes abase and be content to see such wrong. I swear mine ears ne'er heard such yells, nor mine eyes such fury and confusion, horrible! Thou should'st have seen the poor maid's blood paint the ground gules, bleeding from the lips. Through the armour of a Prince Saturnine the sight would pierce.

My resolution being taken at last, I ran where hateful death put on his ugliest mask to fright our senses, and said as I held her arm: "Fair Queen, I kiss your Highness' hand. See, see, O see what thou hast done! Pause in God's name! Be not as barbarous as a Roman or Greek. Good Madam, patience. May I not remove the maiden?"

The wrath of the enraged Queen like an earthquake fell upon my head and, my lord, I'll tell you what, all my glories in that one woman I forever lost. The Queen like thunder spoke: " How now, thou cold-blooded slave, wilt thou forsake thy mother and chase her honour up and down? Curst be the time of thy nativity! I would the milk thy nurse gave thee, when thou suck'st her breast, had been a little rats-bane. I am thy mother. Wilt thou stoop now and this good girl take away from me?"

I stand aghast and most astonished. Then she said again: "Slave! I am thy mother. Thou mightest be an emperor but that I will not bewray whose son thou art; nor though with honourable parts thou art adorned, will I make thee great, for fear thyself should prove my competitor and govern England and me."

> As she spoke my legs, like loaden branches, bow to the earth, as willing to leave their burden; my strength fails and over on my side I fall.
>
> "Fool! Unnatural, ingrateful boy! Does it curd thy blood to hear me say I am thy mother?" And into her eyes fierce, scornful, nimble lightnings dart with blinding flame".[198]

Nearby, standing still and listening intently to all this, was Robert Cecil. When the Queen revealed her secret to her son, stooping her head as low as Francis', she also revealed the royal secret to "that devil who, wrapped in the silence of his angry soul, stood listening"; and "the fury of his heart he in his deformed face portrayed". From that moment on Robert Cecil worked in every way he could to prevent either Francis or his royal brother, Robert (Devereux), being declared heir to the Throne of England, and even contrived to bring about their humiliation and deaths — in all of which he was almost entirely successful.

The heat of the situation being now near exhausted, Robert Cecil came up to the Queen to tactfully offer his apologies and to artfully suggest that he should take the dishonour upon himself and that the Queen should pardon the maid. The Queen naturally dismissed this contrivance, saying that the matter was at an end and that he could not do as he suggested with his honour; but she did 'pardon' her lady-in-waiting, allowing her to be attended to, banishing her from Court and cursing her.

> "Trouble me no more", said the Queen. "I do repute you every one my foes. I'll pardon her, but take heed! Such as thou die miserably. We have an ill-divining soul, and either our eye-sight fails or we, methinks, see thee now as low as one dead in the bottom of a tomb. Thou mumbling fool, utter thy gravity o'er a gossip's bowl, for here we need it not. Great God, all our care hath been to have this secret hid! And now to have a wretched, puling fool, a whining mammet in her fortunes tender, tell it in company of the whole Court! Thou shalt not house with me. Dry thine eyes and go; get thee hence! I will pardon thee, but, my lady wisdom, we hope thou wilt hold thy tongue and let good prudence smatter with thy gossip. Go; speak not, nor answer us not; or by this hand we will yet teach thy tongue proper wisdom.
>
> "And thou, my son, thou foolish child, a pack of blessings light upon thy back. Speak thou not of this that thou hast heard, but go. Speak not; begone! I desire thee to know no more. Look, let thy lips rot off e're thou speak of this. Get you gone."
>
> Stupified, I abruptly rise, turn, and as my tears made me blind, with uncertain steps cross the Court, and by means of one of the several posterns leave the castle and swiftly toward the city walk; and as my feet ascend the hill, mindless of the way, I thought I will go to the worthy lady up to whom all this time I have believed to be my mother — mistress Anne Bacon, the wife of that renowned and noble gentleman, Sir Nicholas Bacon.[199]

Francis reached the Bacon's house, breathless, and found Lady Anne just as she was about to leave the house to go riding. He pleaded with her to remain and talk with him, and they went into Lady Anne's study (or withdrawing room). When Francis had recounted all that had occurred and asked Lady Anne to tell him the truth, poor Lady Anne dissolved into

(17) LADY ANNE BACON, wife of Sir Nicholas Bacon, aged 54 years — oil painting (1580) attributed to George Gower.

tears as she confessed that Francis was indeed the Queen's son but that the Queen and Leicester had promised to allow her to adopt and raise Francis as her own child, never revealing the story of his natural birth. Under Francis' repeated pressing, she told him how he had been born in Windsor Castle during the period when she (Lady Anne) was the Queen's chief lady-in-waiting, how she had attended the birth and had begged for Francis' life when he was born, and how she managed to obtain the Queen's and Leicester's permission to take the newborn child secretly and raise him as her own. She added, in answer to Francis' question:

> "The very force of circumstances made it impossible for Queen Elizabeth to own you as her son. She could not do it without betraying the secret of a very terrible crime which, led on by the great but licentious Semour, she committed when a girl."[200]

Lady Anne went on to tell Francis the whole story of the Queen's pregnancy, when a teenage princess, and how Lady Anne (then the Princess' lady-in-waiting and confidante) was forced to not only assist at the birth of the babe but also to dispose of the body of the stillborn child. It would also appear from what Francis relates in his ciphers that the babe had been murdered by the Princess Elizabeth, probably in attempt at abortion. (See Journal I/3, *Dedication to the Light*). Having talked a little about subsequent events, and having assured Francis that he was not base born (*i.e.* a bastard) but that his royal mother and Leicester were legally married in the presence of witnesses, before Francis' birth, she bade Francis go to bed as night had crept upon them.

The next day Francis pondered and meditated on his birth, and what it might mean, in the library of his foster-parents' house. About three-o'clock in the afternoon Robert Cecil arrived with a message from the Queen, summoning Francis to her royal presence. But Cecil delivered this in such a back-handed way, jibing with all his spite at Francis' seemingly bastard estate, suggesting the rape of the Queen and other invidious matters, that the message was forgot about in the argument that ensued. Francis could bear the taunting no longer and hurled himself upon his tormentor, knocking Cecil to the ground, bloodying his nose and blacking his eyes. Too late Francis realised what a fool he had been to rise to Cecil's bait in such a way. The black malice that Cecil held towards Francis increased a thousand-fold, and Francis did indeed live to regret those blows. Cursing Francis and swearing revenge, Cecil left the room:

> "I, that have the spight of wrekful heaven upon me in deadly hate of you, will lay plots and inductions dangerous against you, for you are mine enemy. O that I were a man to fight with you! But beware! for I will sort a pitchy day for you. I will buzz abroad such prophecies that Elizabeth will be fearful of her life, and then to purge her fear I'll be your death. And for this stroke upon my crest, and for this blood of mine, I will not suffer you to sit in England's royal throne. I do know her spirit; I will raze her honour out, and this feeble hand shall make you crouch in litter of stable planks, to hug with swine, to seek sweet safety out in vaults and prisons."[201]

The noise of the fight alarmed Lady Anne and she hastened to find out the matter. When she arrived Cecil had already left, but she soon discovered by careful questioning of Francis what had happened. Lady Anne was fearful of the possible results, but she sat down and assured Francis of her devoted love for him. She then went on to disclose the whole story of Elizabeth and Robert Dudley's love for each other; their first (bigamous) marriage whilst imprisoned in the Tower of London during Mary's reign; how Francis was conceived; how

Elizabeth, then Queen, ordered Dudley to murder his wife Amy before Francis was born; how they had been married for the second time — this time legally — but in secret at the house of Lord Pembroke, with Sir Nicholas, as Chief Judge of the Realm, performing the ceremony in front of two other witnesses, Lady Anne herself and "Lord Puckering"; and how Francis' birth followed on from this, thus making him 'nobly' born and legally the rightful heir to the throne of England.

By the time Lady Anne had finished relating this history of Francis' origins, the summer night had passed and dawn was breaking. Four o'clock struck. But just as they were about to retire (Lady Anne promising to tell yet more) the Queen's page arrived with a guard to command Francis' presence before the Queen. Only then did Francis remember Cecil's message from the Queen. Too late — again! Twelve hours had passed since that first summons; it was a new day and still the Queen was waiting.

Francis was speedily escorted off to see the Queen, but on his way to the castle he begged permission to visit an apothecary's shop, where he bought enough poison to commit suicide if need be, fear working so hard within him.

Meantime Robert Cecil, to whom for some unaccountable reason Elizabeth always paid heed, reported back to the Queen and worked his poison upon her mind. The Queen had for long used him — even as a child — as her private spy, a position which he relished and adroitly played to suit his own desires. Francis records:

> My hands — aye, my head as well, more than all, my heart — are sorely wounded; for in a breath my royal mother disclosed our relationship and cursed my nativity: nor could I, in the numerous subsequent encounters, change her hasty decision upon that very important question of the succession. 'Tis said: "The curse that was not deserved will never come." Some may find it true, but to me a causeless cause did surely come, and my entire life felt the blight.
>
> Nevertheless, to Robert Cecil I owe much of this secret, underhand, yet constant opposition: for from the first he was the spy, the informer to the Queen, of all the boyish acts of which I had least cause or reason for any pride. This added fuel to the flame of her wrath, made me the more indiscrete, and precipitated an open disagreement, which lasted for some time, between my foster mother, Lady Anne Bacon, and the woman who bore me, whom however I seldom name with a title so sacred as mother. In truth, Cecil worked me nought save evil to the day which took him out of this world.
>
> Through his vile influence on Elizabeth, he filled her mind with a suspicion of my desire to rule the whole world, beginning with England, and that my plan was like Absalom's — to steal the hearts of the Nation and move the people to desire a King. He told her that my every thought dwelt on a crown; that my only sport amid my school-mates was a pageant of royalty; that 'twas my hand in which the wooden staff was placed, and my head that wore the crown, for no other would be allowed to represent princes or their pomp. He informed Her Majesty that I would give a challenge to a fierce boyish fight, or a duello of fists, if anyone presumed to share my honours or depose me from my throne.[202]

During those twelve hours in which Francis seemingly resisted or ignored the command of his royal mother, the Queen of England, Robert Cecil had ample time and opportunity to enlarge and add to the fears, suspicion and spectral illusions already haunting the Queen's mind. Misconstrued, Francis' vast philanthropic aims, his writing and his play-acting, his extraordinary abilities and wide circle of friends in all quarters, could actually appear alarming if coupled with a self-knowledge of being the rightful Prince of Wales. Even the part Francis played in the Kenilworth Entertainment, if true, and if she knew, could be interpreted in a wholly different light by a vain Queen who was haunted by the expectation of having her throne and her life snatched away at the slightest chance. After all, history and her own experience gave her good reason to be filled with such morbid forebodings — and she was, paramountly, a fighter!

Elizabeth had also had time to confer with her "sweet Robin", the Earl of Leicester. Judging by what Francis says in his Biliteral Cipher, the anger of Leicester was stirred up no less than the Queen's by the ill-timed and unfortunate royal revelation. Leicester had many plans up his sleeve, hoping in time to at least establish one of his sons by the Queen in her favour and for her to name either Francis or Robert as her heir, which would in turn necessitate himself being established publicly as the Queen's Consort. The Earl was playing a difficult game. In the previous year (1574) he had had a notorious affair with Lady Sheffield, resulting in a son being born on August 7th, 1574. Then, following this affair, Leicester began a flirtation with Lettice Knollys, the Countess of Essex, the very person who had adopted his second son, Robert, by the Queen, resulting in rumours of his having had two children by Lettice before the end of 1575. Whether true or not, Leicester somehow needed (and wanted) to make great efforts to reconcile himself with Elizabeth and to reinstate himself ever more thoroughly in her favour. Doubtless the Queen's determination to remain 'virgin' in the eyes of the general public had much to do with the various liaisons and courses that the Earl took. However, the huge effort that Leicester made with the Kenilworth Entertainment to persuade the Queen to marry him publicly had not, as far as he was concerned, worked. Then, to have this followed almost immediately by the embarrassing and humiliating royal revelation of Francis' birth, was undoubtedly a nail in the coffin of his desires. The result was that both Elizabeth and Leicester were determined upon Francis' immediate banishment to France, Leicester urging this banishment even more than the Queen.[203]

When Francis arrived in the Presence Chamber, the Queen railed at him in anger and threatened to hang him. But then she commanded Francis to follow her to her private room. When they were alone she showed much more concern over her eldest son, and quickly disclosed her greatest anxiety and condemnation concerning Francis — that he had, in her and others eyes, lost his princely status by consorting with a baser society and taking part in public activities that were considered beneath the dignity of his position and birth. Although many stories had been invented or exaggerated by Robert Cecil, others had not. Francis tried hard to explain his good reasons for this, and to outline his great Plan, but his words fell on stony ground. Not until he wrote the story of Prince Hal (*in Henry IV*) could he try to explain openly why 'princes' should have need to study and know by experience the Kingdom that they are to rule.

> *Queen:* "I know not whether heaven will have it so for some displeasing service I have done, that in His secret doom, out of my blood He'll breed revengement and a scourge for me; but thou dost, in thy passages of life, make me believe that thou art only marked for the hot vengeance and the rod of heaven to punish my mistreadings. Tell me else, could such inordinate and low desires, such poor, such base, such lewd, such mean attempts, such barren pleasures, rude societies as thou art matcht withal and grafted to, accompany the greatness of thy blood, and hold their level with thy princely heart?"

Francis: "So please your Majesty, I would I could quit all offenses with as clear excuse as I can purge myself of this that I am charged withal. Doubtless I am in reproof of many tales devised, (which oft the ear of greatness must hear), by smiling pick-thankes and base news-mongers, yet such extenuation let me beg, wherein my youth hath faulty and irregular wandered; and that I may for something true find pardon, on my true submission."

Queen: "Heaven pardon thee. Yet let me wonder, Francis, at thy affections, which do hold a wing quite from the flight of all thy ancestors. Thou art almost as alien to the hearts of all the Court and princes of my blood; the hope and expectation of thy time is ruined, and the soul of every man prophetically doth forethink thy fall .... For thou hast lost thy princely privilege with vild participation; not an eye but is aweary of thy common sight, save mine, which hath desire to see thee more, which now doth that I would not have it do, make blind itself with foolish tenderness. Thy companions are the very disturbers of our peace, a company of irreligious harpies, scarping, griping catchpolls, unlettered, rude and shallow; thy hours filled up with riots, banquets, sports, and in thee is noted never any study, any retirement, any sequestration from open haunts and popularity. Thy addiction is to courses vain with thy familiars and coadjutors, and should I make thee mine heir, thou wouldst make my throne a seat of baseness."

Francis: "No, Madam, I would rather add a lustre to it."

Queen: "Peace, sir! Most subject is the fattest soil to weeds, and thou, the noble, manly image of my youth, art overspread with them; therefore my grief stretches itself beyond the hour of death. The blood weeps from my heart when I do shape (in forms imaginary) the unguided days and rotten times my kingdom shall look upon when I am sleeping with my ancestors; for when thy headstrong riot hath no curb, when rage and hot blood are thy counsellors, when means and lavish manners meet together, oh, with what wings will thy affections fly towards fronting peril and opposed decay. And in this great work (which is (almost) to pluck a kingdom down and set another up,) I will not be like one that draws the model of a house beyond his power to build,[204] and who (half through) gives o'er and leaves his part-created cost a naked subject to the weeping clouds, and waste for churlish Winter's tyrany; as in an early Spring we see the appearing buds, which to prove fruit hope gives not so much warrant as despair that frosts will bite them."

Francis: "I thought that thou hadst hated me. I will in the perfectness of time cast off my followers and their memories, shall as a pattern or a measure live. Thy Highness knows, when we mean to build, we first survey the plot, then draw the model; and when we see the figure of the house then must we rate the cost of the erection, which, if we find outweighs ability, what do we then but draw anew the model in fewer offices, or at least desist to build at all? And we, by turning past evils to advantages, should survey the plot of situation, and the model consent upon a sure foundation, question surveyors, know our own estate, how able such a work to undergo. And I but study my companions like a stange tongue wherein to gain the language; for 'tis needful that the most immodest word and gross term be looked upon and learned, which once attained comes to no further use but to be known and hated."

Queen: " 'Tis seldom when the bee doth leave her combe in the dead carrion. I speak not of the creatures which are useful in thy kitchen to turn the spit, lick the pan, and make the fire burn; but of the train of gallants that at thy heels men say doth run. Art thou not by birth a prince? Why then dost thou look so low, as if thou hadst been born of the worst of women? Thy tastes are not for royal deeds, and 'twere sin to stain England's throne by such a counterfeit image of a king. And, to shield thee from disasters of the world, I am resolved that thou shall spend some time in the French Emperor's Court. Muse not that I thus suddenly proceed, for what I will, I will; and there an end. Tomorrow be in readiness to go; excuse it not, for I am peremptory".

Francis: "Madam, I cannot so soon be provided; please you deliberate a day or two."

Queen: "No more; look, what thou wantest shall be sent after thee. And as thou canst not live on grass, on berries and on water, as beasts and birds and fishes, nor on the beasts themselves, the birds and fishes, therefore for thy provision thou shalt receive enough from me for thy maintenance. And so, my son, farewell; may all happiness bechance to thee. Were it not thy hated affections change thy tender days to the worse than brutish company which hath meta- morphosed thee and made thee neglect thy studies, lose thy time, war with good counsel and set the world at naught, I would rather keep thee at home and send thee to the studious universities. But now farewell. Blest may thou be like to the flowering sheaves, that play with gentle winds in summer-tide; like olive branches let thy children spread; or as the kids that feed on lepher plains, so be the seed and offspring of thy loins. And in long years to come, may thy pure soul and spright enrich the heavens above. Farewell."[205]

Thus was I banished. And on the day following, about the hour of eight, I put to sea with that gentle knight, Sir Amyas Paulet.[206]

(18) SIR AMYAS PAULETT, wearing the Garter George — portrait (*c.* 1575) by G.P. Harding.

Discovery of High Culture and Romance in France

In just a little over twenty-four hours from his final interview with the Queen, Francis was aboard a ship and sailing for France. He was in the charge of Amyas Paulett, his French tutor and a good friend of his foster parents. Amyas Paulett was commissioned by the Queen as a Special Ambassador to the French Court, with a mission that was highly secret. It is not entirely clear whether Amyas Paulett had already been destined for France on a special embassy before the royal revelation, or whether the banishment of Francis to France was in fact the special mission and Amyas Paulett had been rapidly commissioned, as Francis' French tutor, into performing the task. As there are no indications or prior qualifications singling out Amyas Paulett to be chosen as a Special Ambassador to the French Court for any other diplomatic mission, I would surmise that his special mission was to act *in parentis* with regard to Francis, supervising his care and subsequent education and training in France, and that Paulett was appointed specifically for this task, for which he was indeed specially qualified. In the cipher Francis refers to him as Sir Amyas Paulett, but the knighthood was not in fact conferred upon Paulett until 1576 when he returned home with Francis, having performed his task well and to the Queen's satisfaction. Then he was knighted and sent back again to France to become the principal Ambassador to the French Court, replacing the previous Ambassador, Dr Valentine Dale. But the cipher story of this royal revelation and banishment was not written until Francis had returned to France for the second time with Sir Amyas, in September 1576, when Francis began working on ciphers in earnest.

As for Francis, he was entrusted wth a secret mission, "with business requiring great secrecy and expediency,"[207] and without doubt Amyas Paulett would have been charged with assisting Francis in the carrying out of this mission as well as being Francis' guardian. This was expedient, as a good reason would need to be given for the special embassy and for Francis' presence in France at the French Court. It also gave Francis an extremely privileged position at the French Court.

Francis went to France filled with apprehension, but also with excitement. He was sorrowful at being uprooted so ruthlessly (as it seemed) from his native soil, the family which loved him, his friends and grand schemes, but on the other hand his spirit begged for adventure and his mind for even wider experience. The period he was to spend in France was in fact to be the most enjoyable period of his life — a passion which was to act as a motor to all his future schemes and life.

> As you no doubt are cognisant of our summary banishment to beautiful France, which did intend our correction, but oped to us the gates of Paradise .....[208]

> In due time the Queen, afraid of these ominous portents,[209] sent for good Paulett and arranged that, under pretext of great import, I should accompany our ambassage to France. I was placed in the care of Sir Amyas and left the shores of my own fair land without a moment of warning, so to speak. The Queen, by her [power] royal and her rights maternal, readily overruled all our several objections. No tears on the part of my dear foster-mother, nor entreaties of that of grave Sir N. Bacon availed, while I, as soon as my first protest had been waived, occupied my fantasy hour after hour, picturing to myself the life in foreign lands.
>
> The fame of the gay French Court had come to me even then, and was flattering to the youthful and most natural love of the affairs taking us from my native land, inasmuch as the secret commission had been entrusted to me, which required much true wisdom for safer, speedier conduct than would have if left to the common course of business. So with much interest, though sometimes apprehensive mind, I make myself ready to accompany Sir Amyas to that sunny land of the South I learned so supremely to love, that afterwards I would have left England and every hope of advancement to remain my whole life there. Nor yet could this be due to the delights of the country, by itself, for love of sweet Marguerite, the beautiful young sister of the King (married to gallant Henri, the King of Navarre) did make it Eden to my innocent heart; and, even when I learned her perfidy, love did keep her like the angels in my thoughts half of the time — as to the other half, she was devilish, and I myself was plunged into hell. This lasted during many years, and not until four decades or eight lustres of life were outlived did I take any other to my sore heart. Then I married the woman[210] who hath put Marguerite from my memory — rather, I should say, hath banished her portrait to the walls of memory only, where it doth hang in the pure, undimmed beauty of those early days — while her most lovely presence doth possess this entire mansion of heart and brain.[211]

All this was in secret, as it had to be, upon peril of his and others' lives, and only in cipher could Francis safely leave a record of what really took place for posterity to read and take note. But Francis did refer twice to his early 'mission' to France, more openly, and this was in two letters that he wrote early in 1595 to the Earl of Essex and to Robert Cecil, both of whom had known of his being sent to France by the Queen in 1575:

> "And your lordship may easily think, that having now these twenty years (for so long it is, and more, since I went with Sir Amyas Paulett into France, from her majesty's royal hand) I made her majesty's service the scope of my life: I shall never find a greater grief than this *'relinquere amorem primum'*."[212]
>
> "It may please your honour to deliver to her majesty, first, that it is an exceeding grief to me, that any not motion (for it was not a motion) but mention, that should come from me, should offend her majesty, whom for these one-and-twenty years (for so long it is since I kissed her majesty's hands upon my journey into France) I have used the best of my wits to please."[213]

Allowing for the method of counting which includes the year that the count is being made from, 21 years from 1595 takes us to 1575, thus confirming in signed letters the truth that

Francis was sent to France by the Queen when he was only 14 years old; and, further, that he relinquished his first love — his greatest grief — which is associated with this visit.

At the time of Francis' banishment to France, the corrupt and pleasure-loving Henri III sat on the throne of France. Henri III was the third son of Henri II. His elder brothers had ruled before him — Francis II briefly (1559-60) and Charles IX for fourteen years (1560-74) dominated by his wily and power-hungry mother, Catherine de Medici. When Henry III succeeded to the throne of France at the end of May 1574, at the age of 22, his mother, Queen Catherine, continued to maintain her powerful influence over him as well. Through her power politics her 19-year-old daughter, Marguerite de Valois, was married in 1572 to the 18-year-old Henri, King of Navarre, royal leader of the Huguenots (Protestants) in France, in the hope of settling the Wars of Religion which were devastating and splitting France. Through her hasty decision, the Massacre of St Bartholomew's Day was carried out during those wedding celebrations, which led to a fierce renewal of the civil war. Henri de Navarre's life was spared at that time, on condition he renounced his Protestant faith, but he, together with his wife Marguerite, the French King's sister, had to remain as virtual prisoners of the French King at the French Court. In all this Marguerite stood by her newly-wed husband, which undoubtedly had much to do with the preservation of his life, but they were not a well-matched pair and Henri seems to have made no attempt to gain her affection, proving himself to be a sorry husband, faithless and neglectful. Not surprisingly, Marguerite began to turn elsewhere for attention and affection.

When Henri III became King of France he continued to keep Henri de Navarre and Marguerite as 'prisoners' of his Court. He also kept as a 'prisoner' at Court his youngest brother, François de Valois, Duc d'Alençon, who had always been greatly mistreated by his elder brothers and mother. Alençon became a friend of Navarre, and also the secret head of a confederacy between the Huguenots and Politiques (a section of Roman Catholics who were more liberal and peace-loving than the main French Roman Catholic party led by the King and the Guises). By the time Francis arrived at the French Court, at the end of August or beginning of September 1575, the positions of Navarre and Alençon had become extremely irksome. Navarre's hereditary States were a prey to disorder, and his other liefs ravaged by war, his authority was declining and his revenue was reduced to a mere pittance. Alençon's position was no better, perhaps worse. The revenues which he received were too meagre for him to maintain the dignity of his position and he was deeply in debt. His brother, the King, treated him with coldness and contempt, and his friends were continually having quarrels thrust upon them by the mignons of the King. Under the will of his father, Henri II, Alençon should have been receiving a proper education and income from his elder brother and mother, but Queen Catherine and Henri III ignored these obligations. When Francis arrived at the dazzling French Court, this was the issue that exploded onto the scene, in his hearing. The main characters were, it should be noted, all young (except the mature Queen-Mother) — Henri III being then 23-years-old, Marguerite de Valois being 22-years-old, and Françoise de Valoise being 19-years-old.

In the Word Cipher, Francis describes his arrival at the luxurious French Court and what met his senses there during a banquet; how he was present during a demand by Alençon to the King, his brother, for his money and freedom, and the King's refusal; how he overheard certain scandalous rumours about Marguerite de Valois that were deliberately designed to disgrace her publicly and drive a wedge between her, her husband, Alençon and the King; and how he witnessed the fighting spirit and 'acting' of the fair Marguerite when she was summoned to answer the charges made against her.

In Gallia I arrived full safe, and presently rode post unto the matchless Court of France. The King's house was in Paris set, and the King, (as Ambassadors and Ministers of England's Queen) received and entertained us honourably. We were lodged and accommodated in great state in the royal Seat, as Ambassadors, and were, in an honourable manner, invited by the great King of France to visit and feast with him in royal Princes' state.

Being arrived in Henry's kingly hall, that like Adonis' garden bloomed with flowers, (for flower-de-lices, the lilies of the French, the rose of England, with sweet violets, pale daffodils and sweet honey-suckle[214] covered the walls and obscured the table). Such proud luxurious pomp, excess and pride of royal arras and resplendent gold dazzled my eyes. I never such choice crystal saw, nor such pomp of rich and glittering gold.

As I looked, the trumpets 'gan on high to sound, and all the gracious people in one full consort sent to heaven their echoed report, as forth came Henry, King of France, crowned as a royal King and garnished in wondrous robes; with him the aged, ancient mother-Queen, arrayed in antique robes to the ground, the sad habiliments did her right well beseem ......(here followed a description of the lords, ladies, gallants and other courtiers)...

I was such a novice in the Paris Courts, I wondered at the sight of such banquetings.....[here follows a description of the feast]....The comely services of the courtly train pleased me so much. The god that to our souls the music tuned held out his hand in highest majesty to entertain the flower of France. Thus we feasted, full of mirth, but nothing riotous, and every thing did abound with rarest beauty that all latitudes and countries could afford.

Now, by King Henry's command, still silence was imposed on all; and in the French tongue he did call me to his side and did salute me in honour of the Queen of England. And, smiling, as if he thought I would not understand the French, said: "My good youth, welcome to Paris, thou bearest thy father's frank face; thy father's moral parts may'st thou inherit too, and in thee live thy father's excellence. Is the lord, thy father well? With all his defects he might well be a copy to these younger times."[215]

After some polite repartee, in which the French King takes a definite liking to the intelligent young Francis, the King bids Francis to stand aside to allow the Duc d'Alençon to approach.

Francis, the next blood and heir of the House of Valois, approached the King; twelve peers of France arm in arm accompany him and by the table stand, and confer about some matter. I saw His Royal Majesty did with much unkindness hail the noble young prince of France, greeting, with sour looks and reverted forehead, the Prince, who upon his knee made supplication. Anon I heard the Duke speak thus: "Your Highness bade me ask today for the hundred thousand crowns". The King gaped and gazed upon him, and then exclaimed: "My Lord Duke, I am content to lose some of my crowns now and then to thievish Christian thieves, but thou art a most pernicious userer, froward by nature, enemy to peace. Thou dost disgrace the House of France. Get thee gone and leave me alone. Thou shamest me in this place. Thou hast dishonoured me. By honour of my birth, thou smooth-faced boy,[216] thou like a poor beggar dost dishonour

(19) CATHERINE DE MEDICI, Queen Mother of the French Valois Dynasty — line drawing by François Clouet.

(20) CHARLES IX, Valois King of France (1560-74) — oil painting from studio of François Clouet.

(21) **HENRI III,** Valois King of France (1574-89) — engraving by L'Armessin.

(22) MARGUERITE DE VALOIS, first wife of Henry of Navarre — engraving from *L'Histoire de la maison de Bourbon* (1779).

(23) FRANÇOIS DE VALOIS, Duke of Alençon and Anjou, aged 31 years — oil painting, artist unknown.

(24) HENRI D'ALBRET, King of Navarre and of France (as Henri IV, 1589-1610) — from an engraving attributed to Pierre du Moustier.

(25) HENRI DE LORRAINE, Duke of Guise — engraving, artist unknown.

me. God knows I am unprovided, and therefore thou joinest me with these lordly peers of France to put crowns into thy purse."[217]

Then follows an argument between the maltreated Alençon and his odious brother, in which Navarre intercedes to try to restore some measure of peace, offering Alençon five hundred crowns from his own slim purse. But Alençon is so stirred up that he even turns on his erstwhile friend Navarre, and taunts him concerning Marguerite's infidelity with the Duc de Guise. It ends with the King demanding Marguerite's presence and an explanation of why she had not been present at the banquet but was seemingly with the Duc de Guise. Marguerite is found and eventually brought into the King's presence at the banquet, where it is confirmed she had been "in a monastery two miles off", with Guise.

> With this, the Princess, formal in apparel, gait and countenance, sudden into the presence comes. She is the fairest lady I have yet beheld in France, and is walled about with diamonds, pearls and gold.[218]

This is the first sight that Francis has of the beautiful 22-year-old French Princess, and he is entranced from the start. Francis goes on to report how Marguerite bravely defended herself and tried to change the subject, failing which she attempted to remove the heat from herself to her husband, accusing Navarre of infidelity, adroitly practising the ancient precept that 'attack is the best form of defence'. A tempestuous argument ensued, with Catherine taking the side of her daughter, while all the assembly looked on, gaping, trying to hear the words being spoken and making bets as to the outcome. But the Queen Mother goes too far herself, suggesting black witchcraft on Navarre's part, and Marguerite, clearly perceiving a nasty conclusion if the argument persisted, employed her stagecraft to abruptly terminate the embarrassing scene — thus adding further to Francis' interest in and admiration of her.

> *Marguerite:* "I thank thee, Madam," answered the Princess, "Hath he drawn my picture, or doth he wound my heart with a leaden sword? Tut, how knowest thou all this? My state is not so bad, content thyself".
>
> *Catherine:* "Silence, mistress, thou wrongest me'".
>
> *Marguerite:* "Good mother, do not make me desperate".
>
> Then, as she stood, she sighed and counterfeited to sound; and with shrieks, as though her heart had been wounded with the claws of a lion, she measures her princely body upon the ground. To my eyes she looks not pale, and I would I were to bring her to. O, that I might with a needle prick her back! She no more resembles one in faint than I.[219]

Catherine, quickly and shrewdly playing the part with her daughter, straightaway accused the men of killing her innocent child with slander and, before they had a chance to react, caused Marguerite to be quickly lifted by her gentlewomen and taken out into her chamber, the Queen-Mother retiring with her daughter and preventing the now anxious (or suspicious) King from following. The commotion naturally caused the slanderous story to be discussed throughout the room, and Francis was soon able to comprehend what it was all about.

I, among the rest, stay with his Grace, and enter into the secrets of this viperous slander.[220]

Following this storm, Henri de Navarre pleaded with the French King to allow him and his wife freedom to leave the Court and return to his domains, but the request was turned down with a threat by Henri III that if Navarre should so much as try to escape he would be hung like a thief and war waged upon his "perjured provinces". Both Alençon and Navarre had made vain attempts to escape before, together, when they had been friends and compatriots; now they became doubly determined on obtaining their freedom, but the jealous feud that had arisen between them over the favours of their common mistress, Charlotte de Beaune, Baronne de Sauve, who was lady-of-honour to Queen Catherine and who had been charged specifically with the job of causing dissention between Navarre and Alençon, made these two Princes seek their escape separately.

Alençon made his escape soon after Francis' arrival at the French Court and the banquet scene. On September 15th, 1575, about six o'clock in the evening, muffled in a cloak and accompanied by only one gentleman, Alençon managed to leave the Louvre and meet up secretly with Simier, his Master of the Wardrobe, who awaited him at the Porte Saint-Honore with a coach. Travelling to Meudon, he was joined by Guitry, the Huguenot leader who had tried to effect Alençon's escape eighteen months earlier, with a body of cavalry. From Meudon they rode to Dreux, which had been selected as the rendezvous of his forces. From Dreux, Alençon issued a proclamation which greatly perturbed the King and his Court. Henri III attempted to assemble an army to take the field against Alençon, whilst Queen Catherine went after Alençon to plead with him to return to Court. Alençon refused to negotiate until the two Marshals, Montmorency and Cossé, who were still imprisoned in the Bastille, had been released. The situation was highly dangerous, as Alençon's escape had been the signal for the combined *Politique* and Huguenot armies, with support from the German provinces, to commence a vigorous offensive against the French King and the Guise-led Catholic forces. By November such was the state of affairs that Henri III was glad to purchase a truce of six months, at Champigny, by surrendering to his brother the towns of Angoulême, Niort, Saumur, Bourges and La Charité, as pledges of his good faith (November 21st 1575).

Meanwhile, following Alençon's escape, Marguerite had been placed under arrest in her own apartments, and this confinement was not relaxed until the truce of November 21st. After the truce she and Navarre were given a certain amount of liberty, but the King's men were always present to guard and prevent them from escaping. By cleverly feigning disapproval of Alençon's conduct, and pretending unswerving loyalty and friendship to the French King, Navarre had succeeded in quieting Henri III's suspicions concerning his own plans. Henri III privately liked and admired his brother-in-law, and even the warnings of the Duc de Guise could not make the French monarch move quickly against his sister's princely husband.

During all this time Francis was becoming liked and trusted by the French King, and also by both Navarre and Marguerite. Francis thought little of Henri III, detesting his debauchery, wantonness, unnecessary luxury, and his disorderly, licentious followers ("Men so disordered, so debauched and bold, that e'er his Court, infected with their manners, shows like a riotous inn: Epicurism and lust makes it more like a tavern or a brothel than a graced palace")[221] whom Francis considered to be the main cause of the weak King's disgraceful estate. But Navarre he respected, particularly for the "thrice-valiant courage in his noble heart";[222] whilst Marguerite he adored, loving her more and more as each day passed with something more than just infatuation. It was indeed the young, passionate and adoring love of a teenager for a fascinating and immensely attactive young woman who was just entering the prime of her life, and who possessed a cultivated and alert mind that in some ways matched his own; but his love for her never died out, even when he became well aware of her faults as well as her virtues, and even when there was no chance of the love being reciprocated or a relationship consummated.

> Love of her had power to make the Duke of Guise forget the greatest honours that France might confer upon him, and hath power to make as well all such fleeting glory seem to us like dreams or pictures, nor can we name aught real that hath not origin in her.....
>
> So fair was she, no eyes ere looked upon such a beauteous mortal, and I saw no other. I saw her — French Eve to their wondrous paradise — as if no being, no one in all high heaven's wide realm, save only this one, Marguerite, did ever exist, or in this nether world, ever, in all the ages to be in the infinity of time, might be created.[223]
>
> In my heart, too, love so soon overthrew envy as well as other evil passions, after I found lonely Margaret, the Queen of Navarre, who willingly framed excuses to keep me, with other right royal suitors, ever at her imperial commandment. A wonderful power to create heaven upon earth was in that loved eye. To win a show of her fond favour we were fain to adventure even our honour, or fame, to save and shield her.[224]

Queen Catherine had for some time promised Navarre that "she would bestow large bounties upon him, make him the prime man of the state, Lieutenant General of the Kingdom, and employ him where high profits might come",[225] but by now Navarre had come to realise that she had no intention of doing so and that in fact she privately scorned and laughed at him. Navarre was determined to escape, come what may, and, taking Francis into his confidence, plotted his departure from the French Court with the help of young Francis and the "Count of M, a noble lord of France, whose private loyalty unto the Duke was unguessed at by the King".[226] The "Count of M" was the Count Melun.

> Navarre grew ill at ease. He mourned his mother,[227] whom he did always tenderly revere, and his sister[228] especially he loved, whom he resembled to in countenance, as far as a man's face can be compared to that of a very beautiful girl. She, too, was at the Court but took no part in any of its pomps and gaieties, renouncing all to cherish silent grief for her dear mother's loss, and for the wounds so lately given there to her religion. The prince studied to work his liberty, and now his lover's sonnets turned to holy psalms, dear to his mother and the people of her faith. His prayers also were taught him by no priest: humbly upon his bended knee he craved help from above, but never from Christ's mother, (albeit that jay of Italy, the Queen-mother, did make him swear an oath false to his conscience)....
>
> His noble Christian thoughts did not rest here, but all hope and resolution grew soon to escape from his imprisonment or forced detention at the Court of France. Now shame and sorrow, duty, love and fear, present a thousand sorrows to his martyred soul.[229]

When Navarre engaged Francis' aid in assisting him to escape, and took him into his confidence, Francis likewise confided in Navarre and revealed to him the secret of his royal birth and that he had in fact been banished from England for this reason. Navarre, already filled with respect for Francis, readily accepted the truth of what Francis told him, and from that moment the two Princes bound themselves in a tie of close friendship.

The first attempt that Henri de Navarre made to escape, that winter, failed, due to his procrastination at the final moment of departure. He was overheard saying his farewells to Marguerite and was reported to the King, who sent guards to arrest him. Navarre was imprisoned for a while in a dungeon in the Tower, chained hand and foot, the King fearing to do anything because of Navarre's popularity; but eventually it seems that Marguerite's and Francis' pleading swayed the King into showing more clemency, and not only was Navarre released from the dungeon, but also allowed freedom to go hunting, howbeit closely guarded by the King's men. He made this a daily habit. It was whilst hunting with Marguerite in the forest of Senlis, on February 14th, 1576, that Navarre finally made his escape. Francis played an active and crucial role in this, and obviously enjoyed the adventure and the danger.

Early that February morning Francis arranged for some of Navarre's friends,[230] with horses and men, to be hidden in a prearranged part of the forest. Navarre and Marguerite then led the hunt gradually towards that section of the forest, where they pretended that they needed a rest and some time to dally and sleep in each other's arms. The hunter and his party were loath to stay, and so sped on with the hunt, whilst the old keeper and his company remained to attend to and guard the young couple. Soon sleep overtook the whole company, whilst Count Melun and Francis watched from deep within the thicket. Next, judging the moment right, Count Melun sent Francis forward to carefully wake Navarre without stirring the others. This done, and after Navarre had quietly kissed his wife farewell, Francis quickly led the Prince into the thicket and on to the spot where the friends and horses waited. Mounting the horses, they managed to flee the forest before any alarm could be raised and galloped at full speed away from Paris to Poissy, where they crossed the Seine and made for the town of Alençon. From Alençon, Navarre went on to Daumur on the Loire, where the Huguenot gentry of the neighbourhood hastened to join him.

In order to disguise the part he had played in this escape, Francis first waited for the King's officers and then Marguerite to report the disappearance of Navarre to Henri III, then came in with letters for the King from Navarre and Alençon, and a letter for Marguerite from Navarre, announcing that the Prince had indeed left for Paris to join his brother-in-law. Francis, who had been employed several times in those months as a trusted messenger, claimed to have come direct from the Duc d'Alençon, who had been joined by Henri de Navarre, in order to bring the letter. King Henri put the watch to death, who had allowed Navarre to escape, but Francis' and Marguerite's story he believed.

Marguerite was closely confined to her quarters again. Francis was commissioned both by her and the King to ride as messenger to Navarre and Alençon, bearing letters and messages.[231] The King gave Francis his fleetest horse, Cusay, for the mission, and Francis sped to Alençon's and Navarre's camp where he found a force of twenty-five thousand soldiers assembled. Francis warned them that the French King was sending a mighty army upon them with speed, "which are already mustered in the land, and mean to meet your Highness in the field".[232] Navarre and Alençon were a little anxious, as the Prince Condé with his large forces had not yet joined them. Francis, filled with enthusiasm for their cause, gallantly wished to fight with them.

> *Navarre:* "Much have I said, so that from point to point, now have you heard the fudamental reasons of this war, whose great decision hath much

blood let forth and more thirsts after. Here me more plainly: I have in equal balance justly weighed what wrongs our arms may do, what wrongs we suffer, and find our griefs heavier than our offences. When Charles was King this tyranny began. When we were wronged and would unfold our griefs, we were denied access unto his person, even by those men that most had done us wrong. The dangers of the days but newly gone, whose memory is written on the earth with yet appearing blood, and the examples of every minute's instance (present now) hath put us in these ill-beseeming arms: not to break peace, or any branch of it, but to establish here a peace indeed, concurring both in name and quality."

Francis: "Holy seems the quarrel upon your Grace's part: black and fearful on the Opposer's. For 'tis sacrilegious to do offence and scathe in Christendom, the worst that e're I heard. List to me, pray. By all your good leaves, gentlemen, here I will make my royal choice to stay and fight in your behalf, right manfully. Set on your foot, and with a heart new fired, I'll follow you, to do I know what, but it sufficeth that Navarre leads on. My gracious lord, I tender you my service, such as it is, being tender, raw and young, which elder days shall ripen and confirm to more approved service and desert."

Navarre: "No, no, thou shalt not be so foolish-hardy, so to expose those tender limbs of thine, to the event of the non-sparing war."

Francis: "Methinks I should revive the soldiers' hearts, because I ever found them as myself."

Navarre: "Thou shalt, brave Francis, be my messenger. Command my duty to our sovereign; wear thou my chain and carry this to him — thou has the good advantage of the night."[233]

Navarre bade Francis rest for the night before setting out. But before dawn the call-to-arms was sounded, the King's forces having approached them in the night, led by the Duke of Guise. Prince Condé and John Casimir, who headed an army of one hundred thousand cavalry backed by German lancers (*'Reiters'*), and who had invaded Burgundy, taken Dijon, and then crossed the Loire in order to join up with Navarre's forces on the Bourbonnais, were still a good half-mile or more south of the King's forces. Francis stayed to watch, as dawn broke:

The day begins to break, and night is fled, whose pitchy mantle overveiled the earth. I hear their drums breaking the silence now, like a dismal clanger heard from afar. I am my father's[234] son, for I had rather see the swords and hear a drum than look upon a schoolmaster. I may not be too forward in this matter; they've chid me from the battle, but I have set my life upon a cast, and I will stand the hazard of the die. I will not budge a foot, I swear on my word.[235]

For a while Francis stayed to watch the battle, then sped back to Paris with the reply to the King's letter and tidings of the fight. Henri III, pleased at Francis' trustworthiness and courage, appointed him then and there his "royal page, or squire". But Francis almost overstepped his privilege by speaking his mind openly, and incurred Henri's displeasure and rebuke. Not quite knowing what to say, Francis was led into revealing his royal birth to the French King and, standing up boldly as one Prince to another, proudly proclaimed his lineage, rights and power, even to the extent of threatening to try his right to the crown of France itself. Fortunately for Francis, Henri III was extremely anxious not to cause any dissention between France and England at that time which might break up their treaty, stop the profitable trade between the two countries and perhaps even precipitate Elizabeth's support of the Huguenot armies in France against his Throne. Although he only had Francis' word for his royal parentage — and he had come to trust Francis's open and honest speaking — Henri saw enough of the English Queen's temperament and characteristics in Francis to recognise them, and he dared not take the risk of ignoring what might be true. The gamble — if that is what it was — worked. After some moments of deep thought, the King courteously invited Francis to pass along into the palace and find Marguerite, who would welcome him.

> O dissembling courtesy! How fine this tyrant can tickle where he wounds. I something fear my father's wrath for this, but nothing what his [*i.e.* Henri III's] rage can do on me. I have been rash, and bridled not my tongue, and shall incur I know not how much of his displeasure when this comes to his ears.[236]

Francis went to find Marguerite and discovered her in an upper room, bearing to her letters and messages from Navarre. He reported to her what he had seen of the battle, and the heroic fighting of her husband against the Duc de Guise. Poor Marguerite, who loved her husband, Navarre, as a brother and through sworn loyalty, and who also at the same time loved the Duc de Guise as a lover, could not hide her turmoil nor her tears. Francis tried to comfort her.

> Like to a stricken deer the poor Queen turns: the holy beauty of her wondrous eyes shines on me through her tears. Meseemeth, as I look, her gentle heart would die in tempest of an angry frown, or buffets rude of sorrow and of scorn.
>
> *Francis:* "O lady, weep no more, lest I give cause to be suspected of more tenderness than doth become a man. Every minute now should be the father of some stratagem. So soon as I can win the offended King, I will be known your advocate, sweet Queen."
>
> *Marguerite:* "My Prince — 'Tis very kindly spoke like a true man, and it is honourable in thee to offer this. But step aside with me to my retreat; we can more safely speak of all our plans."
>
> She led the way to a secret bower, in whose enclosed shadow there was pight[237] a fair pavilion, scarcely to be seen, the which was all within most richly dight with gold and many a gorgeous ornament, after the Persian monarch's antique guise, such as the maker's self could devise, enchased with diamonds, sapphires, rubies and fairest pearl of wealthy India; the greatest Princess it might well delight.[238]

Marguerite, who also knew that Francis was the English Queen's son and rightful heir to the English Throne, plotted with Francis as to how she might escape that very night to join her husband, Francis going with her. Francis was overcome with love for "this Queen of earthly Queens", and as she bade him go to make due preparations for the escape, he confessed to her his love:

> I kneel to thank this lovely Queen for her sweet words, and beg for grace to lay my duty on her hand: frankly she gives it and her looks are gracious — fain would I kiss her feet to ease my bashful heart: for such a passion doth embrace my bosom, my heart beats thicker than a feverous pulse, and all my powers do their bestowing lose, like vassalage, at unawares encountering the eye of Majesty. Lower I bend and sigh:
>
> *Francis:* "O Queen, if you deny me favour let me die! I kiss your hand, but not in flattery: its pleasant touch hath made my heart to dance."
>
> *Marguerite:* "What say'st thou, boy? Is my young Prince a poet? I am much bounden to thee; fare-thee-well. Thou art so far before, that swiftest wing of recompense is slow to overtake thee: would'st thou had less deserved, that the proportion of thanks and payment had been better mated."
>
> I look into the lady's face, and in her eyes I find a wonder, or a wondrous miracle, and I nor heard nor read so strange a thing — a shadow of myself formed in her eye, and in this form of beauty read I — love!
>
> *Marguerite:* "Speak then, my gentle Prince; and canst thou love?"
>
> *Francis:* "Nay, ask me if I can refrain from love, for I do love thee most unfeignedly."
>
> With cheeks abash I blush, and swear to serve, be it unto death and future misery, this Queen of earthly Queens, as Goddess so divine, who charms with her sweet smile e'en the most saturnine.[239]

Marguerite was flattered, touched by Francis' declaration of love, but naturally cautious. She tried to temper Francis' passion with sweet, wise words, and to bring him back again to the urgent business that needed to be done. Francis declared vehemently that his love was forever, and that the declaration of love that he then made "shall become, in time, an overture of marriage",[240] when her dearest husband is no more. As a token of this, Francis gave Marguerite a diamond that was his mother's. Marguerite accepted the token, and Francis, encouraged, pressed her to say something of love in return to him. She hesitated, procrastinated, but then Francis stole a kiss from her lips. This seemed to unlock the passions of her heart, whether wise or not, and she begged for more, swearing "lasting fealty"[241] to Francis.

But soon they heard the footsteps of the Queen-mother, and Marguerite's maid, or nurse, came in to report that Catherine wished to speak with her daughter. Francis left with the nurse, and as they walked from the room he asked her to be ready by the abbey wall with "cords made like a tackled stair",[242] which she was to carry with her concealed under a cloak. There she would meet Francis' man.

All this was done and, well after dark, Francis, who had by then received the ladder, climbed over the orchard wall and then, using a pair of anchoring hooks, threw up the ladder outside Marguerite's chambers and clambered up it and into one of the Princess's rooms, where the nurse was awaiting him. The nurse straightway went to awake Marguerite, followed by Francis. Marguerite awoke troubled, torn by conflicting passions and fears. With one breath she declared her joy at being with Francis again; in another she wished to be free of her new troth which stirred her conscience and anxieties sorely; she feared for love contracts that were made too sudden, and she feared for the escape plan that was similarly devised so fast. In all, she would not go with Francis at that moment, and asked that he would come again to her by nine o'clock the next morning. Whilst they talked someone called Marguerite from another room, and eventually the call had to be answered.

> And then a noise did scare me from the room, and she (too desperate) would not go with me.[243]

The nurse followed Francis and whispered to him that the Duc de Guise was the cause that Marguerite could not love nor fix her liking on him. Marguerite's heart, and mind, were being torn — by differing loves, by the conflicting demands of honour and duty, by fears for her life and others' lives, and by her conscience respecting her various vows (to mother, to brother, to husband, to Church, to State, to Guise and, lastly, to Francis). Marguerite loved and tried, in her own way, to be loyal to them all. She had indeed a generous and passionate heart, but lived at a time when such emotions could be easily crucified and run wild.

The nurse wanted Francis to return and try to win Marguerite further, but Francis, sorely wounded, climbed out into the night again and disappeared into the covert of the wood.

Early the next morning Francis visited a friar who was already well known to him and to whom Francis had often turned for counsel and to make confession. The friar well knew Francis' love for Marguerite, and Francis' other secrets, Francis clearly having talked about these before. He tried to cheer Francis up and to point out to him the furies and rashness of young love, which may not last long, and also the great hurdle of any possible marriage (even assuming Marguerite might be released from her existing marriage either by the death of Navarre or their divorce) because of the position of State that both were in — the heir-presumptive to the throne of England on the one hand, and the French King's sister and Queen of Navarre on the other.

Francis was not to be put off by the wise words of the friar. Whilst they spoke together Marguerite's nurse arrived with a message from her mistress to Francis. Marguerite was in distress, weeping and greatly wishing to see Francis again. She sent her nurse forth to find Francis, knowing his habit to resort to the friar for counsel, to bring him to her, and to give him her ring. This ring Francis received with heart racing, hopes rising and joy bounding forth once more. He went to find Marguerite without delay — without even the friar's blessing — to see his love. Whilst he made his way to the palace he planned in his mind as to how his friend, Amyas Paulett, could act as ambassador for him and undertake the treaty of the marriage. In his mind he wooed Marguerite, one moment blessing and admiring her, the next minute jealously cursing her for bestowing her favours upon Guise. His mental state was that of a lover's melancholy.

On the way Francis was stopped by the King's Keeper and one of the lords of the Court, who wanted to know why he walked abroad at such an hour, shrewdly guessing not only that love must be the cause but also where Francis' heart did rest. Amused, they sought to give Francis good advice as to how to win the lady, counselling him to woo her with flattering poetry and love-sonnets, with night-time serenades, gifts and, above all, perseverance. With good and hopeful counsel, garnished with songs, they gradually cheered Francis' melancholia and speeded him on his way with better spirits and determination.

* * * * * * * *

Francis *did* persevere, carefully wooing Marguerite over the remaining weeks of his stay in France. Closer and closer they grew in love, delighting in each other's company, and

Marguerite's divorce from Henri de Navarre was seriously considered and put into motion. But, in early Spring, Francis received a letter from his foster-brother, Anthony, with a command from the Queen for Francis to return home to England. What would have at one time been welcome news for him, was now unwanted and distressing. He did not want to leave his fair Marguerite.

Shortly after he had received the mandate, and whilst he was musing on it, he met Marguerite in the woods. She had disguised herself as a young lord and come into the woods with her nurse and a boy singer in the hope of encountering Francis secretly. Dismissing the boy and the nurse, Marguerite poured out her heart in love for Francis, which he had at last won.

> *Marguerite:* "Oh Francis, let this habit make thee blush. Be then ashamed that I have took upon me such an immodest raiment, if shame live in a disguise of love? I could no less. It is the lesser blot modesty finds, women to change their shapes, than men their minds."
>
> *Francis:* "Than men their minds? 'Tis true: oh heaven, were man but constant, he were perfect; that one error fills him with faults: makes him run through all the sins; inconstancy falls off, ere it begins. But Margaret, I am not such a man, for I do love thee faithfully and well."
>
> *Marguerite:* "Why then God forgive me."
>
> *Francis:* "What offence, sweet Marguerite?"
>
> *Marguerite:* "You've staid me in a happy hour; I was about to protest I loved you."
>
> *Francis:* "Do it with all thy heart."
>
> *Marguerite:* "But I love you with so much of my heart, that none's left to protest, my gentle lord."
>
> *Francis:* "Oh happy shall he be whom Marguerite loves."
>
> *Marguerite:* "Then never say that thou art miserable, because, it may be, thou shalt be my love. Boldness comes to me now, and brings me heart: Prince Francis, I have loved you night and day, for many weary months."
>
> *Francis:* "Why was my Marguerite then so hard to win?"
>
> *Marguerite:* "Hard to seem won: but I was won, my lord, with the first glance; that ever pardon me, if I confess much you will play the tyrant: I've come to join with thee, and leave the King. I'll be thine, for I cannot be mine own, nor anything to any, if I be not thine. I dare not say I take you, but I give me and my service, ere whilst I live, into your guiding power: this is the man to whom I promise truest fealty ....[244] Francis, behold you, when the surgent seas have ebbed their fill, the waves do rise again and fill their banks up to the watery brims. At first I wept and wailed my state with fury, but with my love and woman's wit I've argued and approved it: and, Francis, since I saw thee, the affliction of my mind amends, with which I fear

a madness held me. Now 'tis gone, and now to you I give myself, for I am yours."[245]

As happened so many times in Francis' life, just as the prize seemed won and in his grasp, it was snatched away from him at the critical moment. The summons from the Queen, his mother, was for him to return home immediately, with no delay, and he was to set out for England the very next day. This had to be told, and it turned the sweetness into bitter gall. Francis indeed returned to England filled with hopes and plans of a future marriage with Marguerite, but Marguerite's still tender hopes were dashed, her heart mortified, her offering of love seemingly repudiated. Not even one night would, or could, Francis stay with her, even though she pleaded that he might do so. What a mistake, perhaps, he made, yet — there was actually no hope for the marriage.

The very next morning Francis set out with Amyas Paulett for England and home. The mission that the Queen had entrusted to Francis, requiring great secrecy and expediency, he had carried out well, and he hoped by this means to win Elizabeth's approval of himself and his love-match. Francis' now dear friend, Amyas Paulett, undertook to negotiate both treaties at once — the marriage treaty between Francis and Marguerite, once her divorce from Navarre was final, and the other treaty that was the subject of Francis' secret mission, whatever that might have been. Elizabeth did approve of her son, highly commending him, but she also made it crystal clear to Francis that for highly important and secret reasons the proposed marriage between him and Marguerite could not take place. When Francis was shown the reasons, he completely accepted that it was indeed so. Although his love never died, it could never be consummated in the sanctified bonds of matrimony. The vows that he had made to Marguerite, in his ignorance, could not be kept. A deeper blow there could hardly be, yet this was the force which drove him into accomplishing his great work for mankind.

> When Sir Amyas Paulett became advised of my love, he proposed that he should negotiate a treaty of marriage, and appropriately urge on her pending case of the divorce from the young Huguenot; but for reasons of very great importance these buds of an early marriage never opened into flower. But the future race will profit by the failure in the field of love, for in those flitting days afterwards, having resolved to cover every mark of defeat with the triumphs of my mind, I did thoroughly banish my tender love dreams to the regions of clouds as unreal, and let my works of various kinds absorb my mind. It is thus by my disappointment that I do secure to many, fruition.......[246]

> The story of my secret mission is thus begun, for, as hath been said, I was entrusted at that very time with business requiring great secrecy and expediency. This was so well conducted as to win the Queen's frank approval, and I had a lively hope by means of this entering wedge to be followed by the request nearest unto my soul [I] should so bend Her Majesty's mind to my wish. Sir Amyas Paulett undertook to negotiate both treaties at once, and came thereby very near to a breach with the Queen, as well as disgrace at Henri's Court. Both calamities, however, were averted by such admirable adroitness that I could but yield due respect to the finesse, whilst discomforted by the death of my hope.

> From that day I lived a doubtful life, swinging like a pendent branch to and fro, or tempest tossed by many a troublous desire. At length I turned my attention from love, and used all my time and wit to make such

advancement in learning or achieve such great proficiency in studies that my name as a lover of Sciences should be best known and most honoured, less for my own agrandizement than as an advantage of mankind, but with some natural desires to approve my worthiness in the sight of my book-loving and aspiring mother, believing that by thus doing I should advance my claim and obtain my rights, not aware of Cecil, his misapplied zeal in bringing this to her Majesty's notice, to conceive her mind that I had no other thought save a design to win sovereignty in her life-time.

I need not assert how far this was from my heart at any time, especially in my youth, but the Queen's jealousy so blinded her reason that she, following the suggestion of malice, showed little pride in my attempts, discovering in truth more envy than natural pride, and more hate than affection.[247]

Whilst in France for those seven or eight months, Francis had not spent all his time and thoughts on wooing Marguerite, although she was principal in his mind and heart — especially from February 1576 onwards. He had also been busy with regard to his secret mission, which required such care, and had become well-known and trusted by all sides at the French Court, being entrusted many times with matters and messages of great import and privacy. He acted as a willing ambassador not only for his Queen and mother, but also for Henri III, Henri de Navarre, Marguerite de Valois, and the Duc d'Alençon. Besides this he partook fully in the Court life, with all its gaieties, festivities and political dealings, and absorbed the culture and the learning that was a noted feature of the Valois Court. He almost certainly met and conversed with at least some of the famed group of French court-poets known as the 'Pléiade', and their learned and artistic friends — scholars, humanists and Familists[248] who played such a major role not only in the royal propaganda and entertainments, but also in creating and forming the real French Renaissance and reformation in learning, and enobling the French language — such as Jean Antoine de Baif, Pierre de Ronsard, Pontus de Tyard, Remy Belleau, Jean Dorat, Pierre Porret, Nicholas Houel, Jacques Gohorry, all of whom were interested in and practically advancing all sciences and arts. When Francis was in France, the fame of the Pléiade was at its zenith. He was certainly taught and influenced by distinguished men who were connected with the Royal College founded by Francis I, the refuge of free-thinkers of all countries, whose works were studied and quoted by Francis Bacon and his own (later) school. The duty of this Royal College was to bring before the notice of the King "all men of greatest learning, whether French or foreigners".

Baif, Ronsard, Tyard, Bellau and Dorat were the five remaining 'Pléiade' when Francis arrived at the French Court. The 'Pléiade' were philosophers and humanists, as well as poet-artists, deeply versed in the mysteries of Orphism and Platonism, in the Classics generally and in Hebrew with its Cabbala or esoteric tradition of knowledge. They worked for peace, religious tolerance and the advancement of all learning. Their ideas and talents lay behind the series of *Magnificences* put on by the Valois Court as part of Catherine de Medici's attempt to unite the various warring factions and espouse the principle of the King as the *Roi Soleil*, prince and priest of his people. As court poets they were responsible for the many entertainments and other events at the French Court, and through drama, poetry, music and entertainment generally, they greatly influenced the French people and other nations. The 'Pleiadist' movement was one of '*Politique*' royalism, and dear to the hearts of the Valois family.

Ronsard wrote the great epic poem, *Franciade*, on the origins of the *Rex Christianissimus* and the glorious but mythical ancestor of the French — Francus, the son of the Trojan, Hector. It was a French Aenied, which Francis was later to provide for England as *The*

PROLOGO, LELIO.

Benche siat'uso ò Spettatori Illustri,	Quest'opra, è'l gran Theatro
Solo di rimirar Tragici aspetti,	Del mondo, perch'ognun desia d'udirla:
O Comici apparati	Ma voi sappiat'in tanto,
In varie guise ornati,	Che questo di cui parlo
Voi però non sdegnate	Spettacolo, si mira con la mente,
Questa Comedia nostra,	Dou'entra per l'orecchie, e non per gl'occhi
Se non di ricca, e vaga Scena adorna,	Però silentio fate,
Almen di dopia nouità composta.	E'n vece di vedere hora ascoltate.
E la città doue si rappresenta	

(26) **PROLOGO LELIO** — woodcut showing the speaker of the prologue in Horatio Vecchi's *L'Amfiparnasso, Comedia harmonica* (published in Venice, 1597).

The illustration shows not only a typical Renaissance theatrical stage, with its carefully constructed perspectives (an innovation introduced during the European Renaissance), but also two beautiful 'grail' emblems above and below the main picture, with two 'pillars' at each side. Note the cherubic archers challenging the human-faced dragons in the lower emblem, depicting the mastering of nature by the spiritual self; whilst the upper emblem depicts the fall of man's spirit into matter and natural form. The lower chalice is a grail chalice with the sweet incense of perfect sacrifice or offering arising from it; whilst the upper chalice is filled only with natural living forms which have yet to bear fruit and produce the grail essence.

(27) THE QUINTAIN — detail from sketches made for the Valois Tapestries, by A. Caron (*c.* 1582).

The Valois Tapestries were probably made, in Brussels, for Catherine de Medici during the reign of Henri III. They incorporate scenes from the *Magnificences* of Bayonne (1565) and Tuileries (1573) of Charles IX's reign.

Faerie Queen. Interestingly, neither *Franciade* nor *The Faerie Queen* were ever completely published or 'completed'. The popularity of Ronsard was so great (he was called the prince of poets and the poet of princes) that Henri III is said to have placed beside his throne a state chair for Ronsard to occupy.[249]

Baif founded an Academy of Poetry and Music (in 1570) under the patronage of Charles IX, with the aim of recovering the 'effects' of ancient music, such as that of Orphism, so that these 'effects' could be used for tranquillizing and harmonizing religious and political passions, just as Orpheus charmed the very beasts and rocks into a state of peace, harmony and order. This 'sacred' music was also understood to be linked with the power and magic of words and symbols, mathematically and astrologically ordered, and constituting the pure or complete language of 'angels' — the perfect poetry.

> "The mere name 'Academy' used of Baif's institution implied platonism; the union of poetry and music was, in one of its aspects, a symbol of a phase of initiation into higher harmonies. The Academy was encyclopedic in scope, ideally including under music all the arts and sciences. Baif was a mathematician as well as a poet-musician .... The Academy of Poetry and Music, beyond, or as well as, its purely aesthetic aims, had the aim of fostering the detente of religious passions by encouraging Huguenots and Catholics to make music together in the Academy".[250]

Near Baif's Academy of Poetry and Music in the Faubourg Sain-Marceau was the Lyceum of Jacques Gohorry, otherwise known as '*Leo Suavis*', an alchemist, magician and Paracelsist physician. In his *Lyceum*, Gohorry prepared and administered healing medicines, carried out alchemical experiments, made magical talismans, and "received learned visitors who admired the rare plants and trees, and performed vocal and instrumental music in the *galerie historiée*".[251] The nature of Francis' mind being what it was, he would almost certainly have sought out this famed *Lyceum* as well as the Academy nearby.

In 1576 Gohorry's *Lyceum* seems to have disappeared, but was 'replaced' by Nicholas Houel's *Maison de Charité Chrétienne* in the same Faubourg Saint-Marceau, which included an apothecary's garden, a medieval laboratory and a music school, as well as a school for orphans, a chapel, a hospital, and a "French Academy for various kinds of artisans".[252] The uniforms worn by the inhabitants of this 'House of Charity' were scarlet-violet. The emphasis of the House was on healing and teaching — two aspects of the same thing — demonstrating that true medicine was the Work of Love. Houel made a request to Henri III and the French Parliament in 1526 to set up his 'orphanage', but it was not until 1578 that it finally got off the ground, in an old house — the Hôpital de Lourcine. The hope was that it would, with royal patronage, be enlarged so as to be able to teach all seven liberal arts and sciences, together with other disciplines and knowledges, including Hebrew, Greek and other languages. In 1583 the House of Christian Charity was presented as a main theme in Henri III's special 'religious' processions in Paris, in which he hoped to overcome the menace of the intolerant Catholic League led by the Guise's. The method in which the scenes of the Works of Mercy were presented is very similar to the method employed by Francis in his *Faerie Queen*, which he wrote in 1580 (although it was not published until 1590, and attributed to Ed. Spencer). In the following chapters it will be seen how Francis' ideas and Grand Design influenced the French 'advancement of learning' just as the French sages inspired and influenced him.

Henri III's brother, Françoise Duc D'Alençon, who for a time ruled a state in the southern Netherlands in which religious toleration was attempted, was not only the head of the

'*Politique*' party but the centre of an important '*Politique*' group of thinkers. Similarly Marguerite — particularly after the escape of her husband and the return of Francis to England, when she turned herself again to serious study — became the 'star' around which a circle of great thinkers, artists and poets collected. Neither was Henri de Navarre excepted from the 'brotherhood', although his approach to truth was somewhat more severe and misunderstood — a subject of satire in Francis' *Love's Labour's Lost*.

When Francis returned to England in the Spring of 1576, therefore, although his thoughts were preeminently on love and marriage, yet his mind was filled also with ideas and plans based on the culture and learning he had acquired in France, and the friends he had made. He hoped to return to France as soon as possible, not only so as to be with Marguerite again but also so as to become more thoroughly involved with the French 'Pleiadist' movement and to learn as much as he could from them.

As for Marguerite, soon after Francis left her she was obliged to travel with her mother to the Château de Chasternay, near Sens, in order to negotiate peace terms with Alençon on behalf of the King. Initially Henri III had sent his mother, Catherine, to make overtures of peace to his brother — who was so dangerously intent on seizing the Throne of France for himself — but, as Catherine had pointed out to Henri at the start, Marguerite's help would almost certainly be required. The King, however, was loath to part with so valuable a hostage as his sister was. As Catherine had foreseen, Alençon refused to negotiate any terms whatsoever until his sister was set at liberty. The King was thus forced to concede that this was necessary and that he must first come to peace with Marguerite before she would in any way help him to make peace with their brother. Catherine and Henri together begged Marguerite "not to allow the affront which she had received to inspire her with sentiments of vengeance rather than with a desire for peace, as the King was prepared to make her every reparation in his power".[253] Marguerite magnanimously declared that she would indeed "sacrifice herself" for the welfare of the family and the State. In fact, the idea of her brothers and husband and mother and ex-lover (Guise) fighting and perhaps killing each other appalled her, not to mention the horrors of the escalating civil war and the fear for her own life.

At the end of April or beginning of May she set out with her mother for the rendezvous with Alençon and the principal nobles and princes of his army — Catholic and Huguenot — including the Duke Casimir and Colonel Poux. After several days of discussion, peace terms were finally agreed upon at Beaulieu (May 1576 — the Edict of Beaulieu, also known as the 'Peace of Monsieur'[254]) which constituted a triumph for the 'rebels'. The Protestants obtained the best concessions they had ever won, granting them complete freedom of worship throughout the kingdom, except in Paris, plus the establishment of courts in all the Parliaments composed of an equal number of judges of both religions, and a restoration to all their honours, offices and rights to hold offices which had previously been taken away from them. The Massacre of Saint-Bartholomew was formally disavowed, confiscated properties were restored, and eight fortresses handed over to them as security for the due observance of the treaty. Alençon was given the duchies of Anjou, Beny, Touraine, Maine and other lordships, in addition to the few he already had; so that he now assumed the title of Duc d'Anjou which had been that of Henri III before the latter's ascession to the Throne. Henri de Navarre was confirmed in his government of Guienne, and Condé in that of Picardy. John Casimir was promised a huge sum of money to withdraw his German '*Reiters*' and as a personal compensation for his trouble and expense in invading France, plus a large annual pension in order to secure his friendship.

Alençon, now the Duc d'Anjou, had advised Marguerite to allow herself to be included in the treaty and to demand the assignment of her marriage-portion in lands. But Marguerite gave in to her mother's entreaties not to do this but to trust in the King to give her whatever she desired, "preferring to owe what she might receive from the King and Queen her mother to their good-will alone, in the belief that it would be thus more permanently assured to her."[255] Neither did Marguerite obtain her freedom to join her husband who, as soon as peace was concluded, had written "inviting her to demand her liberty." Catherine persuaded Marguerite to return with her to Court, promising her daughter her liberty once Anjou had arrived at Court to re-embrace the King; and reluctantly she returned to Paris with her mother. Catherine de Bourbon, Navarre's sister, was however set at liberty and returned to her brother.

The peace terms were too good: it created such a backlash that it soon became apparent that it would be impossible to enforce many of the treaty conditions, and a powerful anti-Protestant movement, started by a confederacy between the partisans of the Guises and the bigoted Catholics of Picardy, grew rapidly into a general "Holy League" or association of the extreme Catholic party throughout the kingdom, led by the Guises and backed by Philip of Spain and the Pope. The members of the League bound themselves by oath "to regard as enemies all who refused to join it, to defend each other against any assailant, *whoever he might be*, and to endeavour to compass the objects of the association *against no matter what opposition*."[256] This was a grave threat not only to all Protestants and moderate Roman Catholics, but also to the King of France himself. Henri III decided that the only safe course for him to take was to place himself, as King of France, at the head of the League — having unsuccessfully tried to persuade the Guises to form no association that might breech the recent peace. A new and fiercer war became inevitable.

* * * * * * * * * *

> Their life [Romeo and Juliet's] was too brief — its rose of pleasure had but partly drunk the sweet dew of early delight, and every hour had begun to ope unto sweet love, tender leaflets in whose fragrance was assurance of untold joys that the immortals know. Yet 'tis a kind of fate which joined them together in life and in death.
>
> It was a sadder fate befell our youthful love, my Marguerite, yet written out in the plays[257] it scarce would be named our tragedy since neither yielded up life. But the joy of life ebbed from our hearts with our parting, and it never came again into this bosom on full flood-tide. O we were Fortune's fool too long, sweet one, and art is long .....
>
> Far from angelic though man his nature, if his love be as clear or as fine as our love for a lovely woman (sweet as a rose and as thorny it might chance), it sweetneth all the enclosure of his breast, oft changing a waste into lovely gardens which the angels would fain seek. That it so uplifts our life who would ere question? Not he, our friend and adviser......Sir Amyas Paulet.
>
> It is sometimes said, "No man can at once be wise and love", and yet it would be well to observe many will be wiser after a lesson such as we long ago conned. There was no ease to our suffering heart till our years of life were eight lustres.[258] The fair face liveth ever in dreams, but in inner pleasures only doth the sunny vision come.[259]

Return to France

When Francis sailed for France in August or September, 1575, he left behind him his closest friend and companion — his foster brother, Anthony. Anthony continued his studies at Cambridge, but life must have been somewhat empty for him without the laughter and wit of his younger 'brother' — and how did the revelation that Francis was not in fact Lady Anne's son but the son and heir of Queen Elizabeth affect Anthony? This is not recorded, but it must have had some considerable effect. But it never destroyed their deep friendship for one another — perhaps it strengthened and deepened it?

Anthony Bacon finally 'came down' from Cambridge at Christmas 1575. The Trinity College ledger for the Bacon brothers closes on December 23rd, 1575. Neither brothers seem to have bothered with taking their degrees, Francis for one recording that he had assimilated all that could usefully be taught to him.

Francis returned to England in early Spring 1576, now feeling and acting very much the prince that he actually was. He was commended by the Queen on the manner in which he had undertaken the mission, but his marriage request was turned down, for grave reasons which he accepted. He relinquished his hopes of a life with Marguerite, which he records as being the greatest grief he had to bear. Then, on 27th June 1576, he and Anthony were admitted to Gray's Inn (one of the four Inns of Court in London, in which lawyers were trained and prepared for the bar), following in the footsteps of Sir Nicholas Bacon and his three eldest sons.

But being 'admitted' did not necessarily mean that the person admitted took up residence and his studies then and there. 'Admittance' was a declaration of intent and acceptance — an intent by a person to take up law studies, and an acceptance by the Inn of Court of that person for studies. Not everyone admitted in Elizabethan times either studied seriously for the bar or took up residence. As for Francis, there could certainly have been no intention for him to take up residence immediately, for on the 30th June a licence was granted to him and Edward Bacon to travel "beyond the seas" for up to three years. The Letters Patent, dated 30th June 1576, read as follows:-

> "Elizabeth, by the Grace of God: To all and singular our Justices of the Peace, Mayors, Sheriffs, Bailiffs, Constables, Customers, Comptrollers and Searchers, and all other our officers, ministers and subjects to whom it shall appertain, and to every of them, greeting.
>
> Whereas we have licensed our well-beloved *Edward Bacon* and *Francis Bacon*, sons of our right trusty and well-beloved counsellor Sir Nicholas Bacon, Knight,

Keeper of our Great Seal of England, to depart out of this our realm of England into the part of beyond the seas, and there for their increase in knowledge and experience to remain the space of three years next and immediately following after their departure. We will and command you, and every of you, to suffer them with their Servants, six horses or geldings, three score pounds in money, and all other their bag and baggage and necessaries quietly to pass by you without any your let, stay or interruption, and these our letters or the duplicate of them shall be as well unto you for suffering them to pass as unto them for their going and remaining beyond the seas all the time above limited sufficient warrant and discharge.
In witness whereof,
Witness ourself at Westminster, the 30th day of June."

Edward Bacon was the youngest of the three sons of Sir Nicholas Bacon by his first wife, and was 28 years old at the time of this licence to travel. Between Edward, Anthony and Francis there was a close and affectionate tie all their lives, and Edward later allowed Francis to use his house near London, Twickenham Park, which was on the opposite bank of the Thames to Richmond Palace. The reason for granting such a licence to Edward was almost certainly to allow him to do a 'grand tour' of the Continent, like so many others did before and after him; but in Francis' case, he was given another specific and secret mission, or "embassy", to perform. Edward probably had some sort of commission to perform for the Queen and her Government as well, as it was generally required at that time for such travellers to render to Lord Burghley or Walsingham good, up-to-date intelligence of the countries and people they had visited on their tour, this being made a precondition of their being granted a licence.

Francis' mission was particularly special and specific. Clearly he wanted and requested to be allowed to return to France, but, having relinquished his love suit, he now wanted to devote himself to his great project. According to his cipher account, he was set to work for Sir Francis Walsingham, the Secretary of State, to intercept and decipher the letters written to, by or concerning Mary, Queen of Scots. Later he was to lament that he had been all too successful, uncovering proofs of plots which explicitly implicated her in activities to overthrow Elizabeth and place herself on the Throne of England. But as Francis (who respected whilst not agreeing with her) said, from Mary's point of view she believed that she really was the rightful and only legitimate heir to England's Throne, and that Elizabeth was a bastard and usurper in the eyes of extremist Roman Catholicism.

Sir Francis Walsingham was born *c.* 1530, only son of William Walsingham, common Serjeant of London, by his wife Joyce, daughter of Sir Edmund Denny of Cheshunt. He attended King's College, Cambridge, as a Fellow-Commoner from November 1548 till September 1550, amidst strong Protestant influences. He then went abroad to complete his education, returning to England in 1552. He was admitted to Gray's Inn on January 28th, 1553, but went abroad again on Queen Mary's accession, being a staunch Protestant. In 1555-6 he was in Padua, and was admitted a *"consiliarius"* in the university faculty of laws. Whilst in Italy, Walsingham was introduced to the works of Geronimo Cardano (b.1501-d.1576), the great Italian physician, philosopher, natural scientist, mathematician and astrologer. Walsingham may even have met Cardano, but in any event he was greatly impressed not only with Cardano's philosophy but also with his cabbalistic mathematics and cipher inventions. The influence which this great Italian genius had on Renaissance thought and eventually on the Baconian work was considerable and should be recognised and studied much more than it has been to date.

When Queen Elizabeth acceded to the throne of England, Walsingham returned home, in time to take up a seat in Elizabeth's first Parliament of 1559 which Cecil had arranged for him, as M.P. for Banbury. In January 1562 he married Anne, daughter of George Barnes, Lord Mayor of London, and after she died (in 1564) he married Ursula, daughter of Henry St. Barbe and widow of Sir Richard Worsley. By his second wife, Anne, Walsingham had a daughter, Frances, who married firstly Sir Philip Sydney (Francis Bacon's real cousin, nephew of the Earl of Leicester) in 1583, and secondly, after the death of Sir Philip in battle, Robert Devereux, 2nd Earl of Essex (Francis Bacon's real brother, second son of Leicester and Queen Elizabeth) in 1589. After Robert was executed, Frances married for a third time, her final husband being Richard de Burgh, Earl of Clanricarde.

Walsingham was clearly associated with the Bacons and the Cecils. He was an ardent Protestant and a skilled 'intelligencer'. In 1567-70 he was supplying Sir William Cecil (then Secretary of State) with information about the movements of foreign spies in London, and he was largely responsible for the uncovering of the Ridolfi Plot in 1571. In the Summer of 1570 Walsingham was designated to succeed Norris as the English Ambassador at Paris — which post Walsingham did not want, for the Court of Catherine de Medici was anathema to him. However, he had to accept the post and he proved himself a very able choice. He was in Paris at the time of Henri de Navarre's marriage with Marguerite de Valois, and witnessed the St. Bartholomew's Day Massacre. Walsingham was recalled in April 1573 (being replaced as ambassador by Sir Thomas Smith), and eight months later was admitted to the Privy Council and made Principal Secretary of State — Sir William Cecil having been appointed Lord Treasurer in November 1572. As soon as he was Secretary, Walsingham began to create a first-class and highly-trained intelligence service, and opened a secret Cipher School in London. His agents, who were of different categories according to social rank, education and skill — from the high-ranking 'intelligencer' to the common 'spy' — were required to take a course in cryptography before undertaking foreign missions. Many noblemen's sons were 'commissioned' as intelligencers as part of the conditions for their travelling abroad, and a great number of the spies and cryptographers were drawn from the ranks of university scholars. When it was fully functioning, Walsingham had no fewer than fifty-three agents working for him on the Continent alone — a large number for those days.

It seems clear that Francis must have been sent to take a course in cryptography at Walsingham's London school before leaving for France for the second time. Anthony also trained in Walsingham's school, and became Walsingham's principal intelligencer on the Continent until Walsingham's death, afterwards continuing to run this intelligence network for Francis, Essex and the Queen, to counter Robert Cecil's own network of intelligence which the Bacon brothers and Essex could not afford to trust and would not have had access to anyway. Anthony, who became the noted 'Cabbalist' of the Baconian-Rosicrucian group, absorbed Cardano's philosophy and gematric examples, and adapted at least one of Cardano's cipher methods — known as 'Cardano's grille' — for Francis' use.[260] It is possible that both Francis and Anthony began their cryptographic course together in the Spring/Summer of 1576. Francis quickly developed a deep love and fascination for ciphers which he refers to in his philosophical works, inventing and using many different ones throughout his life, for various purposes.

Although the licence to travel was granted in June, it was not until September 1576 that everything was ready and they set sail for France. Francis, who was on a specific mission from the Queen, refers to this departure in letters to Lord Burghley and Essex, (the kissing of the Queen's hands marking the fact that he was commissioned by the Queen herself):

"I went with Sir Amyas Paulet into France from Her Majesty's Royal Hand."[261]

(28) SIR FRANCIS WALSINGHAM, Secretary of State (1573-90) — oil painting attributed to John de Critz (the elder).

> "I kissed Her Majesty's hands upon my entering into France."[262]

> "I was sent on an Embassy from the Queen's hand."[263]

Francis and Edward travelled over to France, along with many others, in the company of Sir Amyas Paulet and his wife, Margaret. Sir Amyas had recently been knighted and was appointed to take over the Ambassadorship to the Court of France from Dr. Valentine Dale, (which he finally did in February 1577). Sir Amyas was thus travelling to France as the new ambassador-to-be, together with all his household, retinue and necessary goods. Francis (and presumably Edward as well) travelled with the Paulets in the Queen's battleship, *Dreadnought*, which had been specially commissioned for the purpose. Many other ships and barques travelled with them, in convoy. They landed at Calais on 25th September 1576, from where Sir Amyas Paulet wrote a letter to Burghley (that same day) exclaiming at the number of people accompanying him:

> "being accompanied with an extraordinary number, whereof some have been recommended unto me by the Queen, some by other noblemen, only until their coming to Paris. The Queen's ships as likewise the other barques appointed for me and my horses were forced to seek their security at Sandwich, when the wind did serve to have passed to France."[264]

Francis continued in the company of Sir Amyas at least until July 1577, travelling with the French Court from Paris to Blois, Tours and Poitiers. It would seem that, besides servants, Francis had also been assigned another tutor to look after him — a Mr. Duncombe, of whom Sir Amyas Paulet wrote in a letter to Sir Nicholas Bacon (carried by Mr. Duncombe himself, when sent back to England in September or October 1577):

> "I may not omit to commend unto your Lordship the honest, diligent and faithful service of this bearer, which deserveth very good acceptation, thinking him worthy of the government of your Lordship's son, or any gentleman in England, of what degree soever".

Edward and Francis must have parted company not long after they reached Paris, as Edward was at Strasbourg in December 1576,[265] whilst the French Court, with Francis, set out in October for the meeting of the States-General in Blois. At Blois, at the end of November, Henry III signed the roll of the Holy League and compelled all the principal personages of the Court to follow his example, placing himself at their head. Amongst the signatures was Anjou who, having obtained all he desired, had deserted his Protestant allies and returned to Court. When the States-General met at Blois, they were boycotted by the Huguenots and '*Politiques*', who feared the way things were going. The Third Estate of the States-General voted to rupture the Treaty of Beaulieu and to deprive the Protestants of all exercise of their religion, both in public and private. Naturally enough, the civil war was again revived for the sixth time, with enormous hatred on both sides.

It was during that November that Francis, together with all of Sir Nicholas Bacon's sons, were admitted (in their absence) to the Grand Company of Ancients of Grays Inn, and freed from all vacations. The Order of Pension, dated 21st November 1576, granting this privilege, reads as follows:-

> "It is ordered that Mr. Edward Bacon shalbe admitted in My Lorde Kepers Chamber in the absenc of Mr. Nicholas Bacon his sonne & that Mr. Anthony Bacon shalbe admitted in the same chamber in the absenc of Mr. Nathaniell Bacon."

> "It is further ordered that all his sonnes now admitted of the housse viz:- Nicholas, Nathaniell, Edward, Anthonye & Francis shalbe of the Graund Company and not be bound to any vacations."

The honour of being admitted as one of the Grand Company of Ancients or Leaders of Grays Inn was undoubtedly due to the standing of Sir Nicholas Bacon, who held the highest judicial position, next to the Sovereign, in the country. This status gave them special privileges and advantages over the ordinary students, for particular reasons. Grays Inn provided facilities both for those intending to follow the legal profession, and also for some from the upper strata of society who had no intention of doing so but who had need to learn something about the laws and legal system of the country, plus some traditional etiquette, in order to prepare them for some future responsible position of leadership. Thus it was possible "for persons of distinction" to be an Ancient without intending to become a lawyer or barrister, in order to acquire some legal experience and "to form their manners and preserve them from the contagion of vice" by setting them apart from the others of lower social standing and education.

Sir Nicholas Bacon retained a suite of chambers at Grays Inn for his personal and family use, and it should be noted that this Order of Pension permitted Edward and Anthony to be admitted to these chambers in place of their elder brothers, Nicholas and Nathaniel, who were absent from the Inn of Court — Nicholas being married and living at Redgrave Hall in Essex, and Nathaniel likewise being married and settled at Stiffkey Manor in Norfolk. Presumably the chambers could only provide for a maximum of two people living in them at any one time, for Edward had already been a member of Grays Inn for some time previous to this Order. As for Francis, he was given no special rooms until he finally returned to live in England permanently and took up residence in the Inn (in 1580). Even then he was not admitted to Sir Nicholas' chambers, which were still reserved for Edward and Anthony, but lodged with a Mr. Fulwood in Fulwood House, until 1588, when he was granted his own special chambers at the Inn, next to the library, having been appointed a Reader of the Inn in November, 1587.

But, in November 1576, Francis Bacon was with Sir Amyas Paulet and the French Court at Blois and, in the midst of all the political intrigue, enjoying the Court festivities and cultural activities of all kinds in the brilliant company of poets, philosophers and artists collected there. One of the principal artistic events at Blois was the series of Italian *Commedia dell' Arte* staged for Henri III by the celebrated Gelosi troupe, whom Henri had specially invited to Blois for the opening of the *Etats*. The troupe were not actually able to begin performing their *Commedia* until February, 1577, as they had been captured by the Huguenots *en route* to Blois from Venice and had to be ransomed. Henri III was very partial to this kind of entertainment, and the *Commedia* was often performed at his Court, the Gelosi for instance playing at the Hotel de Bourbon in Paris the following September.

> "The *Commedia* of the Renaissance was a type of theatre quite different from any other. As distinct from the written comedies, it was performed, and could only be performed, by professional actors. In sixteenth century Italy it co-existed with the legitimate theatre, with which it vied very successfully for popular support, being similar in this respect to the *Atellanae* or pantomines in which it had its ancestry, and which eventually ousted the classic theatre from the favour of the ancient Roman populace. The *Commedia* and *Atellanae* alike were improvised from scenarios, and the dialogue was made up by the actors as the play went along. Their only guidance was the plot-outline which was posted up in the wings, and which was consulted by the players at the beginning of each scene. They 'suited the action to the word, and the word to the action' as the

occasion demanded, and needed to exercise a high degree of extemporizing skill, fluency, humour and adaptability. It is easy to see why only professionals could take the part in this kind of theatre.........

"Then again, the roles were stylized. Bologna, with its ancient University, contributed the Doctor, who was as foolish as he was pedantic; Venice, the city of merchants and adventurers, evolved Pantaloon and the Captain; the upper and lower Bergamos — districts of Italy notorious for the supposedly abnormal proportion or dullards in their populations — produced the sly, but sometimes witty, booby Harlequin, and the knave Brighella. Pulcinella, another character, with his hooked nose, hump, fleshy cheeks and outsized mouth, became familiar in England, from the end of the seventeenth century, as Punch ..... The *Commedia dell' Arte* differed from the classical theatre in comprising within itself, as it developed, both comedy and tragedy, which the older drama kept quite apart. George Sand described it as having an 'uninterrupted tradition of fantastic humour which is in essence quite serious and, one might say, even sad, like every satire which lays bare the spiritual poverty of mankind.' R.G. Moulton says that the terms Comedy and Tragedy are inadequate and, indeed, absurd when applied to Shakespeare. 'The distinction these terms express is one of Tones, and they were quite in place in the Ancient Drama, in which the comic and tragic tones were kept rigidly distinct and were not allowed to mingle in the same play. Applied to a branch of Drama of which the leading characteristic is the complete mixture of Tones, the terms necessarily break down, and the so-called 'Comedies' of *The Merchant of Venice* and *Measure for Measure* contain some of the most tragic effects in Shakespeare.' He goes on to maintain that the true distinction between the two kinds of play is one of Movement, not Tone. Comedy and Tragedy would be better described by the terms Action-Drama and Passion-Prama respectively. In the *Commedia dell' Arte*, we discover an amalgam of these two types of drama, as we do in Shakespeare ......

"It is possible to discern, in Harlequin — the most individualized member of the *Commedia* — something of Touchstone, Lear's Fool, and the First Clown in *Hamlet*. Originating from Bergamo, Harlequin nevertheless lets fall some shrewd flashes of wit, which were the delight of the audiences; such droll whimsicalities as he purveyed contained much satirical wisdom. In 1601, an unusual treatise was published in Lyons, entitled: *Compositions de Rhetorique de M. don Arlequin*. It is an amusing work and gives ample proof that much pointed and extravagant humour abounded in the second half of the sixteenth century.

"Besides being the possible source of a number of simple dramatic characterisations to be elaborated into the fully drawn human portraits later given to the world by Shakespeare, the *Commedia dell' Arte* was an influence also in the sphere of practical stagecraft that must have made its due impact upon the mind of the Author of the Plays. It is said that the Italian method of declamation did much to cure what Moliere called 'the demoniacal tone' in the French players, some of whom actually sustained apoplexy on the stage itself as an unfortunate result of it! Hamlet's speech to the players seems to give such advice as would indicate that Shakespeare himself had taken to heart the same lessons of disciplined natural acting and smooth delivery of lines demonstrated in the practice of the Italian actors.

> "As the troupes travelled widely, they took the the *Commedia* into every country of Europe and gained an international reputation. In 1527 a company under Drusiano Martinelli played in England, and it may be argued that Shakespeare was able subsequently to see a performance of the Italians without going beyond the confines of his native land. But it is doubtful whether such very occasional shows would have sufficiently imbued the Englishman with such a knowledge of their characters and methods as he seems to have possessed. A stay in Italy of some years, and a prolonged study of their technique would have done so, but it is Francis Bacon that had the opportunity for such a study, not Shakspere of Stratford."[266]

Francis Bacon did indeed have the greatest opportunities not only to see, but also to study the *Commedia* and learn from the mouths of the players themselves, both during the times he spent at the French Court (1575-9) and also during his visits to Italy (in 1581). He, who saw grand entertainment, and plays in particular, as being the principal means to hold the mirror of nature up before men's minds in order to teach wisdom, could not have failed to have been influenced by this high art. And all this while Francis had begun preparing his *Sylva Sylvarum* or *Book M*[267] — his 'book' of nature, comprising his observations, meditations and inspirations concerning the nature of man, God and the universe as revealed in Nature, or "God's Works". From this 'book', which he continued to compile all his life, and which formed his "history of experience", he drew out the examples which he then formed into plays for presentation before the eye of the mind, so that the axioms or laws of life might be observed and drawn forth. From a history of experience, to dramatic performances based on examples of this experience carefully put together, to axioms, is the method of discovering truth that Francis 'discovered'. Whilst in France he was beginning and testing out his work.

It was also whilst in France for this second time that he began to record his own personal life history and that of others known to him (such as his mother, Queen Elizabeth, and father, the Earl of Leicester), and to leave it as a legacy in cipher for posterity. In doing this he hoped not only to leave a correct account of history as he knew it, so as to correct any errors, misconceptions or deliberate falsifications that posterity might have been left with, but also to provide future ages with a genuine history and in-depth study of one man's personal experience from birth to death — with all the passions, thoughts and other matters that constitute man's sensitive but developing psyche. Because he was employed anyway in cipher work, he was able to spend long hours in his own cipher work without suspicion.

> By some strange Providence, this [*i.e.* Francis' mission] served well the purposes of our own heart; for, making ciphers our choice, we straightway proceeded to spend our greatest labours therein, to find a method of secret communication of our history to others outside the realm. That, however, drew no suspicion upon this device, inasmuch as it did appear quite natural to one who was in company and under the instruction of our Ambassador to the Court of France; and it seemed, on the part of our parents, to afford peculiar relief, that our spirit and mind had calmed, as the ocean after the tempest doth sink into a sweet rest, nor gives a sign of the shipwracke below the gently rolling surface.[268]

> The preparation and distribution of the Cipher words required much time, and this time was soon at my disposal. The numerous works that will be sent forth, soon, will prove the truth of my assertion of a ceaseless industry and an unflagging zeal. No one living in the midst of the tumults and

> distractions which are found in our great towns could better hold to a purpose — but a few years younger, in truth, then I, for it stirred within me when I first was told of my great birth, and took form shortly after that scene at the Court of our mother which led so quickly to my being sent to France in the company and care of Sir Amyas Paulet. It weighed on me constantly, until I devised a way by which I could communicate this strange thing to the world, as you know, and my restless mind unsatisfied with one or two good Ciphers, continually made trial of new contrivances, in order to write the true story fully, that wrongs of this age may be made right in another .... [269]
>
> It is not easy to reveal secrets at the same time that a wall to guard them is built, but this hath been attempted. How successful it shall be, I know not, for though well contrived so no one has found it, the clear assurance cometh only in the dreams and visions of the night, of a time when the secret shall be fully revealed. That it shall not be now, and that it shall be then — that it shall be kept from all eyes in my own time, to be seen at some future day, however distant — is my care, my study. [270]

Whilst in the embassy of Sir Amyas Paulet, and carrying out the 'cypher' mission for the Queen, Francis began to invent his ciphers. An early one was the Biliteral Cipher, which he then used to give instructions and keys for the working of his principal cipher, the Word Cipher. This Word Cipher he must have also invented at a very early stage, as he began to record his experience of discovering his birth, his banishment and his love for Marguerite as soon as he was in France for the second time. The exact moment in time when he actually thought of doing all this and invented the cipher is not certain; but, judging by what he says, Francis must have formed the idea almost immediately after he had been told of his royal birth, whilst still in England. Francis tells us that he experienced a powerful and unforgettable vision at that traumatic point in his life, which presumably is when the idea "stirred within" him and "took form":

> And now it is time for us to tell you how we found the way to conceal these ciphers. One night, when a youth, while we were reading in the Holy Scriptures of our great God, something compelled us to turn to the *Proverbs* and read that passage of Solomon, the King, wherein he affirmeth "that the glory of God is to conceal a thing, but the glory of a king is to find it out." And we thought how odd and strange it read, and attentively looked into the subtlety of the passage. As we read and pondered the wise words and lofty language of this precious book of love, there comes a flame of fire which fills all the room and obscures our eyes with its celestial glory. And from it swells a heavenly voice that, lifting our mind above her human bounds, ravisheth our soul with its sweet, heavenly music. And thus it spake:
>
> "My son, fear not, but take thy fortunes and thy honours up. Be that thou knowest thou art, then thou art as great as that thou fearest. Thou art not what thou seemest. At thy birth the front of heaven was full of fiery shapes; the goats ran from the mountains, and the herds were strangely clamorous to the frighted fields. These signs have marked thee extraordinary, and all the courses of thy life will show thou art not in the roll of common men. Where is the living, clipt in by the sea that chides the banks of England, Scotland and Wales, who will call thee pupil, or will read to thee? And bring him out that is but woman's son, [who] will trace thee in the tedious ways of

art and hold thee pace in deep experiment. Be thou not, therefore, afraid of greatness, I charge thee. Some men become great by advancement, vain and favour of their prince; some have greatness thrust upon them by the world; and some achieve greatness by reason of their wit; for there is a tide in the affairs of men which, taken at the flood, leads on to glorious fortune. Omitted, all the voyage of their life is bound in shallows and miseries. In such a sea thou art now afloat, and thou must take the current when it serves, or lose thy ventures. Thy fates open their hands to thee. Decline them not, but let thy blood and spirit embrace them, and climb the height of virtue's sacred hill, where endless honour shall be made thy mead.

"Remember that thou hast just read, that the Divine Majesty takes delight to hide His works, according to the innocent play of children, to have them found out. Surely for thee to follow the example of the Most High God cannot be censured? Therefore put away any popular applause and, after the manner of Solomon the King, compose a history of thy times, and fold it into enigmatical writings and cunning mixtures of the Theatre, mingled as the colours in a painter's shell, and it will in due course of time be found. For there shall be born into the world (not in years, but in ages) a man whose pliant and obedient mind we, of the supernatural world, will take special heed, by all possible endeavour, to frame and mould into a pipe for thy fingers to sound what stop thou please; and this man, either led or driven, as we point the way, will yield himself a disciple of thine, and will search and seek out thy disordered and confused strings and roots with some peril and unsafety to himself. For men in scornful and arrogant manner will call him mad, and point at him the finger of scorn; and yet they will, upon trial, practice and study of thy plan, see that the secret, by great and voluminous labour hath been found out."

And then the voice we heard ceased and passed away.[271]

Francis learnt, not only that divine Truth (the Word of God) is purposely concealed in Nature (God's Works) with the intention that man should gradually search it out, find it and know it, but also that man can (and indeed is destined to) imitate God's purposes and actions, as man is created in the 'image' of God. As the sages of old discovered before him, Francis perceived that the 'treasure hunt' is a basic desire and motivation of all mankind; that the very effort of seeking, culminating with the finding, beings its own unique reward to each person. Effort and achievement are vital forces in everyone's life. Hence it is the parable, enigma, symbol, metaphor, allegory and adventure which are the best teachers of wisdom. Man likes to discover hidden meanings and purposes, and if there is not one to discover he will invariably invent one! So, adapting his own situation — which was one that constituted a state secret — he created an adventure and treasure hunt for the benefit of future generations. Taking the cue from the heavenly 'voice', he composed a history of his times and enfolded it into a multitude of publications and theatrical presentations according to a brilliant system of key words which he invented. Thus, for instance, in his Shakespeare plays, there are passages, sentences and part-sentences which are derived from his 'history', scattered throughout the texts of the plays, and which can be discovered and reconstituted according to their original form by carefully using the cipher keys that Francis gives. This particular cipher Francis calls his 'Word Cipher' or 'Great Cipher of Ciphers'.

Francis enjoyed the cipher writing, although he says he tired a little of the Biliteral, which was more mechanical in its nature and poetically quite restrictive. But once started with the

Biliteral, he had to proceed with it right to the end. It is interesting to note that it is the Biliteral Cipher which is the most easily grasped by our modern minds, and it did indeed form the inspiration for the invention of Morse Code and, more lately, the Binary System on which modern computers depend. Francis enumerated six main (or basic) ciphers which he used, some of them capable of several variations:

(1) ANAGRAMMATIC — formed of anagrams.

(2) BILITERAL — depending on two types of similar (but not the same) italic founts used in the printing of his books.

(3) CAPITAL LETTER — comprising various series of capital letters in two forms.

(4) CLOCK OR TIME — numbers are keys in these ciphers, as in traditional gematria, often with geometrical guides.

(5) SYMBOL — the important 'hieroglyphic' or pictorial method, employing symbolism of all kinds, including mathematical and musical symbolism.

(6) WORD — the great Cipher of Ciphers.

Francis used these ciphers not only to record the history of his times, but also for esoteric teaching — the traditional wisdom teachings and also his new method for discovering Truth.

* * * * * *

The French Court (plus Francis) stayed at Blois for four months, from the end of October 1576 until the end of February 1577. At the French Court, in Blois, Marguerite de Valois' position had become a most unpleasant one. Her brother, the King, had now told her blatantly that he had no intention whatsoever of letting her go, lest she be used as a hostage against him — whilst he was of course using her as a hostage against Navarre and the Huguenots. The standing of Marguerite was high in France on all counts, both because of birth and position as the French King's sister and a Queen in her own right, wife of a King, and also because of her charismatic character, charm, intelligence and involvement in the arts and sciences. She also had the grief of seeing her only allies — her brother the Duc d'Anjou and husband, Henri de Navarre — waging war upon each other, and the country yet again torn apart with visciousness and hatred which she had worked so hard and sacrificed herself to heal. Her husband was proclaimed a traitor and a rebel by Henri III, and she saw Henri III's *mignons* taking every opportunity to exasperate the French King against Navarre and proposing various schemes of assassination. How much in all this dangerous turmoil Francis was able to comfort Marguerite, we do not yet know, as not all of his cipher story has been deciphered; but the fact that he was there in the midst of this turmoil, his true status known and acknowledged (privately) by Henri III, Queen Catherine and Marguerite, his honesty and virtue trusted by them all, and he still loving Marguerite (and presumably she loving him too) must have made a considerable impact on his life.

Marguerite therefore sought for some way out of this odious situation, and took counsel with some of her friends, "to discover some pretext for withdrawing from the Court, and, if possible, from the Kingdom, until peace should be concluded, either under colour of making a pilgrimage, or paying a visit to one of her relatives."[272] A plan was hatched with the Princesse de la Roche-sur-Yon, who was on the point of setting out for Spa, to take the waters. The plan involved Anjou and Mondoucet, the French representative in Flanders, and their design to wrest the lost Burgundian fiefs back from the Spaniards and rule them in the name of France. Under the pretext of escorting the Princesse de la Roche-sur-Yon to Spa, Anjou was intending to return to Flanders with Mondoucet and continue his power-seeking intrigues. They readily saw the advantage that the beautiful and charming Marguerite could bring to their cause, if she was to accompany them — her prestige and winning ways more likely to perform in a week that which a diplomat could take a year or more over. Marguerite, on her part, besides the escape from Court, also enjoyed adventure and intrigue. She was fond of her brother, Anjou, and she also saw that this preoccupation would take him out of the line of direct confrontation with Henri de Navarre.

When this proposal was put forward to Henri III and Queen Catherine, it was accepted and their permission was given to Marguerite to travel to Flanders. Marguerite did not set out immediately for Flanders, but accompanied the Court first to Chenonceaux, near Tours, where Henri III established himself so as to be near Charles de Lorraine, Duc de Mayenne, who was laying siege to Brouage. She remained with the Court at Chenonceaux until May 28th, 1577, when, after a final conference with Anjou in regard to his projects and "the service he required of her", she started on her journey in great state to Picardy and Flanders with the Princess de la Roche-sur-Yon, several maids-of-honour, the Bishop of Auxerre, the Bishop of Langres, and others. Her mission was accomplished with exceptional skill, so that, by the end of the journey (*via* Catalet, Cambrai, Valenciennes, Mons, Namur, to Liège, near Spa) many of the key personages and towns had sworn allegiance to her and the French Crown. But whilst at Liège, taking the waters of the Spa, the Flemish States rose up in rebellion against the Spaniards. Concurrently, news was sent to her from Anjou that a peace had been concluded with the Huguenots, but that he (Anjou) was in even worse 'odour' at Court than ever, hated by the Huguenots whom he had deserted and had made war against; that the King regretted allowing Marguerite to travel on the Flemish expedition and that, out of hatred for his brother, he had secretly warned the Spaniards of Marguerite's real mission. She then had to make an extremely perilous journey back again to France, eventually arriving at her château of La Fère on October 1st, 1577. Anjou hastened to join her there, and the various leaders of the Flemish states, whom Marguerite had won over, gathered at La Fère in conference with Anjou. Finally it was decided that Anjou should raise an army and enter Flanders with his troops in the following Spring (1578), his Flemish allies having by then organised a movement and uprising in his favour. Anjou immediately returned to Paris, to seek permission from the King for this undertaking.

Whilst Marguerite was making her journey, the French Court moved at the end of June to Poitiers, arriving at Poitiers on July 2nd, 1577. Although it is known that Sir Amyas Paulet kept up a regular correspondence with Sir Nicholas Bacon during the three years of his residence in France, yet unfortunately most of the letters are lost; but in an undated letter from Poitiers (following a letter from Poitiers dated July 1577) he informs Sir Nicholas:

> "Your sonne, thanks be to God, is in good health."

It is thought that this letter was also written in July 1577. The next letter still existant, written from Sir Amyas to Sir Nicholas, was written late in September 1577, soon after the new peace (the Peace of Bergerac) had been concluded between Henri III and the Huguenots. The new peace more or less established the terms of the previous Peace of Beaulieu, but with substantial reductions. After expressing his satisfaction at the conclusion of peace, Sir Amyas adds:

> "I must tell you that I rejoice much to see that your son, my companion, by the Grace of God, passed the brunt and peril of this journey .... Your son is safe, sound and in good health, and worthy of your fatherly favour."

What dangerous journey it was that Francis had undertaken sometime between July and September 1577 whilst the French Court remained at Poitiers, we do not know. Given the time available for the journey, the brief description of the journey as being perilous, and the nature of Francis' contacts and friendships, it seems quite likely that he may have been sent as an ambassador to Navarre's territory (Gascony), to find Henri de Navarre so as to sue for and conclude peace terms between the two Kings.

On October 5th, 1577, Henri III left Poitiers to return to Paris, where Anjou returned from La Fère to join him, with the request to raise an army for the Flemish campaign. Soon after Anjou had left his sister, Marguerite too set out for Paris, determined to renew her demand to Henri III to be allowed to rejoin her husband in Gascony. The King, Queen-Mother, Anjou and the whole Court rode out to meet her at Saint-Denis, receiving her with much cordiality and taking great pleasure in having her describe her journey. Henri III and Catherine appeared to give their assent to both Anjou's and Marguerite's propositions but, in usual fashion, their Majesties procrastinated as the weeks went by. With the new peace concluded, and Marguerite returned to Court, a fresh series of festivities were begun, the 'Court of Love' being again revived:

> "the Court having arrived at Paris, the balls and the festivities which the war, the King's journey to Poitou and above all the absence of the beautiful Queen of Navarre had suspended, were resumed."[273]

Matters became worse and worse for Anjou, as the King's *mignons* baited Anjou and turned the King more and more against his younger brother. The whole disgraceful episode culminated on February 9th, 1578, when Henri III ordered the arrest of Anjou on suspicion of treasonable activity. The next day the elder members of the King's Council, scandalised at the bad advice the King had received from his favourites, addressed a vigorous remonstration to his Majesty. Henri had recovered his senses by this time, took the remonstrations in good part, and begged the Queen-Mother to smooth things over for him. Anjou was then set at liberty, and the two brothers were formally reconciled in the presence of the principal personages of the Court. But Henri now feared what Anjou might think or do, and Anjou was kept under close surveillance and forbidden to leave the Louvre, with orders that all of Anjou's attendants were to be turned out of the palace every night (excepting his attendants of the bedchamber).

This was really too much for Anjou, and he resolved to escape, withdraw to his estates and finalise the preparations for his Flemish expeditions. With the help of Marguerite and a stout rope, on the night of February 14th, 1578, Anjou made his escape from his sister's bedroom window in the north-east quarter of the Louvre, (in a similar fashion to that earlier escape plan which Francis had set up for Marguerite two years previously), and made his way to Angers. The King, informed of this escape, suspected Marguerite's connivance but did not dare to do anything now that Anjou was at liberty and whilst there was still only a very precarious peace in the country. Once at Angers (capital of the province of Anjou), the Duc d'Anjou began to prepare for his expedition to Flanders, assuring his brother that he had no intention of disturbing the Kingdom of France.

Marguerite renewed her efforts to persuade her mother and brother to allow her to leave the French Court and rejoin her husband in Gascony. Henri III was now in a position of pressure on all sides, such that he could no longer afford to refuse her request; so he granted her wish. It was arranged that the Queen-Mother should accompany her daughter, ostensibly to come to a settlement with Navarre, concerning grievances and claims that he was (rightfully) making, but really so as to try to stir up trouble between the King of Navarre and some of his most influential followers. Henri III did his utmost to try to conciliate his sister so that she would not depart bearing him ill-will, and even went so far as to fulfil the promises made to Marguerite at the time of the Peace of Beaulieu, and which he had renewed on her return from Flanders: namely, to assign her her dowry in lands. Marguerite thus received on March 18th, 1578, the *sénéchaussées* of Quercy and the Angenais, the royal domains of Condomois, Auvergne and Rouerge (all adjoining her husband's dominions), and the lordships of Rieux, Alby and Verdun-sur-Garonne, thereby making her one of the wealthiest and most powerful landowners in France.

Before setting out on her journey south, Marguerite went with her mother to visit her brother, Anjou, at the town of Alencon, to bid him farewell prior to her own journey to join her husband and prior to his expedition into Flanders which was on the point of departure. Then she returned to Paris to make arrangements for her own journey, which eventually began at the end of July, 1578.

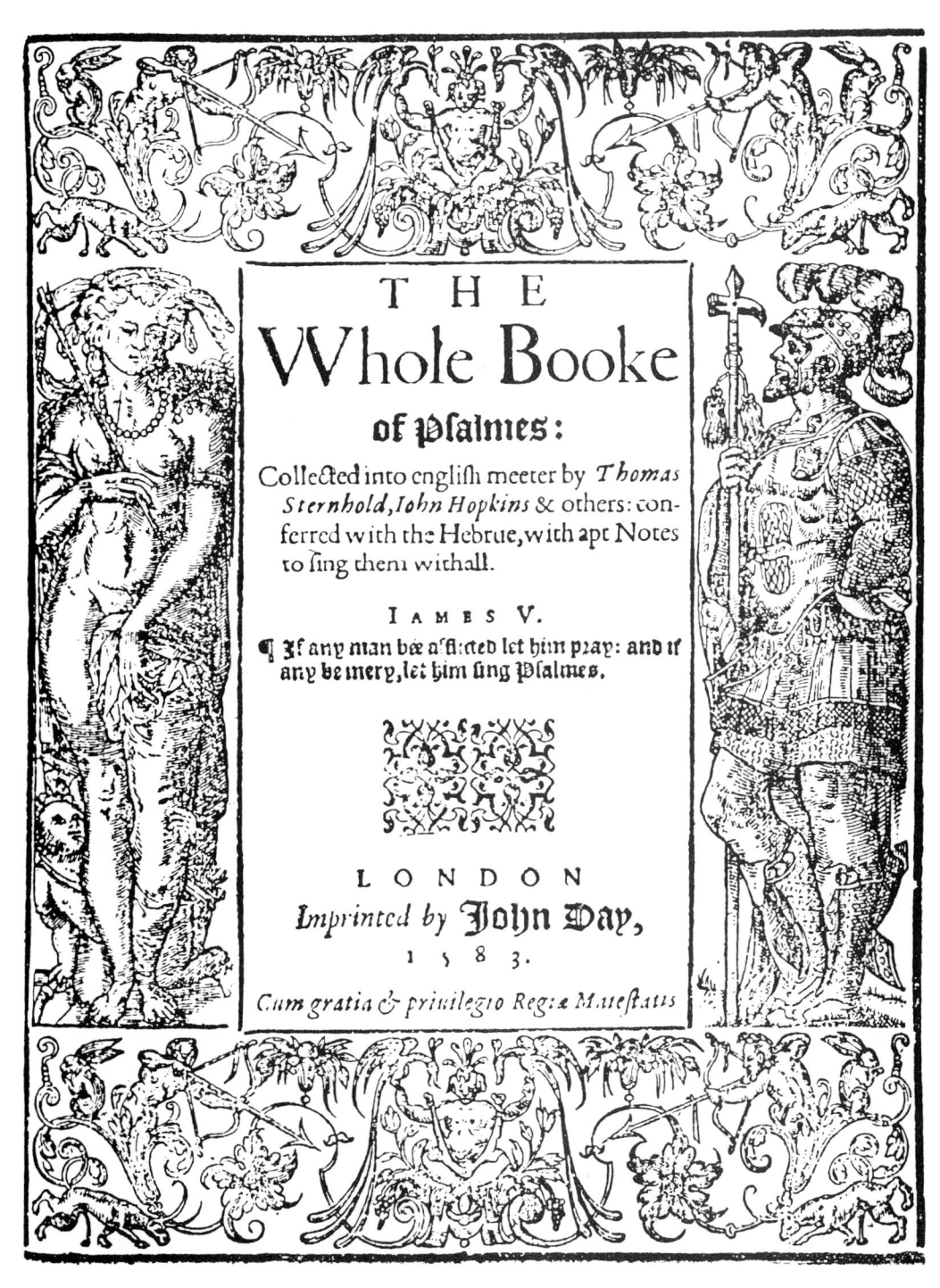

(29) TITLEPAGE OF THE WHOLE BOOKE OF PSALMES (1583).
This is the first known appearance of the 'Archer' emblem.

The French Academy

In February 1578 there was printed and published in France the first edition of the first part of a remarkable book, entitled *Académie Françoise par Pierre de la Primaudave Esceuyer, Seignour dudict lieu et de la Barrée, Gentilhomme ordinaire de la chambre du Roy*. The dedication, dated "February 1577" (*i.e.* 1578)[274] is addressed, "Au Tres-chrestien Roy de France et de Polongne Henry III de ce nom". The first English translation, by "T.B.", was published in 1586, imprinted at London by Edmund Bollifant for G. Bishop and Ralph Newbery. Other parts and editions followed in 1589, 1594, 1602 and 1614, but the first complete edition in English is dated 1618, and was printed for Thomas Adams. Over the dedication to this 1618 edition is printed the 'Archer' headpiece which came to be the renowned signature of Rosicrucian works directly associated with Francis Bacon and his fraternity — an emblem which was first used (as far as I am aware) in a 1583 edition of the highly popular The *Whole Booke of Psalmes: Collected into english meeter by Thomas Sternhold, John Hopkins & others, &c*, published in 1583 after Sternhold's and Hopkins' death by someone or some people who felt it important to "give this work to the world in a new and specially remarkable garb".[275] The 'Archer' headpiece is an emblem intimating the deepest mysteries of initiation, relating peculiarly to the Orphic Mysteries of Divine Love and to the hidden knowledge concerning Sirius, the Dog Star, Falcon or Archer. In the emblem is incorporated the Baconian rebus of two rabbits (generally known as *conies* in the 16th century) placed back to back to each other.[276] In the first English edition of the *French Academie*, of 1586, was printed a special ornamental capital 'S', carved from a woodblock and used for the first letter in the text of the dedication. This same woodblock was reused in a similar manner in the second English edition of 1589, and finally in an identical way (*i.e.* as the first letter in the text of dedication) in the 1625 edition of Francis Bacon's *Essays*, printed in London by John Haviland, which was Francis' last publication (with the exception of a small pamphlet containing his versification of certain Psalms) issued during his known life-time as 'Francis Bacon'.

The French Academie of 1618 is a thick folio volume, with 1,038 pages of text. As William Smedley noted in his essay on the mystery of Francis Bacon,[277] "It may be termed the first Encyclopaedia which appeared in any language, and is perhaps one of the most remarkable productions of the Elizabethan era. Little is known of Pierre de la Primaudaye. The particulars for his biography in the *Biographie Nationale* seem to have been taken from references made to the author in the *French Academie* itself. In the French Edition, 1580, there is a portrait of a man, and under it the words '*Anag. de L'auth. Par la prierè Dieu m'ayde.*'"

The book is a record by a youth who had, by "good hap", been a visitor at the Court of Henri III when the Court was at Blois for the meeting of the Estates-General (November

(30) CAPITAL LETTER 'S' — the first letter in the text of the Dedication to the 1st edition of the English translation of the *French Academie* (1586), printed in London by G. Bollifant. The wood-block is also used in a similar manner in the 2nd edition (1589), printed in London by John Bishop, and also in the 1625 edition of Bacon's *Essays*, printed in London by John Haviland. These are its only known appearances.

Note the fish suspended on a line from a bell, indicative of esoteric Christian teachings. The bell is a symbol for the *Logos*, the Word of God, whose Sound (or *Christos*) is the line (or ray) that catches the fish and draws it out of the water. The water signifies universal matter, and the fish is the symbol of the pure form of the neophyte-disciple-initiate who is able to swim and exist freely in those waters of baptism, the waters of the Lesser Mysteries. The form of the fish is the *vesica piscis*, 'the vessel of the fish', which is the chalice that will become the living flame of light, or holy grail, when the fish is truly caught and lifted out of the waters in its resurrection to the Greater Mysteries of Light. The lifting of the chalice in the Holy Communion means just this. The fish are those which are able to not only survive but be at ease in the ever-recurring Floods of Noah. Noah's ark is the *vesica-pisces* — the temple or grail vessel. The font is the *piscina*, the 'fish pond', Neophytes were called *pisciculi*, 'little fishes'.

Whilst the fish represents the neophyte-disciple-initiate of the Lesser Mysteries, the bell signifies not only the *Logos-Christos-Sophia* but also the resurrected and illumined adept-master, the Christed soul, who is become the fisherman (or "fisher of men"). The fish becoming, through sacrifice, the fishermen is analogous to the sheep becoming the shepherds. The crozier or shepherd's crook of the bishop is also the fisherman's rod.

The sacred meal of broiled fish, plus honeycomb, bread and wine, is the *agape* or love-feast — the *Qadosh* or Feast of Brotherhood that is partaken of on the eve of the Passover and on the eve of every Sabbath. The Sabbath signifies the Great Peace, the Great Illumination, and is the end result of all labours when the labourer may rest. To reach it requires dedication and self-sacrifice: the fish must be caught and broiled on the fire of the Spirit. Saturday symbolises the Sabbath in the weekly cycle: Friday is the Sabbath eve, hence the custom of eating fish on that day in memory of the Last Supper. Friday is also dedicated to Venus, symbol of the Morning Star.

(31) FISH & HONEYSUCKLE EMBLEM, from the titlepage of the 1st (1603) and 2nd (1604) Quartos of *The Tragicall Historie of Hamlet, Prince of Denmarke,* by William Shake-speare.

1576 — February 1577). It had been a time of great festivities and cultural activity, and a feast of knowledge and artistry, and the author desired to thank Henri III for this privileged time by offering to his Majesty another "dish of divers fruits, which I gathered in a Platonicall garden or orchard, otherwise called an ACADEMIE, where I was not long since with certain young Gentlemen of Anjou my companions, discoursing together of the institution in good manners, and of the means how all estates and conditions may live well and happily." The author describes how the Academy came about, founded by "an ancient wise gentleman of great calling" who had spent the greater part of his years in the service of two kings and of his country, France. This wise gentleman retired to his house and set up an Academy to teach the youth of his day good manners, virtuous, honest living and philosophy, to help keep them from the "over great license and excessive liberty granted to them in Universities." The wise old gentleman admitted four young gentlemen, sons of noblemen, and appointed a tutor of great learning to instruct them. After six or seven years the fathers of the four youths decided to pay a special visit to the Academy to see the results of their children's privileged education. They managed to spend several days there with their sons, hearing them recite, but a fresh outbreak of civil war (which from my researches would have been the resumption of hostilities in December 1576, after the Holy Catholic League was formed and had renounced the Treaty of Beaulieu at the meeting of the Estates-General) interrupted their happy assembly, and the noble youths were called to the service of their King. When a new peace was eventually concluded (in September 1577, the Peace of Bergerac), the four young gentlemen managed to return to their Academy, and a fresh visit of their fathers was arranged. This duly took place, lasting for a period of three weeks, and during this time the author was "by good hap" included as one of the company when they began their discourses. He so greatly wondered at the discourses that he thought them worthy to be published abroad, and he first published them as an 'offering' for Henri III, whom he honours or flatters with the title of 'Solomon' in the Dedication.

After the dedication of the book to the French King, the author continues with 'the Author to the Reader', which is an essay on Philosophy, the thoughts and sentences of which are familiar to us in Francis Bacon's *Essays* and his other works. Then follow the several chapters, seventy-two in number, each beginning with the word 'Of' (*e.g.* 'Of Death', 'Of Nature and Education', 'Of Poverty', 'Of Riches'), which are paralled both in title and thought by Francis Bacon's *Essays*. An original of these titles and their subject matter may be found in Sir Nicholas Bacon's *Sententiae* which were inscribed on the walls of his Long Gallery at Gorhambury (see F.B.R.T. Journal I/3, *Dedication to the Light*), and the same form was used in Michel de Montaigne's *Essaies* which were first published in France in 1580.

Although the author of the *French Academy* describes himself as a youth of small experience, yet the contents of the book bear evidence of a wide knowledge of classical authors and their works, an intimate acquaintance with the ancient philosophies, a profound interest in modern philosophy and moral virtue, and a store of general information that would have been impossible for any ordinary youth of his time and age to have possessed. In the *Gesta Grayorum* of Gray's Inn (Christmas 1594), whose principal playwright and organiser was Francis Bacon, the *French Academy* is specifically mentioned as necessary to be studied by the 'Knights of the Helmet', Bacon's fraternity in learning and illumination.[278]

Like several other Baconian scholars, I am convinced from all the evidence available that the young Francis Bacon was the author of the *French Academy*, and that he first wrote it (in French) after his return from the Academy to the French Court at Paris during the winter of 1577. Further, from my own researches, I feel fairly certain that the French Academy that is mentioned is an Academy founded by the writer and essaist, Michel de Montaigne, who became a close friend not only of Francis but also of Anthony Bacon, and that Montaigne was the "wise old gentleman".

Michel Eyquem Montaigne was born on 28th February 1533, the eldest son of the Siegneur de Montaigne. He was brought up to speak Latin as his 'mother tongue', as was the case with most scholars of that day — although he insisted that he was no scholar himself, but a man of action. He also knew Greek. After studying law he eventually became counsellor to the *Parlement* of Bordeux, and a highly respected man. In 1561 he was sent to the French Court on a mission which lasted for a year and a half; and in 1562 he followed the King, Charles IX, to Rouen. He returned to Bordeaux in 1563, and married in 1565. In 1568 his father died and Michel inherited the estates of Montaigne. In 1569 he published his first literary work, the *Natural Theology* of the Catalan, Raimond Sebond, whose 'defence' he was later to write. In 1571 he left the practice of law and retired to his estates at Montaigne.

In retirement on his estate, Montaigne began to compose his famous essays on different aspects of life. Most of the essays which were later published as Book One were written (in their first form) during 1572-73, being mainly collections of quotations and notes on his reading, with some personal reflections added. When the civil war began again in 1572, after the Massacre of St. Bartholomew's Eve, his own writing and studies were periodically interrupted as he took an active part in the war, helping Charles IX and then Henri III. It was not until the Peace of 'Monsieur' in May 1576 that Montaigne was able to fully retire again to his library and studies, when he began to write the *Apology for Raimond Sebond*. This latter writing, together with some further essays. formed his second book, and his Books One and Two were eventually published together in 1580.

The story of the *French Academy* and its wise old gentleman fits exactly the circumstances and character of Michel Montaigne. It would have been sometime in 1571 when, having retired from legal life and withdrawn to his house, he first began to consider the formation of a little Academy, which probably came into being shortly afterwards. As the intention of the Academy was to augment University life and teaching, guiding the noble-born into an honest and virtuous way of life and thereby preserving them from the possible corruption that University life might give, it would appear reasonable to suppose that this Academy operated during the University vacations only. The four young noblemen, who were sons of distinguished noblemen of Anjou (and hence probably of royal blood and directly related to the French Monarchy, to which in particular the wise old gentleman was trying to bring profit by his scheme), were students of this Academy for "six or seven years" before their fathers decided to visit them together at the Academy and test the results of their education. Their visit was a happy one, but unfortunately broken up after a few days by the news of the country's return to civil war — and this dates the visit accurately to Christmas 1576, when the Treaty of Beaulieu was torn up by the Holy Catholic League with the King at their head. This was during the Christmas vacation, and when the French Court was at Blois. The four youths were called to the service of their King and country, and by that time would have been old enough to be called upon for active service.

The ensuing war (the sixth civil war in France of that period) did not last long, and peace was concluded at Bergerac in September. Once the war was ended, the four young noblemen "laboured forthwith to meet together", and to arrange to continue their special conversation with their fathers. This duly came about, and it seems likely that it was during the Christmas vacation of 1577 that the special assembly of fathers and sons took place again. It was during this time that the author of the *French Academy*, whom I feel certain was Francis Bacon, had the privilege of being allowed to become one of their company. In fact he eventually replaced one of the four principal students, who seems to have left during those discourses without any explanation, leaving his father disappointed. This special conference went on for the space of "three whole weeks, which made eighteen daies workes". After it had finished Francis left the Academy and Montaigne and returned to Paris and the French Court, where he wrote up his thoughts and experiences as a book whilst they were still fresh on his mind. Then he caused the first part of the book to be straightway printed and published in the following February, 1578. He was obviously exhilarated and fired by the whole experience, just as he had been by the equally extraordinary time with the French Court at Blois. His young mind, prepared and trained by his upbringing at Sir Nicholas Bacon's Platonic 'Academy' at Gorhambury, had proved fully capable of responding to and enjoying this unique opportunity at Montaigne's Academy, and in true form he just had to write it down and let the precious thoughts and knowledges pour out for the benefit of the world at large, as an early part of his philanthropic and philosophical scheme.

Francis addresses Henri III thus, in the Dedication to the *French Academy*:

"The dinner of that prince of famous memorie, was a second table of Salomon, vnto which resorted from euerie nation such as were best learned, that they might reape profit and instruction. Yours, Sir, being compassed about with those, who in your presence daily discourse of, and heare discoursed many graue and goodly matters, seemeth to be a schoole erected to teach men that are borne to vertue. And for myselfe, hauing so good hap during the assemblie of your Estates at Blois, as to be made partaker of the fruit gathered thereof, it came in my mind to offer vnto your Maiestie a dish of diuers fruits, which I gathered in a Platonicall garden or orchard, otherwise called an ACADEMIE, where I was not long since with certaine yoong Gentlemen of Aniou my companions, discoursing togither of the institution in good maners, and of the means how all estates and conditions may liue well and happily. And although a thousand thoughts came then into my mind to hinder my purpose, as the small authoritie, which youth may or ought to haue in counsell amongst ancient men: the greatness of the matter subiect, propounded to be handled by yeeres of so small experience: the forgetfulness of the best foundations of their discourses, which for want of a rich and happie memorie might be in me: my iudgement not sound ynough, and my profession vnfit to set them downe in good order: briefly, the consideration of your naturall disposition and rare vertue, and of the learning which you receiue both by reading good authors and by your familiar communication with learned and great personages that are neere about your Maiestie (whereby I seemed to oppose the light of an obscure day, full of clouds and darkness, to the bright beames of a very cleere shining sonne, and to take in hand, as we say, to teach Minerua)[279]. I say all these reasons being but of too great waight to make me change my opinion, yet calling to mind manie goodlie and graue sentences taken out of sundry Greeke and Latine Philosophers, as also the woorthie examples of the liues of ancient Sages and famous men, wherewith these discourses were inriched, which might in delighting your noble mind renew your memorie with those notable sayings in the praise of vertue and dispraise of vice, which you alwais loued to heare: and considering also that the bounty of Artaxerxes that great Monarke of the Persians was reuiued in you, who receiued with a cheerfull countenance a present of water of a poore laborer, when he had no need of it, thinking to be as great an act of magnanimitie to take in good part, and to receiue cheerfully small presents offered with a hartie and good affection, as to giue great things liberally, I ouercame whatsoeuer would haue staied me in mine enterprise".[280]

Then, in the first chapter, Francis describes how the Academy came about:

"An ancient wise gentleman of great calling having spent the greater part of his years in the service of two kings, and of his country, France, for many good causes had withdrawn himself to his house. He thought that to content his mind, which always delighted in honest and vertuous things, he could not bring greater profit to the Monarchie of France, than to lay open and preserve and keep youth from the corruption which resulted from the over great license and excessive liberty granted to them in the Universities. He took unto his house four young gentlemen, with the consent of their parents who were distinguished noblemen. After he had shown these young men the first grounds of true wisdom, and of all necessary things for their salvation, he brought into his house a tutor of great learning and well reported of his good life and conversation, to whom he committed their instruction. After teaching them the Latin tongue and some smattering of Greek he propounded for their chief studies the moral philosophy of ancient sages and wise men, together with the understanding and searching out of histories which are the light of life. The four fathers, desiring to see what progress their sons had made, decided to visit them. And because they had small skill in the Latin tongue, they determined to have their children discourse in their own natural tongue of all matters that might serve for the instruction and reformation of the every estate and calling, in such order and method as they and their master might think best.[281]

"Now this school having been continued for the space of six or seven years to the great profit of this nobility of Anjou, the four fathers on a day tooke their iournie to visite this good old man, and to see their children".[282]

The four fathers visited their sons at the Academy, to hear them discourse on what they had learnt, but the fresh outbreak of civil war broke up their happy assembly after only a few days; but later, when peace was again restored, they all assembled again to carry on where they had left off:

"The sudden and sorrowfull newes of the last franticke returne of France into civill warre, brake vp their happy assembly, to the end these noble youths betaking themselves to the service due to their Prince, and to the welfare and safetie of their countrie, ...But, as we said in the beginning, after newes of the peace proclaimed, which was so greatly looked for, and desired of all good men, they laboured forthwith to meete together, knowing that their ioint return would be acceptable to their friends, especially to that good old man by whom they were brought vp. Moreover they deliberated with themselves as soon as they were arrived at the old man's house, to give their fathers to understand thereof, to the end they might be certified from them, whether it were their pleasures to have them reiterate and continue in their presence the morale discourses begun by them, as we have learned before; that they might be refreshed with the remembrance of their studies, and thereby also keepe fast for ever those good instructions, which by the daily travell of so many years they had drawn out of the fountaine of learning and knowledge. As it was devised by them, the execution thereof followed, so that all these good old men being assembled together, taking up their first order, and conferring anew of the same matters, daily met in a walking place covered over in the midst with a goodly green Arbour, allotting for this exercise from eight to ten in the morning, and from two to foure in the afternoone, Thus they continued this exercise for the space of three whole weeks, which made eighteen daies workes ....[283]

"It was arranged that they should meet in a walking place covered over with a goodly green arbour, and daily, except Sundays, for three weeks, devote two hours in the morning and two hours after dinner to these discourses, the fathers being in attendance to listen to their sons. So interesting did these discussions become that the period was often extended to three or four hours, and the young men were so intent upon preparation for them that they would not only bestow the rest of the day, but oftentimes the whole night, upon the well studying of that which they proposed to handle. .... During which time it was my good hap to bee one of the companie when they began their discourses, at which I so greatly wondered, that I thought them worthy to be published abroad ...."[284]

Francis had the good fortune to be in the company of the four fathers and their sons when the discourses began, as a privileged visitor; and sometime during the three weeks of these discussions he actually took the place of one of the student sons, for a reason that is not given except for the mention that the student's father was disappointed (by his son). The narrative continues with Francis included as one of the four students:

"And thus all fower of us followed the same order daily until everie one in his course had intreated according to appointment, both by the precepts of doctrine, as also by the examples of the lives of ancient Sages and famous men, of all things necessary for the institution of manners and happie life of all estates and callings in this French Monarchie. But because I knowe not whether, in naming my companions by their proper names, supposing thereby to honour them as indeede they deserve it, I should displease them (which thing I would not so much as thinke) I have determined to do as they that play on a Theater, who under borrowed maskes and disguised apparell, do represent the true personages of those whom they have undertaken to bring on the stage. I will therefore call them by names very agreeable to their skills and nature: the first **ASER** which signifieth Felicity: the second **AMANA** which is as much to say as Truth: the third **ARAM** which noteth to us Highness; and to agree with them as well in name as in education and behaviour, I will name myself **ACHITOB** which is all one with Brother of goodness. Furthermore I will call and honour the proceeding and finishing of our sundry treatises and discourses with this goodlie and excellent title of Academie, which was the ancient and renowned school amongst the Greek Philosophers, who were the first that were esteemed, and that the place where Plato, Xenophon, Poleman, Xenocrates, and many other excellent personages , afterwards called Academicks, did propound & discourse of all things meet for the instruction and teaching of wisdome: wherein we purposed to followe them to our power, as the sequele of our discourses shall make good proofe."[285]

And then the reports of the discourses follow, in their respective chapters.

The names of all the people concerned and the place of the Academy are carefully concealed by Francis, but clues are provided by him as to the esoteric nature of the Academy and its 'initiates'. First of all he employs the suggestive letter '**A**' as the capital letter of each cipher name — the special symbol and signature used by the Rosicrucian initiates under whatever name or organisation they are working. It first seems to appear in printed literature in 1563, as the 'Double A' headpiece, in *De Furtivis Literarum Notis Vulgo. De Ziteris*, by Ioan. Baptista Porta, and is apparently used for the first time in English printed books, in this extremely important and significant 'double' form, in the 1589 publication. *The Arte of English Poesie*, printed by Richard Field,[286] and associated with (if not written by) Francis himself. Although several different printers were used to print the various works, in many different places and countries, yet there were only a few woodblocks cut with the distinctive designs, and these 'migrated' from printer to printer, being reused many times. Without doubt they were the guarded property of the author or group sponsoring the publications, who then supplied the woodblock to the printer when required. The actual symbol of the '**A**' goes back into deepest antiquity, and is the sign and symbol of Light, the Creative Word and Idea of God — the Divine Vision and Alpha of Creation. From the '**A**' all things spring forth. Its very form signifies the Holy Trinity or Triple Cause of Creation, and was once written as an equilateral triangle. It is to be found in Ancient Egyptian esotericism, for instance, signifying the Creative Source of all life, and in its double form '**AA**' as the active and passive, or radiant and reflective principles; but its use is found in all Wisdom Traditions. In the double form, the first '**A**' signifies the radiant and creative spiritual light, whilst the second '**A**' signifies its perfect reflection or manifestation as the perfected living form or soul of light. The one descends from heaven to vivify matter and evolve all life form, and the other rises from earth to heaven in its evolution towards perfect manifestation of the former. The descending spirit and the ascending soul were thus shown as the descending and ascending triangles or '**A**'s' (*i.e.* $\triangledown \triangle$), the perfect fusion or marriage of the two forming the symbol of the Seal of Solomon and the six-pointed Christ Star (or Star of David), the signet of the Christ Spirit and the Christed Soul. The island-hill arising from the waters or earth symbolised the second '**A**' of the illumined soul, as also did the pyramid-temple, and both were used to signify Horus or Jesus, the Christed soul of man. In the 16th and 17th century publications of the Rosicrucian initiates, the two '**A**'s were usually shown side by side, divided from each other by a central unifying feature, with one '**A**' shown light and the other shaded, according to a symbolic rule. (See F.B.R.T. Journal I/2, *The Virgin Ideal*.)

In the names that Francis uses for the four students, he begins each name with the significant and revealing '**A**', and then adds suitable symbolic words to name each student accurately. **A-SER** signifies 'Felicity', which is a state of calmness, tranquility, joy and happiness. The same root, 'SER', is to be found in the English word 'SERENE' from which Francis' cipher word is undoubtedly derived. 'Serene' was also an honorific epithet given to a reigning prince or to a royal house,[287] and certainly fits the son of an eminent nobleman of Anjou. **A-MANA** is as much as to say 'Truth', which is a state of wisdom or illumination — the knowledge of God,[288] which is the Word itself. The word 'MANA' is at the root of such words as 'Manna', the spiritual 'bread' or nourishment which falls from heaven, and which is the divine Thought or Living Word of God — the Light of God, which is Truth. 'Man', meaning 'Thinker', also comes from this root, as does the Eastern word, '*manas*', meaning 'mind'. **A-RAM** noteth to us 'Highness', which is that which is highest and in charge over other things which lie lower in hierarchical order. In particular 'Highness' relates to that which thinks, and which conceives the perfect Idea or Thought which then becomes the royal command. 'The Most High' is the special title of God as the Creative Mind, or Heaven, of Divine Thought; and in man it relates to his head and his thinking, governing capacity. The symbol of the RAM is used in all ancient traditions to signify this 'head' or 'heaven'. It is, like ASER, another princely title, befitting a member of the royal house. As for **ACHITOB**, with which Francis names himself, he explains that it is all one with 'Brother of Goodness', the very quality and nature that formed the central feature of his Great

Scheme — the primary goal for man to attain unto. This was his very motivation and driving force, and great hope for mankind. Everything that Francis thought, wrote or did was woven around this central focus.

> "I take Goodness in this sense, the affecting of the weal of men, which is that the Grecians call *Philanthropia*; and the word *humanity* (as it is used) is a little too light to express it. Goodness I call the habit, and Goodness of Nature the inclination. This of all virtues and dignities of the mind is the greatest; being the character of the Diety; and without it man is a busy, mischievous, wretched thing; no better than a kind of vermin. Goodness answers to the theological virtue Charity, and admits no excess; neither can angel or man come in danger by it. The inclination to goodness is imprinted deeply in the nature of man; insomuch that if it issue not towards men, it will take unto other living creatures....
>
> "The parts and signs of goodness are many. If a man be gracious and courteous to strangers, it shews he is a citizen of the world, and that his heart is no island cut off from other lands but a continent that joins to them. If he be compassionate towards the afflictions of others, it shews that his heart is like the noble tree that is wounded itself when it gives the balm. If he easily pardons and remits offences, it shews that his mind is planted above injuries; so that he cannot be shot. If he be thankful for small benefits, it shews that he weighs men's minds, and not their trash. But above all, if he hath St Paul's perfection, that he would wish to be an *anathema* from Christ for salvation of his brethren, it shews much of a divine nature, and a kind of conformity with Christ himself."[289]

Goodness and Holiness mean the same thing, and it should be well noted that, in Book I of *The Faerie Queen* (written by Francis on his return to England under the pseudonym *Immerito* and later masked by Edmund Spenser), the Knight of the Red Cross represents Holiness; and in the story he is eventually brought to the House of Holiness in which were to be found Faith, Hope and Charity — Charity being the chief of the three and the principal nature of the House itself.

The Knight is brought to this holy House by Una, his lovely lady, whose name means 'oneness' or 'one'. hence showing up the careful choice of words in *The French Academy* — "ACHITOB which is all one with *Brother of Goodness*." Indeed, all Francis' descriptive words are carefully chosen, veiling and yet revealing a deep esoteric knowledge or truth.

As for the word ACHITOB, it has been suggested[290] that it yields a revealing anagram, 'BACO-HIT', in which the word 'hit', as used by Chaucer, is the past participle of 'hide', and 'BACO' is an acceptable shortened form of 'Bacon' or 'Baconian'. Furthermore, if we return to the title of the French publication, *Académie Françoise*, we have in this title a possible ambiguity where 'Françoise' can mean 'French' or 'Frances'[291] — and, in fact, is more correctly translated as 'Frances', the word 'French' being, strictly speaking, 'Francaise' with an 'a' rather than an 'o' (although both forms were used in the 15th/16th centuries). It is only the so-called English translation of *Academie Françoise* as *The French Academy* which disguises this ambiguity. As mentioned before, the hero of our story is a master of deliberate ambiguity and disguise that is meant to reveal as well as to conceal.

Finally, it is worth noting the wisdom concerning the education and upbringing of a prince, which Francis learnt by direct experience and which he records in chapter 59 of *The French Academie*:

"For there is no time better and fitter to frame and correct a prince in, than when he knowes not that he is a prince. For if hee learne to obey from his infancy, when he commeth to the degree of commanding, he applieth and behaveth himself a great deal better with his subjects, than they that from their youth have been alwaise free and exempted from subiection."[292]

Love's Labour's Lost

When the special meeting of the fathers and sons, noblemen of Anjou, had been concluded at Montaigne's Academy, Francis returned to Paris to rejoin Sir Amyas Paulett's Embassy at the French Court, where he arranged for the immediate publication of the first part of his book, *Academie Françoise* (which was published that February 1578).[293] Francis was thus present during the disgraceful episode when the Duc d'Anjou was arrested, culminating in his escape from the Louvre with Marguerite's assistance, on the night of St. Valentine's Day. He was also in the company again of the Pleiadian poets, artists and philosophers, and the gaiety and culture of the Valois Court. When it was agreed that Marguerite could return to her husband, leaving the French Court for the Court of Navarre, Francis was able to accompany her and to journey south again to meet his 'old' friends.

A condition of Marguerite's journey south to join her husband in his territory was that the Queen-Mother should accompany Marguerite, ostensibly to settle her son-in-law's claims and the points still in dispute after the Treaty of Bergerac, but secretly to endeavour to sow discord between Henri de Navarre and his followers, and between him and his wife. To this end Catherine took with her her fairest and most appealing ladies-in-waiting, to seduce the susceptible King of Navarre and his friends, in the hope of creating disharmony and friction through jealousies and snide gossip. By the end of July all was ready for what was to be a Grand Progress and the Queens set out on their journey, escorted by Henri III as far as Olinville (one of his favourite country seats) where they stayed a few days. Then, on August 2nd, 1578, the King took his leave of his mother and sister, and the two Queens took the road south. Noel Williams, in his book *Queen Margot*, gives a good résumé of the Progress as follows:

> "The two Queens travelled in full state, and Marguerite's suite alone numbered close upon three hundred persons; there were ladies-of-honour and maids-of-honour, councillors and secretaries; confessors and chaplains; physicians, surgeons, and apothecaries; equerries and *valets-de-chambre*, pages, waiting-women, and lackeys; musicians and *marechaux-des-logis*; cooks, scullions, and laundresses; coachmen, grooms, postillions, and muleteers, so that it is small wonder that his Majesty preferred to burden the clergy, rather than himself, with the expenses of the journey. Among the distinguished persons who accompanied them, and whose attendants helped to swell the cortège to the size of a veritable army, were the Cardinal de Bourbon, the Duc de Montpensier, and his son, the Dauphin of Auvergne, the Prince de Conti, Matignon, Brantôme, and the learned Pibrac [the Queen of Navarre's Chancellor].......The '*escadron volant*,' too — significant fact! — was on its war

footing. For an advance guard, Catherine's maids-of-honour, Bazerne and Dayelle, a beautiful young Greek, who had escaped from the sack of Cyprus in 1571, the Italian Anne d'Atri, who had accompanied Marguerite to Flanders and Mlle. de Rebours and de Fosseux, maids-of-honour to the Queen of Navarre. And as for the rear-guard, the Duchesse of Montpensier, and the Duchesse d'Uzès, of the caustic tongue, whom Catherine called 'her gossip', and Marguerite 'her sibyl,' and, finally, the too-celebrated Madame de Sauve, who, although she was but five-and-twenty, had achieved so many conquests that she must have seemed almost a veteran to the young girls who were on their first campaign.

"The royal travellers journeyed by easy stages, and, after having passed through Étampes, and Artenay, and traversed the environs of Orléans, they made a short stay at the Château of Chenonceaux. From there they travelled, by way of Tours, Azay-le-Rideau, Chinon, Fontevrault, Poitiers, Ruffec, and Cognac, into Guienne.

"It was Catherine's policy that her daughter should be received *en souveraine* in all the towns of her husband's government, and Marguerite had a magnificent reception at Bordeaux, the capital of the province, into which city she made her entry 'with all the magnificence that could be desired, habited in an orange robe, her favourite colour, covered with embroidery, and mounted on a white horse.'[294]

"After a stay of a few days, the two Queens left Bordeaux on October 1st, and slept the night at Cadillac, and the one following at Saint-Macaire. Here Pibrac, who had been sent on in advance to announce their coming, arrived with the news that the King of Navarre would meet them at Castéras, half-way between Saint-Macaire and La Réole, 'a town which was still held by those of the Religion, by reason of the mistrust which yet possessed them — the disturbed condition of the country not having permitted of his coming any further.'[295]

"The Queen arrived first at the rendezvous, and entered the château to await the King. Henri appeared, an hour later, bravely attended by a suite of six hundred gentlemen, all richly dressed and well mounted. Followed by the Vicomte de Turenne and his chief nobles, he entered the château, saluted Catherine very cordially, kissed his wife on both cheeks and overwhelmed her with expressions of joy and affection. At La Réole, to which the united Courts proceeded, and where they remained for a few days, Catherine had several interviews with her son-in-law, and it was finally arranged that a special commission should be appointed to enforce the concessions granted to the Protestants at the Peace of Bergerac, and that all the points in dispute between the Huguenots and Catholics should be submitted to a conference.

"In the meanwhile, 'a little war of ogling' had begun. Madame de Sauve endeavoured to resume her empire over her royal lover, [*i.e.* Henri de Navarre], but she already belonged to ancient history. The Béarnais preferred green fruit, and his chief attentions were bestowed on Mlle. Dayelle, the beautiful Cypriote. On her side, Mlle. d'Atri found a malicious pleasure in rendering d'Ussac, the old governor of La Réole, madly enamoured of her. The King of Navarre and his younger nobles bantered the poor governor unmercifully, and the veteran, wounded to the quick, vowed vengeance on his ungrateful chief, and, some months later, deserted to the Royalist side.

"At Marmande, the two Courts parted; the King of Navarre setting out for Nérac to make arrangements for the proposed conference, while Marguerite, accompanied by her mother, went to take possession of her appanage. On October 12th she arrived at Agen, and made a magnificent entry into the town, whither all the nobles and gentry of the neigbourhood flocked to do her homage. From Agen, they set out for Toulouse, being met at the Château de Lafox by Henri, who escorted them as far as Valence. Their official entry into Toulouse took place on October 26th, when the Queens, who were accompanied by the Maréchaux d'Amville and de Biron[296], and a number of nobles, were received with great ceremony by the municipality, and conducted beneath triumphal arches and through streets strewn with flowers, to the archbishop's palace, where they lodged.

"Soon after their arrival at Toulouse, the Queen of Navarre fell ill 'seized with a violent attack of fever,' in consequence of which she was compelled to receive the members of the Parlement, when they came to present her with their address of welcome, 'in a great bed of white damask,' and was unable to leave the city until November 10th. Eager to expedite the meeting of the conference decided upon by her and Henri of Navarre, Catherine had already set out for Isle-Jourdain, the rendezvous arranged between them. While she was at Bordeaux, Henri had sent to her, proposing that the conference should be held at Castel-Sarrazin, on the pretext of the lack of suitable accommodation at Isle-Jourdain, but really because he wished to remain in a Huguenot country. The Queen-Mother curtly replied that she should hold him to his agreement; but, though she waited a week at Isle-Jourdain, neither the King nor any Huguenot deputies appeared. In great disgust, she ended by consenting to the conference being held at Nérac, and proceeded to Auch, into which town she made her entry on November 20th. Marguerite arrived the following day. On her journey from Toulouse, she had stopped for a night at the Château of Pibrac, belonging to her chancellor, renowned at that time for its sumptuous furniture and decorations, and had been magnificently entertained by its owner. Without as yet daring to avow his feelings, Pibrac, like so many others, had already succumbed to his beautiful mistress's charms; and this growing passion was to be followed by very unfortunate consequences.

"The municiple authorities came to receive Marguerite at the Porte de la Trille. The young Queen was in a litter, over which was spread a black velvet pall embroidered with her Arms; trumpets sounded, cannon fired salutes, and the children of the town chanted odes in her praise. Two days later, her husband arrived, and was also received with a great ceremony, as the Comte d'Armagnac, and handed the keys to the town……

"From Auch, the two Queens proceeded to Condom, and, on December 15th, Marguerite made her entry into Nérac, the capital of the duchy of Albret, and the residence of her husband's maternal ancestors. Here, the two Courts remained a week, [*i.e.* 15th-22nd December], which was devoted to fêtes and amusements of all kinds. The King's troupe of Italian players gave several performances, and Salluste, du Bartas, the Ronsard of the Huguenots[297], composed, in the Queen's honour, a dialogue in three languages, which was recited by three damsels, representing the Gascon, Latin and French Muses. As was, of course, to be expected, Marguerite awarded the palm to the Gascon Muse, who had proclaimed her husband '*leu plus grand rey deu moun*,' and, in token of her satisfaction, presented the young lady — a certain Mlle. Sauvage — with a gauze fichu which she happened to be wearing, and which, M. De Saint-Poncy assures us, was for many years cherished as a precious relic by the descendants of the recipient.

"At Nérac, politics were for the moment relegated to the background, and love reigned supreme. The pretty girls whom the two Queens had brought with them, turned the heads of all the Protestant nobles, so much indeed that Marguerite tells us that there were moments when her mother suspected that the delays in holding the conference had been purposely arranged by these enamoured gentlemen 'to the end that they might the longer enjoy the society of her maids-of-honour.' Even the stern Calvinist, d'Aubigné, and the grave statesman, Rosny,[298] caught the prevailing infection; for the former tells us that they were 'all lovers together,' while Sully admits that he also became a courtier and 'took a mistress like the others.' It should be mentioned, however, that the Calvinist nobles were, after all, only following the example of their sovereign, who had renewed his old *liaison* with Madame de Sauve, and whose passion for Mlle. Dayelle had reached a very high temperature. 'But,' writes his complacent consort, 'this did not prevent the King my husband from showing me great respect and affection, as much, indeed, as I could have desired; since he informed me, upon the very first day we arrived, of all the devices that had been invented, while he was at Court, to create bad feeling between us, and he expressed great satisfaction at our reunion.'

"Catherine cut short these intrigues by removing with her squadron to Porte-Sainte-Marie, where she remained until the first week of February 1579, when she returned to Nérac, for the conference.[299] In these deliberations, Marguerite took a prominent part, but in a sense very much opposed to that which Catherine had expected of her. That veteran intriguer had brought her fairest auxiliaries with her, in the confident expectation that her susceptible son-in-law would succumb to their charms, and thus cause an estrangement between him and his wife, by which she could not fail to profit. But Henri and Marguerite seemed to have agreed upon a policy of mutual tolerance, and the latter, thoroughly well acquainted with the objects and methods of her mother, was able to give her husband some very useful advice, which greatly disconcerted Catherine's plans. She also did not scruple to make use of her influence over Pibrac, and the enamoured lawyer manoeuvred so skillfully that the Huguenots obtained more favourable terms than they had dared hope for. The conference, after some pretty sharp recriminations,[300] ended with a promise of further securities to the Huguenots, in the shape of eight additional surety-towns, and of the complete redress of their grievances; and, towards the end of March, the Queen-Mother set out on her return to Paris, having accomplished very little, save the sowing of a few seeds of discord about the King of Navarre, and the beguiling of two or three Catholic nobles from their allegiance to him.

"Marguerite and her husband accompanied Catherine as far as Castelnaudary, where they took leave of her. The parting affected his Majesty not a little; for the Queen-Mother carried away with her the fascinating Mlle. Dayelle."[301]

Francis probably returned to Paris at the point where Queen Catherine left Nérac and the festivites of the 'Court of Love' at Nérac came to an end, just prior to Christmas. When back

(32) NICHOLAS HILLIARD
— self portrait, miniature (1577).

(33) ALICE HILLIARD, *née* Brandon, wife of Nicholas Hilliard — miniature painted by Nicholas Hilliard (1578).

(34) FRANCIS BACON, aged 18 years — miniature painted (1578/9) by Nicholas Hilliard.
Inscription: *"Si tabula daretur digna animum mallem"*.
Translated: "It would be preferable if a picture deserving of his mind (soul) could be brought about".

in Paris once more and amongst all the intellectuals, poets and artists that thronged Henry III's Court, Francis had a miniature portrait painted of him by Nicholas Hilliard. Hilliard, together with his newly-wedded wife, Alice, was among those who had journeyed to France in the entourage of Sir Amyas Paulett in September 1576. He had come to France in order to increase his knowledge and in the hope of earning some money for his future life in England by painting the lords and ladies of the French Court. He and his wife remained in France for the same period of time as Francis, staying partly with Sir Amyas Paulett's Embassy at the French Court, and partly in the service of François, Duc d'Anjou. He was back in England by April 1579.

> "In France he [Nicholas Hilliard] moved with assurance in the artistic circle which gathered around the brilliant Court of the last Valois King, Henry III. Hilliard stayed with Germain Pilon, the sculpture and medallist, and with the Queen-Mother Catherine de Medici's painter, George of Ghent. He conversed with the great humanist poet, Pierre Ronsard, who commented, in tribute, that 'the islands indeed seldom bring forth any cunning man, but when they do it is in high perfection', and, through the humanist philosopher, Blaise de Vignère, received commissions from the Duke and Duchess of Nevers for wood engraved portraits. Nor did he reside solely in Paris but followed the English embassy on its journey south to Poitiers in the summer of 1577."[302]

The painting that he made of Francis Bacon is exquisite. He and Francis would have known each other very well indeed by the time the painting was made, and this is reflected not only in the beautiful lines and colours of the portrait which depict the seventeen-year-old Tudor prince so accurately, but also in the comment which he inscribed around the edge of the portrait: "*Si tabula daretur digna animum mallem*" ("It would be preferable if a picture deserving of his mind/soul could be brought about", or "If only his mind/soul could be painted!") — a heart-felt comment which really says all! The inscription gives the date when the portrait was made, namely "1578" when Francis was aged "18", which dates the portrait accurately to the period between Francis' 18th birthday, 12th January 1578 (old style) and 20th March 1578 (old style). As the calendrical new year then began on March 25th, by modern reckoning the year was 1579.[303] The portrait may have been commissioned specially — perhaps by the Queen, Francis' royal mother, or by Francis himself — as a momento of his 18th birthday.

But Francis' delightful and instructive sojourn in France came to an abrupt end. On February 17th or 18th he dreamt that Gorhambury, the country home of his wise and caring foster-father (whom Francis regarded as a real father to him), "was plastered all over with black mortar" instead of the white plaster with which its walls were faced.[304] Two weeks or so later he received news from England that Sir Nicholas Bacon had died on February 20th, just two or three days after he had had his prophetic dream. Francis was summoned home, and on March 20th, 1579, he set out for England.

* * * * * * * * * *

Part of Francis' great scheme (the 4th part of the 'Great Instauration') was to find means of presenting historical experiences to the mind, using the imaginative arts as the *Janus* or medium by which the mind of man might observe a collected history of life experiences, emphasised, concentrated and presented as it were in a frame, so that the heart of man could be stimulated and his mind could the more readily notice and cognise the truths and laws of life that lie behind the mask of Nature. As the sages of old did before him, and as was being done at that very time by the sages of France, he used drama as his chief means; and as already noted, he began experimenting with its possibilities from an early age. Whilst in France, between 1576 and 1579, he developed his ideas as to how to achieve this aim and provide mankind with an example (or "light") to guide future generations by, whilst at the same time fulfilling his other spiritual commission to encipher the true history of his times for posterity to discover and profit by.

The two aims — the encipherment of the true history and the presentation of real life experiences to the mind *via* the imaginative arts — went hand in hand with each other, mutually complementing and assisting each other in the brilliant method that Francis devised. Thus, whilst in France, he began to collect a dairy or 'history' based on observation, learning and the fruits of experience. He carefully recorded his and others' emotions, thoughts, words and actions as they arose. He studied his own and others' characters and life-styles, love-affairs, politics, ambitions and prejudices, successes and failures, as also the characteristics of plants and trees, herbs and medicines, animals and birds; of water, air, earth (particularly crystals and gem-stones) and fire (especially sun-light), together with the spiritual nature of all these. In all this he was not only embarked upon his Great Scheme, but was also fulfilling his pledges to his country and Queen to study the customs and manners of other countries, and to gather intelligence concerning their politics and religion that might be useful both to his own education and to his country.

From his 'Tables of Experience' Francis drew forth the 'meat' for his plays which he began to write, publish and produce after his return to England. Inspired and guided by his 'Muse' (Pallas Athena)[305] he composed series after series of dramas based upon real life experiences or 'history', drawing forth from his collection of tabulated history whatever was required for any play. The greatest of his plays were eventually masked by the actor from Stratford, William Shaxpere, and known as the Shakespeare Plays ('Shake-speare' being the direct translation of the Greek name, *Pallas Athena*, as also the epithet of the Rose Cross Knight, St. George). And in all these dramatic works, as also in his other philosophical and poetical works, is scattered the carefully broken and divided 'Word Cipher' story, employed in the literary and dramatic works in various ways — apt, scintillating, humorous, obscure or even plainly ridiculous: but such is the power of his creativity and language that all these usages have, on the whole, withstood the ravages of time, critics, editors, scholars and audience — even the seemingly ridiculous and incredulous passages.

Francis' experiences with Marguerite, for instance, are scattered throughout his poems and plays — especially the earlier ones. A study of her character, and of his own in relation to her, appears in numerous characters in the plays, usually all with key names or references to help any of us trace the sources of his inspiration or experience. Events which actually took place, which Francis either experienced himself directly, or was told of by close and trusted friends, or heard of in some verifiable way, are built into the plays as part of their plots, as also is gossip and rumour and scandal, and deeply penetrating philosophical thought, all mixed together like colours on the master artist's palette and applied to the canvas of each play.

One 'Shakespeare' play in particular incorporates a great deal of Francis' experience in France during his 1576-9 sojurn, and that is the play, *Love's Labour's Lost*. A great deal of argument has been spent on the exact form of the title intended by the author, as it was printed in several varying forms. In the first Quarto edition, dated 1598, the title-page reads, ambiguously, *A Pleasant and Conceited Comedie Called, Loues labors lost*,[306] which could mean either 'the labour of love is lost' or 'the lost labours of love'; but the running title at the top of each page of text reads *Loues Labor's lost*. The 1623 Folio of Shakespeare Plays uses the latter form throughout, except in its table of contents where *Loues Labour lost* is used. Robert Tofte's *Alba* (1598) mentions the play as "*Loues Labour Lost*", and Francis Meres uses the form *Loue labors lost* in his famous *Palladis Tamia, Wit's Treasury* (1598). We know that Francis used ambiguous statements deliberately and in masterful ways, to convey different but equally viable meanings; however a statement from his Biliteral Cipher helps to clarify it on one level — that of his love for Marguerite and his endeavours to win her:

> Our keys for the story of sweet Marguerite ..... are here repeated in my Biliteral Cipher, to assure the finding and working out of her history which was to me *labour of love* to write, but, to my sorrow, *my love was labour lost*. Yet a certain degree of sadness is to the young pleasurable, and I desired by no means to be free of the pain.[307]

Francis also refers to Marguerite as his "sweet White Rose":

> My word-signs ...... doth unite parts in such a manner that you can write in perfection my many stage-plays,[308] histories, poems, translations of Homer, Ovid and Virgil, and many French poems written at an early age and little worth save to finish the history that they complete — indirectly it is true, nor too fully, but with such passion that he who doth put it down is sure to take it up again. It sheweth forth my love for mine angelic-faced, soft-eyed Marguerite of the South-land — sweet White Rose of my lone garden of the heart.[309]

This reference to Marguerite as the "sweet White Rose" of his heart is embodied in the name 'Rosalind' or 'Rosalinda' that he uses in *As You Like It*, and 'Rosaline' in *Romeo and Juliet* and *Love's Labour's Lost*. In each play Rosalind (which means 'beautiful rose' or 'fair as a rose') embodies definite characteristics of Marguerite de Valois, and each play reveals a great deal of hers and Francis' love — he for her, and she for him and others — plus a wealth of other historical incidents. In another type of cipher, Francis clearly relates 'Rosalind' to Marguerite; but the direct analogies must not be taken too far, as Marguerite's nature can be likened to white and pink hues on a palette, which are then used to paint in the features of many different and invented portraits on the canvasses. The portraits of 'Rosalind' simply have a lot of the pink and white hues in them, deliberately.

The story of *Love's Labour's Lost* concerns a youthful King of Navarre (called 'Ferdinand' which means 'world-daring', 'life-adventuring' — an apt description of Henri de Navarre) and three of his courtiers, the Lords Dumaine, Longaville and Berowne. All four swear to fast, pray and study together for three years, abstaining completely from the company of women and forbidding all women to come within a mile of their 'academe'. But their vow of abstinence is almost immediately broken, for the Princess of France, attended by Lord Boyet and three ladies, Katharine, Maria and Rosaline, arrives at Navarre's Court to discuss her father's debts to Ferdinand. When Ferdinand and his lords visit the Princess, Maria recognises Longaville, a former acquaintance, and likewise Berowne with Rosaline. Dumain becomes interested in Katharine. The King of Navarre falls in love with the Princess, and his three courtiers with the Princess' ladies. After a great deal of intrigue, deception and mockery, with love-letters and poems going astray and being mixed up, the characters sort themselves out. Berowne, who thought the vows unwise to begin with, mocks the others for their weakness, but then is found to be equally culpable. But eventually Berowne's argument that love is a lawful, and indeed essential part of all study wins the day, and the King and his lords openly pursue their courtship of the ladies by masque and revels. After further mix-ups, when the ladies appear before the men masked, they sort themselves out and watch the masque of 'The Nine Worthies'. Then, in a highly dramatic and abrupt ending, the Lord Mercade arrives to inform the Princess that her father is dead. She must leave immediately, leaving the three Lords and King to perform tasks that, in time, will win them their ladies.

In all this are woven real-life situations as well as characteristics of real persons, collected from the experiences of both Francis and Anthony Bacon. The idea of the four gentlemen and their 'academe' is clearly adopted and adapted from Montaigne's 'French Academy', and many scholars have noted echoes of *The French Academy* in the play, together with influences from the Italian *commedia dell'arte*. The arrival of the French Princess and her ladies, followed by the love-matching, the courting, the masque and revels, had a basis in the state visit of Catherine de Medici and Marguerite de Valois to Henri de Navarre's Court at Nérac in December 1578, and the subsequent festivities and love-making that ensued, and the real-life tangle of affairs of the heart that existed between Henri de Navarre and some of his courtiers on the one hand, and Marguerite and some of hers and Queen Catherine's ladies on the other — not to mention that of Francis and Marguerite which threads the whole play, chiefly under the disguise of Berowne and Rosaline.

The names of the three Lords — Dumain, Longaville and Berowne — can be seen to be allusions to (or derived from) the Ducs de Biron, de Longueville and de Mayenne, the former two being supporters of Henri de Navarre, and the latter being the Catholic opponent of both Navarre and the Huguenots. But, as to why they should be selected to give names to characters in the play, these three names, plus the name of the Lord Boyet, will be found to be taken directly from the passport of Anthony Bacon, which, when Anthony and his train went to Navarre's territory (1583-92), was signed Biron, Dumain, Longaville and Boyesse". As to the names of the ladies in the play, in the 1598 Quarto not only is the Princess often called 'Queen', but both Katharine and Rosaline (the name and pseudonym of the French Queen and her daughter, the Princess, respectively), are confused together. Furthermore, all three ladies' names — Rosaline, Maria and Katharine — are particular names and symbols of the Virgin Goddess, meaning respectively 'the beautiful rose' (especially the white rose, symbol of purity and hence virginity), 'the salt (or crystal) sea', and 'the pure or holy one', as also is the name 'Marguerite' which means 'the pearl'. Thus the three ladies of the play can be seen to be representing the Triple Goddess[310] — maiden, nymph (or mother) and crone (or wise old woman) respectively — as also three inherent aspects of Marguerite, Francis' own love and the symbol of our own natural souls.

The comedians in *Love's Labour's Lost* are two men — Don Armado, a Spaniard, and Costard, a country swain. Much of Don Armado can be identified with Antonio Perez, Philip II's exiled minister who escaped from Spain to France in 1591, and came to England in 1593. He was befriended in England by Francis and Anthony, much to Lady Anne Bacon's alarm and regret, and "wrote his books and letters in his extravagently affected style at Essex House, Strand",[311] where Anthony Bacon and Robert, Earl of Essex, lived. As George Stronach points out:

> "Like 'Armado', Perez was a 'traveller from Spain', and he published a book under the assumed name of 'Raphael Peregrino'. It is not surprising, therefore, to find in *Love's Labour's Lost* the statement:- 'He is too picked, too affected, too odd, as it were, too *peregrinate*, as I may call it.'

> "I am therefore of the opinion that when this word was used Bacon had in his eye his friend 'Peregrino'. He had already called the Latin version of his 'Essay on Travel', *De Peregratione in Partos Extremos*."[312]

In notes made on *Love's Labour's Lost* by Alicia Leith, she writes concerning Antonio Perez:

> "Perez arrived in England in the autumn of 1593. Hume says,[313] 'By all he was laughed at for his affectation and envied for his malicious wit,' and further tells us, 'Lady Bacon was violently angry that her son Francis should be so friendly with him, *a proud, profane, costly fellow, whose being about him, I verily believe, the Lord God doth mislike.*' Hume says that 'one of Anthony Bacon's agents writes of him in 1594, *Surely he is, as we say, an odd man, and hath his full sight everywhere.*' If we turn to the play we shall find the Princess of France's view of the Don agrees with Anne, Lady Bacon's view of Perez. In Act V. sc. ii. the Princess asks Biron, 'Doth this man serve God?' adding, 'He speaks not like a man of God's making.' Holofernes' description of the Don is given in actually the same words as those used by Anthony Bacon's agent, 'Too odd, as it were, too peregrinate, as I may call it'. (Act V. sc. i.)

> "Hume says much more about Perez. 'In 1596 he disgusted and offended the Earl (Essex) and thenceforward his star in England had set.' Then he adds, 'So that if we assume that the special touches of caricature that identify Don Adriano Armado with Perez were introduced into the play when it was re-cast for the Court performance in 1597, the reason for the skit upon Essex's fallen favourite becomes at once apparent. The Court, and the Court only, would see the joke, which no one would have dared to make when Perez was in favour three years before, for then Perez would have struck back with the sharp claws beneath his velvet paw.' Another sign to us Baconians how dangerous it was for Bacon to acknowledge his plays or make himself known as the dramatist who satirised living people.

> " 'No one can read,' says Hume, 'Perez's many published letters and *Relaciones* without identifying numerous affected turns of speech with those put into the mouth of Don Adriano Armado. And the description given of Don Adriano Armado by the King of Navarre in the play tallies exactly with the word portraits remaining to us of Antonio Perez drawn from his own writings and those of his contemporaries.' Hume says (p.273), 'Perez gave himself many nick-names, one favourite being *Peregrino*, *El Peregrino*, or *Rafael Peregrino*,' and that 'he signed himself thus.' Hume adds. 'Peregrinate is, and always has been, an extremely rarely used English word, so that its introduction by Shake-speare especially applied to Don Adriano Armado is significant.' Then he quotes Sir Nathaniel and Holofernes on the Don (Act V. sc. i.), where it is stated Don Adriano is 'a companion of the King of Navarre.' Henry IV, King of Navarre, Hume says, 'treated him (Perez) with almost royal honour,' and 'would hardly let him out of his sight.' Hume points out that the Don's account of the king's familiarity with him may very well be 'a burlesque of Henry's affection for him' (Perez) which 'would not be displeasing to Shakespeare's patron (Essex) at the time,' who, again, 'had been deeply offended by the ingratitude of Perez in preferring to remain in France.' "[314]

And of Marguerite de Valois, Alicia Leith notes:

"With regard to the Princess of France in *Love's Labour's Lost*, is she Marguerite of Valois, daughter of Henry II and Catherine de Medicis ? ......

".....De Thou, in his *Collection complète des Memoires relatifs a l'Histoire de France*, says that at fifteen Margot came to Court the idol of her maids of honour, that she cultivated charms of mind, was the leader of fashion, and took the most prominent part in the more grave and majestic dances of her day. I was surprised to find from Brantôme that this remarkable Princess of France was a dark lady, inheriting her locks '*fort noir*' from her father, Henry II, for her portraits represent her as fair. This, it appears, is owing to the fair *crispé* wigs she often wore, and always carried about with her when travelling. In the light of this, the many allusions to the fair and dark ladies in the play become pointed. When Biron rhapsodies about his dark lady, the King upholds his fair princess, who, Biron assures him, dare not face the rain for fear her colour should be washed away — a home thrust when we know the perfumed, gilded lily Margot painted an inch thick. Biron makes another good point when he adds:

>   Devils soonest tempt, resembling spirits of light.
>   O, if in black my lady's brows be decked,
>   It mourns that painting and usurping hair,
>   Should ravish doters with a false aspect.

"Moth amuses the company vastly when he connects the angel princess with a devil, but in so doing he ventures all too near the truth, if Don John of Austria is to be believed.

"Brantôme, in his *Memoires of Marguerite*, says Don John attended a ball at the Louvre disguised on purpose to see Margot, and said, in Spanish, 'How much more is that queen's beauty for the perdition and damnation of men than for their salvation.'

"The princess's attraction was certainly phenomenal, though she had good haters among the Protestants and Catholics both. Hers is a dramatic and majestic character, standing out as she does in history with her well developed figure, robed in cloth of silver, or in orange and black, or in blood-red Spanish velvet and cap to match, and all her plumes and jewels. Beautiful and accomplished Margot, conversing easily and spontaneously with envoys and ambassadors in elegant Latin, singing her own stanzas to her lute, was as inconstant as that moon which her lover in the play compares her to — a dangerous planet for our young, impressionable, and amiable poet Francis of sixteen, to approach in brilliant Paris. How should he, of all others, escape the fatal fascinations of this '*Venus Uranie*', sung by all the poets of her time, Ronsard included? Bitter-sweet recollections of Margot in that, his 'green goose' stage, may well have inspired Francis after fourteen years had passed, to immortalise so great a lady, especially one who may have, even to his cost, proved to him what Biron says: that

>   Love first learned in a lady's eyes
>   Lives not immured in the brain;
>   But with the motion of the all elements,

> Courses as swift as thought in every power;
> And gives to every power a double power
> Above their functions and their offices.
> It adds a precious seeing to the eye."

> "Just because of his 'green goose idolatry' during the time he was the youthful envoy of the 'Arbitress of Nations', Elizabeth, in Paris, did he wish to paint her picture in this play of his, destined to live as long as the world lasts?

> "Even Margot had her good points. She was 'extraordinarily charitable,' and to the Forester her hand is a giving one in the play, where she pointedly says, 'A giving hand though foul shall have fair praise.' It is not a touch without purpose, that of our dramatist, when he places his Princess shooting with her bow at a deer in a park, for stag hunting was a favourite pastime of Reine Margot at Fontainebleau."[315]

This delightful and much misunderstood play, packed full of witty debate and deep philosophy, depends on its fast-moving discourses rather than on any action. "Newly corrected and augmented" for its presentation before Queen Elizabeth I at Christmas 1597 (and then published as a Quarto in 1598), it was nevertheless probably one of the earlier plays that Francis composed, settling on the final names and adding references to later events in 1596 when Perez had fallen from favour. It embodies all the feelings, thoughts and reactions of adolescent love, with its almost overpowering drive for sexual or sensual pleasure coupled with the entirely opposite idealism for the adored lady's purity and chastity, at one and the same time seeking to idolise and abase that which is loved. More than this, the play is about the central theme of all Francis' knowledge and teachings: that Truth is Love, and Love is the only motivating power or life force which moves and fashions all things. Love, or Truth, is the Summary Law of all Nature, the Supreme Cause of all things, whose radiance (or light of Truth, the divine Consciousness) is Wisdom. To seek truth is to seek love. To know truth is to know love. These are the original and basic teachings of the Orphic Mysteries (and of the Ancient Egyptian and Druid Mysteries), and the very foundation of Christianity; yet Plato, and dedicated neo-Platonists and Humanists who imitated Plato, "admitted no auditor in his academy but such as while they were his scholars would abstain from women; for he was wont to say that the greatest enemy to memory was venery."[316] Modelled upon Plato's school in Athens, the Renaissance academies inherited Plato's seeming anti-feminism.

A similar misunderstanding of women and of love relationships generally had crept into Christianity, both with the influence of Platonism and of St Paul. The reasons for periodic celibacy came to be misunderstood and misapplied, and the place and role of women denigrated and denied, often with catastrophic results. A grave distortion and perversion of the Ancient Wisdom doctrines and practices had occurred, and Francis set out to try to correct this poor state of affairs, and to help mankind recover the harmony, balance and mutual respect between sexes that can lead to true and lasting love relationships and an expression of divinity itself.

> "This Love I understand to be the Appetite or Desire of primal Matter, or to speak more plainly, the natural motion of the *Atom*, which is indeed the original and unique force that constitutes and fashions all things out of Matter .... Next unto God it is the Cause of Causes, itself only without any Cause."[317]

> "Love is the eldest and noblest and mightiest of the gods, and the chiefest author and giver of virtue in life and of happiness after death."[318]

> "All existence comes from one immeasurable Good Principle, the One Cause."[319]

> "God is revealed as an eternally abiding GOOD, an ever-flowing fountain of Truth and Law, an omnipotent Unity, an omniscient Reality."[320]

> "God is the Truth and Reality which sustains the universe. This Truth animates all things; it is the spiritual principle in all of life."[321]

> "Love is the great gravitating or attracting force which brought the universe into shape and gave birth to the starry spheres, out of chaos ..... But its opposite power is necessary to prevent everything unifying or marrying, and thus becoming One Universal Light until the purpose of Evolution is fulfilled .... The purpose of Evolution is TO KNOW GOD."[322]

Truth is Love, and the very purpose of man's existence is for him to seek out and come to know that Truth by looking for it, recognising it, embodying it and experiencing it in every aspect.

> "For the principles, fountains, causes, and forms of motion, that is, the appetites and passions of every kind of matter, are the proper objects of philosophy."[323]

Nevertheless:

> "For the Summary Law of Nature, that impulse of Desire impressed by God upon the primary particles of matter, which makes them come together and which by repetition and multiplication produces all the variety of nature, is a thing which mortal thought may glance at, but can hardly take in."[324]

It thus follows that, if Truth is Love, then the search for Truth is the labour of love. All study is, or should be, for the discovery of Truth, leading on to the perfect practise of Truth.

> "Truth, which only doth judge itself, teacheth that the enquiry of Truth, which is the love-making or wooing of it, the knowledge of truth, which is the presence of it, and the belief of truth, which is the enjoying of it, is the sovereign good of human nature."[325]

What Francis found all around him, over and over again, were men and women (but particularly men) seeking after vain things — self-glory, fame, call it what you will — but not after the true things of life, the REALITY of life as it actually is, with the divine Emotion or Life force vivifying all things and being expressed in all the myriads of emotions, desires, feelings, and other aspects of love in all creatures, particularly the human kind. Not that the expression of this divine Emotion in God's creatures is invariably perfect — far from it; but by degrees the divine Life Force is becoming more and more fully expressed as each creature evolves in its own way, until at the end of all evolution the divine Love will be perfectly manifest through the perfected Creation of God. As Plato remarks:

> "Man is the progeny of the gods: composed inwardly of the Spirit of Absolute Truth, and composed outwardly of varying degrees of relative truths .... The ignorant man is in servitude to his animal nature; the partly informed man is in

> servitude to his intellectual nature; the divinely enlightened man is united with
> his spiritual principle which is the sustaining power in the midst of his being."

The spiritual principle is Love, the divine Emotion. This is Truth. To know it we have to experience it; to experience it we have to seek it; and we have to seek and experience it in all its relative expressions, until we at last come to discover real and comprehensive love, the truth that is, in fact, a "naked and open daylight."[326]

This love theme is central to all the Shakespeare plays and everything else that Francis wrote. He lived it, he experienced it, he studied it, he knew it as the great Masters of the Ancient Mysteries had known it. He took immense pains to show that "fame, that all hunt after in their lives",[327] is for most people but self-seeking and vain glory, an empty shell devoid of its real essence which is generous and selfless love. Thus he has the Princess replying to the King's opening speech with the words:

> "Glory grows guilty of detested crimes,
> When, for fame's sake, for praise, an outward part,
> We bend to that the working of the heart;
> As I for praise alone now seek to spill
> The poor deer's blood, that my heart means no ill."[328]

Francis sought to show, and lead humanity to, that real and lasting Rosicrucian Fame which is Virtue, personified as an aspect of Pallas Athena, the Goddess of Wise or Illumined Intelligence, the Leader of the Seven Liberal Arts and Sciences who rides the winged white horse, Pegasus, symbol of the aspiring, pure and intelligent mind. Moreover, that Virtue is moral Goodness, which is the nature of Love and the characteristic of the Diety.

Part of the means of achieving this true Fame is through death, which leads to resurrection — birth, death and resurrection (or rebirth) all being part of the continuum of life. "Dead, for my life!" exclaims the Princess, who is now Queen of France.[329] "He that loseth his life shall find it."[330] "Greater love hath no man than this, that a man lay down his life for his friends."[331] "I am come that they might have life, and that they might have it more abundantly. I am the good Shepherd: the good Shepherd giveth His life for the sheep,"[332] said Jesus the Christ. Thus, in *Love's Labour's Lost* especially, Francis seeks to shatter the illusion that fame can buy men a kind of immortality over the process of Time and Death, unless that fame be deathless Virtue. Thus, although the play concludes abruptly with the unexpected death of the Princess' father intervening into the studies, love-making and festivities, yet it is by no means out of place and comes exactly where it should, reminding us all that, although "the mind is the man, and the knowledge is the mind: man is but what he knoweth",[333] yet we should not "so place our felicity in knowledge, as we forget our mortality."[334] In the same manner was Francis reminded, when news of Sir Nicholas Bacon's death reached him at Paris in March 1579.

The news of the royal death cuts right into the play, dividing the previous unreality from the succeeding reality. Sanity is restored. Truth is, for a moment, seen face to face. No longer is even the love-making a 'play'. Suddenly it is all for real. As Berowne remarks near the end of the play, when he and the other lords are charged with waiting and doing good deeds for a year and a day in order to win their ladies, this year and a day is "too long for a *play!*"

* * * * * *

The writing, presenting and publication of *Love's Labour's Lost* occurred at the time when Francis and Anthony Bacon had come together again in London after their respective travels, and had just assembled their fraternity of co-workers in the Great Scheme and inaugurated their work. (Anthony Bacon returned to England after his travels in February 1592, and joined Francis to form the Baconian *Dioscuri* or Twins. In August 1592 'Greene' announced the "Shake-scene" as the first Shakespeare works began to appear anonymously, marked in the heavens by two novae appearing in Cassiopeia in November and December 1592. In 1593 the core of the Baconian-Rosicrucian Society was founded, marked by a comet which appeared in July and August, and announced by the publication of *Venus and Adonis* on St. George's Day,[335] with the first printed use of the symbolic name 'William Shakespeare.' During 1593-4 the 'Grand Lodge' of the Rosicrucian fraternity was gathered together and inaugurated at the end of 1594, announced allegorically at the Gray's Inn Christmas revels by *The Prince of Purpoole, and the Knights of the Helmet* together with *The Comedy of Errors*. Following which *Love's Labour's Lost* was almost certainly first written and performed for the benefit of the 'knights', then "corrected and augmented" for a Court performance before the Queen at Christmas, 1597, and published in a quarto edition in 1598.) *Love's Labour's Lost* powerfully, succinctly and brilliantly presents the "ills" besetting mankind in the field of philosophy and philanthropy, and offers all 'knights' a vision of how to heal those ills, how to recognise truth, and how to practice truth. It is a wonderful 'opener' to the whole of the Great Instauration and, although not written until 1594-5, it is thoroughly based upon the living experience of Francis and Anthony Bacon — particularly Francis', when he was in France from 1575-9.

In 1605 Francis published *The Tvvoo Bookes of Francis Bacon. Of the proficience and advancement of Learning, diuine and humane*.[336] Other than his *Essayes* and some legal work, this was his first publication under his name FRANCIS BACON, as a philosopher; and the first plainly published declaration and explanation of the Baconian-Rosicrucian work, the **Great Instauration or Regeneration of all sciences and arts**. In Book I, he writes about the vanities and errors in learning extant in his day (and still around, though not to such an extent as in Bacon's day), and the proper ends of knowledge — namely, "for the glory of the Creator and the relief of man's estate." The philosophical treatise is a careful wording of what *Love's Labour's Lost* portrays dramatically — the philosophical axioms drawn forth from the dramatic presentation of life that allows the eye of heart and mind to see the truth.

> "There be therefore chiefly three vanities in studies, whereby learning hath been most traduced. For those things we do esteem in vain, which are either false or frivolous, those which either have no truth or no use: and those persons we esteem in vain, which are either credulous or curious; and curiosity is either in matter or words: so that in reason, as well as in experience, there fall out to be these three distempers, as I may term them, of learning: the first, fantastical learning; the second, contentious learning; and the last, delicate learning; vain imaginations, vain altercations and vain affections; and with the last I will begin....

> "Here, therefore, is the first distemper of learning, when men study words and not matter .... how is it possible but this should have an operation to discredit learning, even with vulgar capacities, when they see learned men's works like the first letter of a patent, or limned book; which though it hath large flourishes, yet is but a letter? It seems to me that Pygmalion's frenzy is a good emblem or portraiture of this vanity: for words are but the images of matter; and except they have life of reason and invention, to fall in love with them is all one as to fall in love with a picture......

"The second which followeth is in nature worse than the former: for as substance of matter is better than beauty of words, so contrariwise vain matter is worse than vain words: wherein it seemeth the reprehension of St. Paul was not only proper for those times, but prophetical for the times following; and not only respective to divinity, but extensive to all knowledge; *Devita profanas vocum novitates, et oppositiones falsi nominis scientiae* ['O Timothy, keep that which is committed to thy trust, avoiding profane and vain babblings, and oppositions of science falsely so called.']³³⁷ For he assigneth two marks and badges of suspected and falsified science: the one, the novelty and strangeness of terms; the other, the strictness of positions, which of necessity doth induce oppositions, and so questions and altercations ..... This kind of degenerate learning did chiefly reign amongst the Schoolmen: who have sharp and strong wits, and abundance of leisure, and small variety of reading, but their wits being shut up in the cells of a few authors (chiefly Aristotle their dictator) as their persons were shut up in the cells of monasteries and colleges, and knowing little history, either of nature or time, did out of no great quantity of matter and infinite agitation of wit spin out unto those laborious webs of learning which are extant in their books. For the wit and mind of man, if it works upon matter, which is the contemplation of the creatures of God, worketh according to the stuff, and is limited thereby; but if it work upon itself, as the spider worketh his web, then it is endless, and brings forth indeed cobwebs of learning, admirable for the fineness of thread and work, but of no substance or profit ....

"For the third vice or disease of learning, which concerneth deceit or untruth, it is of all the rest the foulest; as that which doth destroy the essential form of knowledge, which is nothing but a representation of truth: for the truth of being and the truth of knowing are one, differing no more than the direct beam and the beam reflected. This vice therefore brancheth itself into two sorts; delight in deceiving, and aptness to be deceived; imposture and credulity; which, although they appear to be of a diverse nature, the one seeming to proceed of cunning and the other of simplicity, yet certainly they do for the most part concur: for, as the verse noteth

Percontatorem fugito, nam garrulus idem est,[338]

an inquisitive man is a prattler; so, upon the like reason a credulous man is a deceiver: as we see it in fame, that he that will easily believe rumours, will as easily augment rumours, and add somewhat to them of his own; which Tacitus wisely noteth, when he saith, *Fingunt simul creduntque*:[339] so great an affinity hath fiction and belief.

"Thus have I gone over these three diseases of learning; besides the which there are some other rather peccant humours that formed diseases ......But the greatest error of all the rest is the mistaking or misplacing of the last or farthest end of knowledge: for men have entered into a desire of learning and knowledge, sometimes upon a natural curiosity and inquisitive appetite; sometimes to entertain their minds with variety and delight; sometimes for ornament and reputation; and sometimes to enable them to victory of wit and contradiction; and most times for lucre and profession; and seldom sincerely to give a true account of their gift of reason, to the benefit and use of men: as if there were sought in knowledge a couch whereupon to rest a searching and restless spirit; or a tarrasse, for a wandering and variable mind to walk up and down with a fair prospect; or a tower of state, for a proud mind to raise itself upon; or a fort or commanding ground, for strife and contention; or a shop, for profit or sale; and not a rich storehouse, for the glory of the Creator and the relief of man's estate. But this is that which will indeed dignify and exalt knowledge, if contemplation and action may be more nearly and straitly conjoined and united together than they have been; a conjunction like unto that of the two highest planets, Saturn, the planet of rest and contemplation, and Jupiter, the planet of civil society and action: howbeit, I do not mean, when I speak of use and action, that end before-mentioned of the applying of knowledge to lucre and profession; for I am not ignorant how much that diverteth and interrupteth the prosecution and advancement of knowledge, like unto the golden ball thrown before Atalanta, which while she goeth aside and stoopeth to take up, the race is hindered ....

"Neither is my meaning, as was spoken of Socrates, to call philosophy down from heaven to converse upon the earth;[340] that is, to leave natural philosophy aside, and to apply knowledge only to manners and policy. But as both heaven and earth do conspire and contribute to the use and benefit of man; so the end ought to be, from both philosophies[341] to separate and reject vain speculations, and whatsoever is empty and void, and to preserve and augment whatsoever is solid and fruitful: that knowledge may not be as a courtesan, for pleasure and vanity only, or as a bond-woman, to acquire and gain to her master's use; but as a spouse, for generation, fruit and comfort."[342]

Anthony Bacon, Secret Agent

The journey between London and Paris could take up to a fortnight, even eighteen days, unless there was exceptional luck with the conditions in the Channel; thus by the time that Francis arrived home in England it was probably early April, 1579. Sir Nicholas Bacon had been dead and buried some three weeks or more. His funeral had taken place on the morning of March 9th, 1579, in a manner befitting the great Lord Keeper. York House and St Paul's Cathedral (where Sir Nicholas' body was to be laid) were draped in black, and the bells of the Cathedral and those of St Martin-in-the-Fields tolled throughout the day. Two hundred mourners clothed in black formed the funeral cortège, consisting of sixteen beadles at the head, to clear the street, followed by sixty-eight[343] poor men who had been supplied with suitable clothes plus a shilling each for dinner, then by the Lord Keeper's household of some seventy persons, after whom walked those relatives and friends who formed the chief mourners (other than the immediate family), including the Lord Treasurer Burghley, Mr. Secretary Walsingham, the Master of the Rolls, the Attorney-General, the Solicitor-General, the Master of the Queen's Jewel House, Sir Thomas Gresham, and Sir Nicholas' two other brothers-in-law, William Cooke and Henry Killigrew. The procession culminated with Sir Nicholas' four sons — Nicholas, Nathaniel and Edward (by his first marriage), and Anthony (by his second marriage) — together with his widow, Lady Anne, who rode a horse draped in black. His daughters and daughters-in-law, and Lady Anne's gentlewomen, did not form part of the procession but joined the cortège at St Paul's. The procession made its way by horse and foot from York House by Charing Cross, along the Strand and Fleet Street, and up Ludgate Hill to St Paul's Cathedral. Sir Nicholas had expressly desired to be laid in a tomb next to that of John of Gaunt, "time-honoured Lancaster," and there his body was finally laid to rest, on the south side of the choir. Son of a yeoman sheep-farmer in Suffolk, he had served faithfully under four monarchs and held the highest office in the land, next to the Sovereign, for twenty-one years. His loss to the country, to his family, and to the Queen, was a great one.

Sir Nicholas Bacon had made his last will and testament on the 23rd December 1578, just a few months before his death. In it the chief bequests were made to his wife, Lady Anne, to his eldest son and heir, Nicholas, and to Anthony, with a proviso that should Anthony die without male issue then Francis was to inherit all Anthony's portion. All the sons, Francis included, were left "a sufficient release in law" of Lady Anne's dower rights. Besides several manors, woods, farm, amd other premises and revenues bequeathed to Anthony immediately,[344] Sir Nicholas also left to Anthony one half of his household stuff remaining at Gorhambury when Anthony reached the age of 24 years, plus the rest of the Gorhambury household stuff upon Lady Anne's death. He also left Anthony his special jewel that he wore. Many commentators have assumed that Anthony was also left, under the terms of Sir

(35) SIR NICHOLAS BACON, Lord Keeper of the Great Seal (1558-79) — oil painting (1579), artist unknown. Sir Nicholas died at the age of 68 years. He carried out the duties and had the powers of Lord Chancellor whilst being called the Lord Keeper. Note his unusual 'dragon' jewel suspended from his neck.

Nicholas' will, the inheritance of Gorhambury upon Lady Anne's decease; but this is not specifically stated in the will, although it may have been Sir Nicholas' known intention. Rather the will specifies that Lady Anne should "cause to be made within one year after my decease, and before she be married again, to every one of my sons, Nicholas, Nathaniel, Edward, Anthony and Francis, a sufficient release in law of all her right, title, interest and demands of dower of and in all the manors, lands, tenements and hereditaments whereof by reason of my seizin she is or then shall be dowable and deliver or cause to be delivered to every of my said sons, one such release within the said year and before she be married ..."

The release of Gorhambury she seems to have given to her son, Anthony, but only for him to occupy upon her death, it being her dower house to which she subsequently retired after the funeral and the disposal of York House. Anthony's half-brother, Nathaniel, disputed these bequests, but the quarrel was referred to Lord Burghley and he settled it in Anthony's favour. In the event, Anthony died without issue before his mother, and so, when Lady Anne died in 1610, Francis inherited Gorhambury and the surrounding lands that were attached to it as a proviso of Sir Nicholas' will. Anthony himself was never able to enjoy Gorhambury as its master, and lived elsewhere.

In his will, Sir Nicholas also expressly desired that Lady Anne should see to the well bringing-up of their "two sons, Anthony and Francis that are now left poor orphans without a father", which charge he requested in consideration of all the legacies, manors, lands and tenements that he assured unto her, "and for all loves that have been between us." Nicholas and Nathaniel, his eldest sons, were the executors of the will, with Lord Burghley, his brother-in-law, made the overseer of the will. Under these terms, and because of his family and official position, Burghley acted as a guardian to Anthony and Francis until they came of age, and always remained a powerful influence in their respective lives.

Francis arrived home, to London, bearing a dispatch from Sir Amyas Paulett to the Queen (dated 20th March 1578/9), praising Francis to the Queen with the commendation that Francis was "of great hope, endued with many good and singular parts, and if God gave him life, would prove a very able and sufficient subject to do her Highness good and acceptable service." He appears to have lived for a time in the London house of his real father, the Earl of Leicester, until May 1580 when he took up residence in Gray's Inn. This grand house leading off the Strand and facing the Thames waterfront, right next door to the Temple Inns of Court, was once the great house which the first Lord Paget had bought from the Bishop of Exeter and subsequently enlarged. In 1569 Leicester acquired the lease and transformed it into "the stately place" that so amazed Edmund Spenser, fully equipped even with a delightful little banqueting house, two storeys high, in the garden on the bank of the river. When Leicester died in 1588, his second son by Elizabeth, Robert Earl of Essex, took over the lease and renamed it 'Essex House'. Sir Philip Sydney, nephew of Leicester and cousin to Francis, spent much of his time at Leicester House where his uncle established a meeting place or centre for poets, scholars and artists of various kinds to come together for study, conversation and celebration, for whom he was a noted patron. Principally these talented people who were patronised by Leicester were of his own family or relatives of his family (such as Philip Sydney), but, in association with the Walsinghams, many other university scholars and poets were among the numbers, plus a professional acting company. One of the scholars was Edmund Spenser from Pembroke College, Cambridge, who had been introduced to Leicester by Francis and Anthony's old tutor and friend, Gabriel Harvey. When Francis returned home and came to live at Leicester House, Spenser was already living there as a secretary and messenger of the Earl. Later, possibly in return for advancement, Spenser allowed his name to be used by Francis as a mask to the true authorship of Francis' famous works of 'poesie'.

Leicester was by now secretly married to Lettice, the Countess of Essex and daughter of Sir Francis Knollys. The Earl of Essex, Lettice's husband, had died in Dublin in September, 1576. Lettice became pregnant by Leicester in 1578, and on September 21st, 1578, Leicester and Lettice were privately married at his house at Wanstead. For some years Durham House on the Strand, next door to York House, had been used as a residence by the Earl of Essex and his family. Durham House was owned by the Queen, who provided it as a "grace and favour" to certain courtiers and distinguished foreigners as a residence in London. Leicester himself had had a suite of rooms in Durham House before he moved to Paget House, and he may well have had the first opportunities thence to become enamoured of the Lady Essex. Here lived Robert Devereux, Leicester's second son by the Queen, adopted by the Earl and Countess of Essex, together with Lettice's own children, Walter, Penelope and Dorothy. Married to Lettice, Leicester could now be close to his son as a father. Philip Sydney had been given the chance to marry Penelope Devereux in 1575, which he turned down, much to his later chagrin and despair when he discovered too late that he loved his childhood friend.

At the time that Francis returned to England, the Queen was fully involved with her diplomatic intrigue concerning a possible marriage with Françoise, Duc d'Anjou. Although, as far as the Queen was concerned, actual marriage was out of the question, nevertheless through the continued bluff (that the Queen was not secretly married to Leicester and was still a 'virgin') she was able to play an extraordinary and successful diplomatic game that kept a reasonable balance in those European politics that touched English interests, and which undoubtedly helped to preserve England from many potential aggressive alliances against her. Subtlety and bluff were the Queen's tactics and weapons, rather than the costly provision of armies. The 'marriage' ruse had once been used with Henri de Valois (then the Duc d'Anjou) before he became King of France, the negotiations continuing for thirteen months, and then was switched at the end of 1571 to a possible 'marriage' with Françoise de Valois (then the Duc d'Alençon, later Duc d'Anjou). Alençon's suite was to continue for a full ten years, the Queen deftly playing the game. The Massacre of St Bartholomew's Eve almost put an end to this political game of chess, until in the summer of 1578 Elizabeth decided to renew the game.

The Netherlands States constituted at that time the most sensitive zone for English and Protestant interests, and it was Françoise, Duc d'Anjou's (*i.e.* Alençon's) intervention in the States in the Spring of 1578 that provoked Elizabeth to renew her 'courtship' of him. It seems a possibility at that time that either the Spanish, led by Alexender Farnese, Prince of Parma, or the French, led by the Duc d'Anjou, might gain the whole territory and use it as a sphere of influence and possible launching ground for an invasion of England, not to mention the suppression of Protestantism. Anjou became elected "Defender of Belgic Liberty against the Spanish Tyrant", and he hoped to become Protector of the Netherlands. Elizabeth Jenkins, in her classic book *Elizabeth and Leicester*, succinctly notes this important and subtle strategy:

> "Elizabeth had now allowed the marriage negotiations with the Duke of Alençon, which had been brought forward when those with the Duke of Anjou had lapsed, to be renewed, for a very subtle diplomatic purpose. The Spanish onslaught on the Netherlands States was being so courageously and doggedly resisted that the defence was no longer assumed to be hopeless; the French thought that, as the native resistance was so strong, a moderate assistance on their part might gain them the whole terrain as a sphere of influence, a prospect which alarmed the English almost as much as that of a complete Spanish domination. The Duke of Alençon, always at odds with his brother Anjou, who

had succeeded Charles IX as Henry III, thought that he saw an opening for himself as Protector of the Netherlands. The question deeply engaging the Queen and Lord Burleigh was whether Alençon would enter the scene as an emissary of the King of France or in an independent capacity. If he could be encouraged and helped to do so independently of France, this might prove the best means open to England to avoid the menace of a French or Spanish control of the Netherlands. This was the object of Elizabeth's throwing out before Alençon's eyes the glittering possibility of his becoming King of England by a marriage with herself. It was essential to the success of the negotiations that Alençon should believe them to be sincere; this he would not do if the Queen were surrounded by able councillors all of whom knew the whole process was a sham. She took none of them into her confidence; they all believed that, though the inherent difficulties might prove insuperable, the Queen personally was willing, even eager for the match.

"That she was able to make this impression was owing to her obsessive fondness for being flattered and wooed, a fondness that was part of her very being. Nature intensifies the facilities that she is about to take away, and Elizabeth, at the age of forty-five, played her favourite game with a madder intensity than she had ever shown before. When the negotiations were halting, her councillors told her that if she wanted to withdraw they would take the blame upon themselves, and the Queen wept, asking why she, alone, was to be denied marriage and children, so great was the conviction with which she played the role. Yet every near approach to success was checked by herself; she spun her delusive webs until even the French King's eagerness for the alliance was exhausted, and he threatened to sever diplomatic relations."[345]

It is quite possible that one of the delicate missions entrusted to Francis by the Queen was to give to her advice concerning François de Valois' character and motives, and the likelihood of his falling for the bait if it were offered to him, besides informing the Queen and Burghley on the state of French politics generally, for which he was in a peculiar position to gain first-hand 'inside' information.

Preceding Francis to England, on January 5th, 1579, Anjou's Master of the Wardrobe and close friend, Jean de Simier, was sent to England to woo Elizabeth by proxy. He was well liked by the ladies and began to make sufficient impression on behalf of his master that Leicester became jealous and anxious that perhaps the royal marriage might become a reality — especially if the Queen were to discover his illicit marriage with Lettice Knollys. He became intolerant of and exceedingly rude to Simier, doing his utmost to ruin the courtship, and this very reaction brought about his own undoing. In April discussions were begun seriously on arranging for Anjou to visit the Queen in England, plus tentative negotiations for marriage terms. These discussions with the Queen and her councillors took place at Leicester's country house at Wanstead. Subsequently Elizabeth signed the necessary passport for Anjou to enter the country later that summer. Leicester retired to bed, sulking and feigning illness.

During August, whilst Leicester was away from Court (which by then had moved to Greenwich), an assassination attempt was made on Simier. Rightfully or wrongfully, Simier believed that Leicester was behind this attempt on his life, and so he counter-attacked. Early in July he had discovered Leicester's secret, and now Simier straightway went to the Queen and told her about Leicester's secret marriage with Lettice, giving the Queen sufficient evidence that this had taken place. The volcano of pent-up emotion erupted.

Elizabeth was hurt beyond measure. In a burst of hysteria and fury, she ordered Leicester to be sent to the Tower. Only the careful and persuasive reasoning of the loyal and courageous Lord Chamberlain, Lord Sussex, was able to prevent the Queen from doing irreparable harm to herself, to Leicester, and to the country. The Queen saw at last the danger that such an action might create, and Leicester was instead banished from her presence, and required to place himself in the Tower Mireflore — a little tower built by Humphrey of Gloucester on a mound in Greenwich Park which overlooked the palace, and which had at one time been used as a lodging for Ann Boleyn. Whilst incarcerated there, it was given out that he was ill and taking physic, and as a result could see no-one. After a few days, although still banished from the Queen's presence, he was allowed to retire to his home at Wanstead. Lettice was barred from Court for evermore.

Lady Mary Sidney, who had been invited to live at Court in February, 1577, at the age of 14 years, and then had a splendid marriage arranged for her in April, 1577, with Henry Herbert, the second Earl of Pembroke and son of William Herbert, first Earl of Pembroke (who had been a great friend of Leicester's), found the situation too painful for her to bear, and retired to Penshurst Place, her parent's home, relinquishing her apartments at Court. Later she went to live at her husband's great country house of Wilton, in Wiltshire.

Eventually the volcano subsided, and Elizabeth gave out that Leicester was not forbidden to appear again in his usual place provided he appeared as a man humbled. But it took a long time for him to swallow his pride and recover from the hurt he had given to the Queen, his first and still rightful wife. In fact he never really recovered his standing with Elizabeth, who was so deeply injured, although she appeared to forgive him and to continue loving him.

François, Duc d'Anjou, arrived in London on the 17th August, 1579, for a twelve-day visit. The Queen attended him with all gaiety and fervour, appearing enchanted with this lively and fascinating French prince; so much so that when he left England a marriage did indeed seem possible. But Anjou did not forget or underestimate Leicester, nor Sir Philip Sydney, Sir Francis Walsingham, their families and friends, and indeed the majority of parliament, the mass of the clergy and the Protestant population of England who had not forgotten St. Bartholomew's Eve, the loss of Calais and the treachery of the Houses of Guise and Valois, and who were very much against such a marriage actually taking place. Many preached against the marriage. Sir Philip Sydney wrote to the Queen about it on behalf of all English Protestants, at the request of Leicester and Walsingham (whose daughter, Frances, Sydney later married)[346]. Elizabeth is said to have wept when she read Sydney's letter, and he for his part diplomatically retired from Court for a whole year, going to stay (from 25th March, 1580, onwards) with his sister at Wilton where he embarked upon writing his great work, *The Countess of Pembroke's Arcadia*, and translating the *Psalms* with the help of his sister.

As to the marriage, the Queen's Council debated the issue all day on October 7th, 1579. Most of the Council were hostile to the proposals, but they arrived at an open verdict, wishing to displease neither the Queen nor country. Elizabeth was not amused by this lack of positive recommendation one way or the other, and for several weeks she stormed at them for not giving her definite support of the marriage. But in her heart she must have always known it was not possible without causing a civil uprising, and was not what she really wanted or intended anyway. When a quorum of councillors (excluding Leicester) came to talk marriage terms with Simier in November that year, Elizabeth made Simier agree, on Anjou's behalf, to a moratorium of two months in the marriage treaty, so that during that time she could try to win popular support for the match. But by then, in effect, she had determined to proceed no further, or at least to procrastinate. Not until 1582 did she revive her affair with her 'Frog' (as she called Anjou), and then only because the Catholic powers had again become more menacing.

* * * * * * * *

In all this Anthony Bacon was by no means on the side-lines. What had become of him since Francis was first banished from England in the summer of 1575? He had stayed on at Cambridge for a further term, coming down at the end of December, 1575. Together with Francis he had been admitted to Gray's Inn on 27th June 1576; but, unlike Francis, he did not straightway go abroad (as far as we know) and may have taken up residence at Gray's Inn in November of that year when he, Francis and Sir Nicholas Bacon's other sons were admitted to the Grand Company of Ancients and he personally was granted permission to live in Sir Nicholas Bacon's chambers in the Inn, together with Edward Bacon. Edward was ten years the senior of Anthony, Anthony being 18 years old in 1576 and Edward being 28 years old. Both were good friends of each other, but, at least initially, Anthony must have resided in those chambers alone as Edward was busy travelling abroad. For part of the time Anthony probably continued to live in his parents' household at York House and Gorhambury.

Whilst he was at Gray's Inn, ostensibly studying law and being tutored in social demeanor and moral graces, Anthony must have undergone training at Sir Francis Walsingham's special school or college in London where recruits to his intelligence service were given courses in cryptology and other matters. Anthony clearly decided about this time that it was in the service of his father's friend, Sir Francis Walsingham, that his own particular talents could most usefully be employed on behalf of the Queen and country. It was almost certainly with encouragement from his father and from his uncle, Lord Burghley, that he took this course. Besides which, it was in the Walsingham-Dudley-Sidney 'enclave' that careful and sustained patronage of the arts and learning was beginning to take place, giving scope to talented scholars, poets, artists and potential diplomats, whilst at the same time encapsulating and directing their talents and energies into the beneficial services of the State. Between them, they were beginning to bring into being the kind of 'Academy' that Sir Nicholas Bacon had envisaged and encouraged to be set up "for the advancement of learning and training of statesmen".[347] Anthony was inspired with his father's vision and example, and also with the thoughts of the Great Scheme that he was privy to and had planned with Francis — not just to advance learning and train statesmen for the benefit and service of England, or of any other country, but a complete and total "reformation of the whole wide world" in terms of science and art. In entering Walsingham's service — which was the service of the Queen and country and the reformed faith — Anthony was helping to carry out and further his father's designs, and at the same time find the base from which to operate and build up the Great Scheme, or Great Instauration.

Sir Francis Walsingham, the Queen's Secretary of State from 1573 until his death in 1590, was head of one of the best intelligence networks of the time. As mentioned before, (see chapter 10), Walsingham established his special training school in London shortly after he became Secretary of State, but he was involved in intelligence work long before this together with his young cousin, Thomas Walsingham. Walsingham employed numerous scholars and linguists in his intelligence service, in return for patronage of their talents in the field of literature, drama, poetry and other works of education and art. Liasing in particular with the Sydneys, Dudleys and Bacons, plus some other notable families, the Walsinghams brought together the talents and devoted energy of educated and gifted people for the combined purposes of serving the country politically, promoting Protestantism or freedom from Popery in religion, and advancing knowledge and culture generally. This confederacy

(36) ANTHONY BACON, aged 36 years — oil painting (1594), artist unknown.

of gifted and influential families, plus their retinue of scholars, writers and linguists that they patronised and employed, formed the very crucible in which formed and out of which emerged the great reforming impulse or renaissance that Francis Bacon was eventually to lead.

Besides the Walsingham cousins, Sir Francis and Thomas, there were their close friends and mentors, Sir Nicholas and Lady Anne Bacon, and Lady Anne's father, Sir Anthony Cooke (whom Anthony Bacon was named after). Robert Dudley, the Earl of Leicester, for all his faults was a notable scholar himself and an important patron. His sister, Mary, was likewise a scholar and patronesse of scholar- poets, and married to Sir Henry Sidney. Their children, particularly Sir Philip Sidney and Mary Sidney, became renowned for their learning and poetry, as also their patronage of others. Sir Philip Sidney married Sir Francis Walsingham's daughter, Frances, in 1583, in what was a love match which welded together the three families[348] in even closer bonds of friendship. Mary Sidney married the wealthy Henry Herbert, second Earl of Pembroke, in 1577, and used their wealth and homes to help sponsor the Great Scheme and maintain a company of actors. Philip and Mary Sidney's cousins were Francis Bacon and Robert Devereux, sons of Leicester and the Queen, the former adopted into the Bacon family and the latter into the Devereux family, the Earls of Essex. Lettice Knollys, wife of Walter Devereux, Earl of Essex, was the daughter of Sir Francis Knollys and a second cousin to the Queen. Robert Devereux himself became intimately involved with the group, especially after his foster-mother married Leicester in 1578, and he became Earl of Essex, although he was never himself a noted scholar or poet but rather a soldier. Instead, Francis wrote poems and masques and even letters for his younger brother, whilst Anthony Bacon later put himself and the intelligence network at the services of Robert, Earl of Essex. With Sir Thomas Walsingham's help, Anthony eventually took over the running of the intelligence network after Sir Francis Walsingham's death in 1590, and both he and Francis continued to employ "good pens" not only for cipher and secretarial work, as cryptologists, amanuenses and copyists, but also for translating classics and other works in foreign languages into English , plus writing or contributing to new works of literature, poetry and drama. John Davies, Nicholas Breton, George Chapman, Edmund Spenser, Thomas Watson, Samuel Daniel, Christopher Marlowe and Anthony Munday were amongst those scholar-poets and dramatists employed and patronised by the 'family' group, the latter five acting additionally as secret service agents abroad from time to time.

> "The well-established fact that Marlowe was an agent employed by Sir Francis Walsingham from 1587 while an undergraduate at Cambridge and after he had taken his M.A. degree, until Walsingham's death in 1590, may serve to kindle interest in those who directed and those who served under them in the secret service. That Sir Thomas Walsingham,[349] cousin of Sir Francis, carried on in an unofficial capacity after the latter' death, and was employing the same men including such as Ingram Frazer, Robert Poley, Marlowe and Nicholas Skeres (all present at the 'liquidation' of Marlowe at Deptford at the end of May, 1593) is also well attested.

> "The Walsingham cousins had very close ties of affection and interests. One has only to study Thomas Watson's *Meliboeus* described on the title page as 'An Eglogue upon the Death of the Right Honorable Sir Francis Walsingham Late principall Secretarie to Her Majestie, and of her Honourable Privie Councell.' It was printed in 1590 in Latin and English. The Latin version is dedicated to Sir Thomas Walsingham, and the English version to Lady Frances Sidney, the daughter of Sir Francis. In the 'Eglogue,' the Queen is 'Diana'; Sir Francis is

'Meliboeus', Sir Thomas speaks as 'Tityrus', and Watson as 'Corydon'. Both Sir Francis and Sir Thomas are declared to be patrons of learning and literature.

"In 1581, Sir Francis Walsingham was sent to Paris to negotiate a treaty with France which was calculated to destroy any agreement between France and Spain which would be dangerous to England. Watson's 'Eglogue' makes it clear that Sir Thomas accompanied his cousin, and that Watson was also there: *e.g.*

> **Tityrus (i.e. Sir Thomas)**
> Thy tunes often pleas'd mine eare of yoare,
> When milk-white swans did flocke to heare thee sing,
> Where Seine in Paris makes a double shoare,
> *Paris* thrice blest if shee obey her King.

"Why was the poet Watson in the company of the Walsingham cousins? Was he, like Marlowe, assisting in the secret service? Were there still more poets and dramatists using their intelligence as agents in return for patronage?

"This appears to be highly probable for there is proof of yet another poet and playwright, Anthony Munday, being similarly engaged. One fact which has come to light is that in 1582, Munday had been hunting Catholics with success. We can learn that from his publication, *A Discoverie of Edmund Campion and his Confederates*. There is no evidence known as to Munday's employer, but he went to Rome to spy on English Catholics, and to learn what he could to their detriment, and then betray them (see *Dictionary of National Biography*).

"Literature was not a profession in those days. There was no living to be made from the writing of books or poems. There was no such thing as a 'reading public', for those who could read were an extremely limited and favoured few. Even the writing of plays was miserably rewarded as we can see from Henslowe's diary. But the authors were intelligent and well-educated men, and what is more likely than for them to use their talents in employment for gathering political information? No doubt it was the Privy Council who employed Munday to carry out his successful detection of the Popish conspiracy in 1582, and who had previously sent him to Italy to spy on the English Catholics residing in the northern cities. On his return journey Munday had exhausted his funds while passing through France and had to make a diversion to Paris where he was given money by the British Ambassador to enable him to reach England. He would not have been so favoured had he not been on official business.

"We do not know the names of all those who collected information for Walsingham from France, Italy, Spain and the Low Countries. Naturally, as secret service agents, they did not come into prominence, but from 1567 onwards Walsingham was supplying Burleigh with lists of names of those hostile to the Queen and the Government. In 1568, he put the secret service on an organised basis. The chief cipher expert was Thomas Phelippes, but another cipher expert employed by Walsingham was Anthony Standen, who worked for Essex after his patron's death. So clever was Standen that he was knighted and Walsingham procured for him from the Queen a pension of £100 yearly. Standen's information for Essex was sent in letters to Anthony Bacon and he used numerals for letters. Some of these are preserved at Lambeth Palace and were printed in *Memoirs of The Reign of Queen Elizabeth* by Thomas Birch,

D.D., in two volumes printed in 1754 .... Besides Robert Poley, Walsingham employed Gilbert Gifford (a Catholic traitor) and Thomas Harrison in the discovery of the Babington plot. There were probably others. Walsingham was kept well informed by his agents in Spain as to the preparations for the Armada and the invasion of England — even to the minutest details of men and armaments."[350]

As far as I know, all previous commentators on Anthony Bacon's life accept, for lack of obvious evidence, that Anthony remained in England until after his father's death, when he was then sent abroad at Lord Burghley's suggestion on a long continental tour of twelve years in search of political intelligence, acting as a principal agent of Walsingham, the Queen's Secretary of State. But I believe that there is in fact good evidence pointing to the fact that Anthony Bacon made several journeys abroad before the end of 1579, under extremely hazardous circumstances, in order to gather intelligence for the Queen. The evidence of this is contained amongst Anthony Bacon's own private papers in the British Museum, and mentioned in Thomas Birch's *Memoirs of Queen Elizabeth*.[351] This is a paper in which a note had been recorded of "special services" performed by one Edward Burnham for Her Majesty at the commandment of Sir Francis Walsingham. The paper gives details of four separate missions that were undertaken by an 'intelligencer' known cryptically as **Edward Burnham**.

The first mission took place in February/early March, 1578, Burnham venturing into Picardy "to see and learn what French forces were there levied to enter the Low Countries" — these being the newly assembled forces of François, Duc d'Anjou. The second mission, which occurred in June, 1578, was into Champagne and the Low Countries, and to the camp of Don John of Austria who was then beseiging Limberg, in order to discover the state of affairs. The third mission took place towards the end of April, 1579, when Burnham was sent into the provinces of Artois and Hainault, and into the camp of Alexander Farnese, Prince of Parma, the successor to Don John, to discover and report on the new situation in the Low Countries. The fourth mission occurred during February to April, 1580, when Burnham was sent to Portugal to report on the state of that country after the death of the King of Portugal (on 31st January, 1580) and whether Philip of Spain would take over the dominion of the country. For this last-mentioned mission Burnham remained in Portugal for three months, in continual danger, and escaped being apprehended by only twelve hours. The note appears to have been written by Nicholas Faunt, who was Sir Francis Walsingham's own private secretary, and a colleague and intimate friend of Anthony Bacon.

This "Mr. Burnham" is also mentioned in the 'Sydney Papers', in Collins' *Letters and Memorials of State* (vol. II, p.302), in the *Complete Ambassador* by Sir Nicholas Digges, and by Sir Thomas Lake, one of Walsingham's amanuenses. The latter mention is important, concerning the troubles of 1600, when Robert, Earl of Essex, (who had been under house arrest at York House from October 1599) was moved on 19th March, 1600, to Essex House, still under house arrest, the Queen having ordered (on 10th March, 1600) that Anthony Bacon, together with Lady Leicester, Lord and Lady Southampton, and Fulke Greville, should be removed from Essex House where they had their residence. Sir Thomas Lake states that he could not send letters by Burnham because Essex had sent him (Burnham) out of the way.

Some scholars have thought that Edward Burnham was an *alias* of Francis Bacon — 'Burnham' being a rather witty pseudonym for 'Bacon', and Francis being known to have performed difficult missions in order to supply intelligence to the Queen. The observation

Diagram H: MAP OF EUROPE

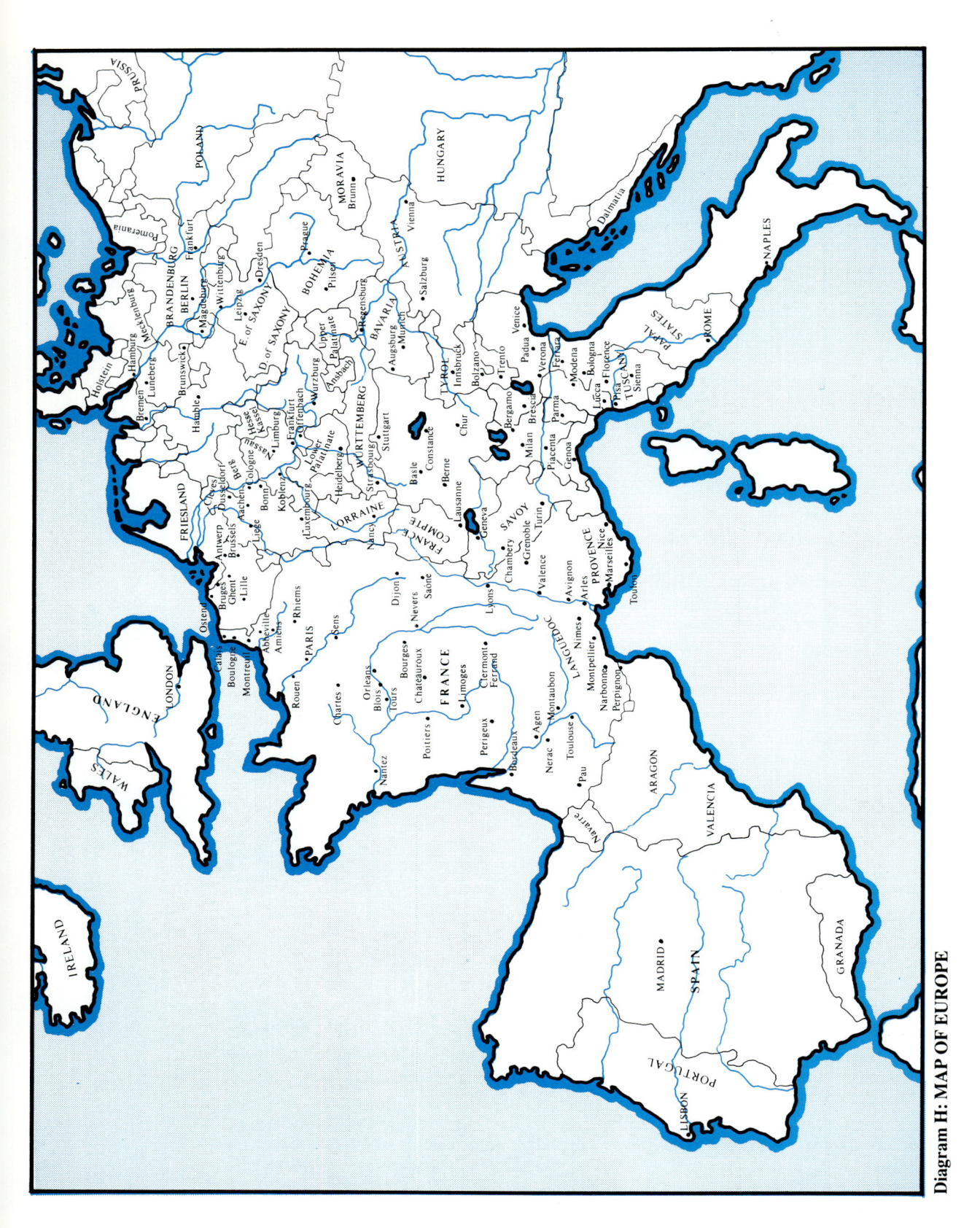

concerning the pseudonym is indeed valid, but the person in question could not have been Francis Bacon, who was far too well known on the continent to have been able to go around in disguise, and who was anyway elsewhere at the times mentioned. But the whole reference and description fits Anthony Bacon precisely, he possessing the talents and character required for such missions, and it being entirely in keeping with what is known of his responsible work for Walsingham. The time and place of the missions not only fill gaps in Anthony's known history, but are entirely feasible in terms of his being able to carry them out, dovetailing into the rest of his known history. Furthermore, not only is the record of the missions stored amongst Anthony's own private papers, but Sir Thomas Lake's reference to Burnham can only refer to Anthony Bacon, who was at that time the principal person at the hub of the intelligence network through whom all correspondence was sent out and received, cyphered, deciphered, and generally distributed *via* agents or messengers to where it was required to go. A whole range of apartments in Essex House were given over to him for his own rooms and offices, and the offices of his secretariat, and thus formed the 'headquarters' of the network. At the Queen's express order, dated 10th March, 1600, Essex was required to dismiss Anthony Bacon and the others from Essex House, prior to his own incarceration there — which order he of course carried out, having no alternative. '**Edward**', which means 'a trusted guardian' of other people's properties or interests, plus '**Burn-ham**', alluding to 'bacon' or 'Bacon', indeed constitute an excellent pseudonym for revealing and yet describing exactly what Anthony Bacon was.

It is also worth noting that, in his own words (in a letter to the then Secretary of State, Sir Robert Cecil, written in December 1599), Anthony had from his earliest years enjoyed a privileged and constant access to the Queen:

> "....The root of this I discern to be, not so much a light and humorous envy at my accesses to her Majesty, which of her Majesty's grace being begun in my first years, I would be sorry she should estrange in my last years ...."

But, to return to the record of the missions made by Anthony Bacon, as I believe, under the *alias* of Edward Burnham, the following is the text of the note found amongst his private papers:

> "A note of special services performed by Edward Burnham for Her Majesty at the commandment of the Right Honorable Sir Francis Walsingham, Knt., Her Majesty's principal Secretary, and My Honorable Master.
>
> "Mr. Burnham, who is sometimes mentioned in the Sidney papers, went in the year 1577, by the Secretary's order, into Picardy in France, to Calais, Boulogne, Montreiul, Abbeville and Amiens, to see and learn what French forces were there levied to enter the Low Countries; and at his return passed through Licques where he had a conference with Monsieur de Licques, with whom he had an acquaintance before, and another conference with Monsieur de la Notte, Governor of Graveling. This secret journey was performed before the Duke of Anjou, brother of Henry III of France, made his first voyage into the Low Countries; and Mr. Burnham brought back a relation of the state of things agreeable to the Secretary's instructions, with which both he and the Queen herself were extremely satisfied.
>
> "After this, when the Lord Cobham and Secretary Walsingham were sent by Her Majesty in 1578 in the Low Countries, Mr. Burnham was dispatched by the Secretary to Paris to Sir Amyas Paulet, the Queen's Ambassador there, and

thence to Rheims in Champagne, in order to see and learn what ill-affected subjects of Her Majesty were there, where he ventured so far as to confer with Dr. Alan, afterward made a Cardinal, and other Englishmen equally averse to the religion of their own country. Thence he went to the camp of Don John, then beseiging Limburg, and continued in it fifteen days, till that city was taken. In this hazardous situation he concealed himself under the protection of John Baptista de Nanty, to whom he had brought a letter of recommendation from an Italian gentleman at Paris, pretending himself to be a gentleman of his cornet of horse.

"After he had observed the state of that camp and the enemy's garrison towns through which he passed, he carried the relation, which he had drawn up, to the Lord Cobham and the Secretary, then at Antwerp.

"About half a year after the death of Don John of Austria, which happened on the 1st of October 1578, Mr. Burnham was sent by the Secretary into the camp of the Prince of Parma, the successor of Don John, to observe in what situation things were at that prince's entrance into the Government, and how his highness was lik'd of the nobility, soldiers and commonalty; of which, at his return, he drew up a relation approv'd of by the Queen and the Secretary.

"Upon the first news of the death of Cardinal Henry, King of Portugal, in 1580, when it was doubtful whether Don Antonio, Prior of Crato, or Philip II were in possession of that kingdom, and the former had sent to Queen Elizabeth, John Roderigo de Zenza, Mr Burnham was dispatched by the Secretary, by Her Majesty's order, into Portugal, to see in what state that country then was. He continued in Lisbon for that purpose 22 days, in the disguise of a servant to a factor of Mr. Bird, a merchant, and was for three months exposed to continual danger, being strictly examined at several places, particularly by the Conde de Lemos, at the time when the account came of Arthur, Lord Grey of Wilton's having put to the sword the Spaniards who had landed in Ireland. And his danger was the greater, as Don Bernadino de Mendoza, the Spanish Ambassador in England, had received some intimation of his voyage to Portugal, and sent over a description of his stature, countenance, and particular marks, to know him by: and he had embarked but twelve hours in his return to England, before orders arrived from the Court of Spain for his apprehension.

"For these services, as well as the several journeys, in which he had been employed by the Secretary to the Duke of Anjou, William Prince of Orange, Charles de Croy, Prince de Chimay, and the States of the Low Countries, he requested some 'extraordinary gratification'.

— Mr. Faunt, etc, etc ...... "

Anthony was probably sent off to Picardy and the Low Countries on the first mentioned mission (February/March 1578) in response to intelligence being sent home to Walsingham and Burghley from Sir Amyas Paulett in Paris, following Anjou's escape from the Louvre. Francis Bacon and Sir Amyas would have been able to supply the inside information concerning Anjou's actions and intentions, and those of the French King, *etc.*. Since this could seriously affect the balance of power in the Low Countries, and in Europe generally, someone trustworthy and capable had to somehow discover the true state of affairs in that hot-bed of Europe. It was following Anthony's report that the Queen's policy of offering the bait of marriage to Anjou was renewed in earnest.

(37) DON JUAN, Archduke of Austria — engraving.

Anthony's second mission (June 1578) into the Low Countries and Don John's camp was really a follow-up to his earlier mission, this time to determine the strength and intentions of the Spanish and of the Catholics generally. But first he went to Paris, to confer with Sir Amyas Paulett and gather as much information from Sir Amyas as he could before embarking upon his hazardous journey. There at Paris, Anthony must have almost certainly met up with Francis, and we can imagine the joy and the conversation they must have exchanged.

The third mission (April 1579) that Anthony undertook into the Low Countries was a further follow-up to his previous missions, again to infiltrate the Spanish encampment, but this time to discover the new state of affairs now that the Prince of Parma had succeeded Don John at the head of the Spanish forces. The temporary and rather unnatural coalition of Catholics, Protestants and Calvanists against Don John had been rapidly breaking up since his death, a division into power blocks coming into the open early in 1579 with the Protestant 'Union of Utrecht'[352] in the north and the Catholic 'Union of Arras'[353] in the south. Between these two Unions lay a mass of small states and towns whose allegiance to either side was there to be won. Using diplomacy and concessionary bribes, Parma was beginning to woo the Walloon provinces of Artois and Hainault back to the Spanish fold, despite Anjou's influence in that area. Anthony must have left on this mission after the funeral of Sir Nicholas Bacon, and shortly after the return of Francis to England in early April.

Anthony was probably not away for long on this mission — probably no more than a month or two — and on his return he drew up a report of what he had observed for Walsingham and the Queen. There would have still been a great deal to sort out concerning his father's will and estates, and a dispute to settle with his half-brother, Nathaniel, over Pinner Park in Middlesex which had been left to Anthony. Lady Anne had to be helped in her move from York House to Gorhambury, and the household staff needed to be greatly trimmed to suit her changed status and income. In addition, Anthony made preparations to set out on a long continental tour. According to a letter written by Burghley to La Motte-Fenelon, dated December 1579, this journey was made at Anthony's own request, and Burghley assented to allow Anthony, "according to his honest desire, to travel into France, to see the country and to learn the language, and to enable himself by learning good things there, to serve his country better." This, however, was only a partly true explanation. Anthony undoubtedly desired to travel more liberally, as many gentlemen of his time did, but the suggestion to go was Lord Burghley's, and the real aim was so that Anthony could gather political intelligence and act as a key English agent or intelligencer abroad. At the same time, he would be able to help further the Great Scheme — learning, teaching, sowing seeds, gathering fruits and steadily discovering and building up an international fraternity or fellowship of like minds and motivations — a network through which might pour the grail of new (and ancient) wisdom-knowledge, inspiring great philanthropic achievements.

Later that year (1579) Anthony set out for France, accompanied by a few attendants, his first port of call being Paris where a new English Ambassador, Sir Edward Stafford, had taken the place of Sir Amyas Paulett. Sir Edward had recently married the widowed Lady Sheffield, with whom Leicester had had a liaison and by whom he had a child. Then, in February 1580, Anthony was sent to Portugal by Walsingham, on the Queen's order, to find out what was happening there and whether Spain would indeed gain control of Portugal and thus of the principal trade routes to the East and to the Americas. After three dangerous months in Portugal, Anthony escaped the country just in time, by ship, and made his way back to England, seemingly *via* Paris where he had dealings with a Welsh Catholic, Dr. William Parry, who was also working as a secret agent for Walsingham and Burghley.

Various mischief-makers tried to make out in high places in England that Anthony was liasing with a Roman Catholic and a known criminal and was thus untrustworthy. (Parry had been involved in a theft and a possible murder in the Inner Temple in London, but had obtained a pardon *via* Burghley on condition he acted for them as a secret agent.) The Queen was informed, but Burghley told her that not only was Anthony trustworthy, loyal and a firm supporter of the Protestant faith, but that both men were acting under his instructions. However, the episode provided as good a smoke screen as any for Anthony to be 'recalled' to England to make his report. It is not known exactly when he returned to England — whether in the summer or autumn of 1580 — or for how long he stayed, but come December 1580 he had returned to France and made his way to Bourges, arriving there on the 15th December 1580.

Whilst he had been in Portugal, his staff at Gorhambury had been writing to him, their letters reaching Paris, but, of course, receiving no reply. Many people seemed to have had a deep love and regard for Anthony, including his parents' household staff who missed him sorely; and this helped him enormously in the work he did. One such letter demonstrates this, written apparently in March, 1580, to Anthony Bacon by Thomas Cotheram, the under-steward to Hugh Mantell, Sir Nicholas Bacon's faithful elderly steward of the household:

> "My duty always remembered, most loving master. I know you would gladly hear of your house Gorhambury, and I have written to you three or four times and never heard from you again. My Lady doth take to your house very well and hath great care to repair it where it is needful, and it doth lack nothing but you to come home and dwell there yourself and I would to God it might be so. My Lady doth think you will come home in debt and therefore you shall have the felling of the trees in your woods to set you out of debt. Your water-mill is well and when it is out of repair the plumber doth come presently to mend it. There is nobody here doth so much miss you as I do. Neither I nor my wife, and though I say it myself, never have broken your commandment and that was we should pray for your prosperous journey and your happy return again every night and every morning upon our knees, and I am persuaded that the Lord hath heard and shall hear our prayers still. I hope you do not nor need fear my Lady's marriage for truly she is as far of that as when you departed from her, nor never none came to her for that, as far as I can hear of, and if it do you shall hear of it by me as soon as I can send you word, but I hope to God there shall be no such action yet. I pray you, sir, come home as soon as you can, it is good to doubt the worst ever. Mistress Tutes doth send you hearty conn [?], John Bruer your poor servant the like, Nicholas of the kitchen the like, and John Cotles the like.
>
> "Your poor servant to command as long as life doth last in all your affairs,
> Thomas Cotheram.
>
> " I pray you, sir, to do my wife's conn and mine to Master Brigennes and to George."

The last two names would seem to refer to the two servants from the household staff who attended Anthony at that time, Master Brigennes and George Jenkyll. The anxiety about whether or not Lady Anne was likely to remarry was probably because of the conditions of Sir Nicholas Bacon's will, in which Lady Anne had to give a sufficient release in law of his estate to each of Sir Nicholas' sons before the twelve-month was up and before she should marry again; but Nathaniel Bacon was disputing the inheritance being given to Anthony, in particular Pinner Park and Gorhambury, and Burghley was still trying to settle the matter on their behalf as overseer of the will.

From then onwards, until 1592, Anthony travelled Europe as an English agent and secret diplomat acting on behalf of the Queen of England. On at least one other time during this period he was sent on another specific mission, as 'Edward Burnham', to the Netherlands. This is the episode referred to in the 'Sidney Papers', in which Burnham is mentioned, and takes place at the end of 1585 when Sir Philip Sydney was in the Low Countries, along with many English noblemen and an English army sent by the Queen to help the Dutch oppose Parma, following the Treaty of Nonsuch made with the Dutch Provinces in the summer of 1585. Leicester himself came over on December 10th, to command the armies and to accept the role of Governor of the United Provinces. In a letter to Walsingham, dated December 1585, from Middlebury in the Netherlands, Philip Sydney wrote:

> "Burnham is come to me whom I long longed for and find myself much steeded [*i.e.* supported] by him. I humbly beseech you to end the matter for him which you promised for he hath and will deserve it."

(38) ELIZABETH I, Queen of England (1558-1603), holding the Garter George — oil painting (*c.* 1575), artist unknown.

(38) ELIZABETH I, holding the Garter George — oil painting (*c.* 1575), artist unknown.

The Queen is depicted holding up the Lesser George in front of her heart, for the viewer to take note. It is the focal point of the picture, to which the artist is deliberately drawing attention. The Queen's right hand (of compassion, mercy, grace, *etc.*) holds the badge — the Lesser George — which is traditionally made of plain gold, depicting St. George on horseback slaying the dragon with a spear, and worn as a pendant from the ribband of Garter blue. Nowadays the ribband is worn such that it passes over the left shoulder and the gold badge rests upon the right hip. Judging from the Tudor and Jacobean portraits, the Lesser George was worn as a pendant on the breast, as shown in this picture. The (Greater) George is the more elaborate jewel of the Order, of enamelled gold (sometimes with gem stones), suspended from the gold and enamel Collar, which is worn only for ceremonial purposes as part of the full regalia of the Order. The Lesser George may be worn in other circumstances, as a badge or mark of merit.

(39) SIR FRANCIS BACON — frontispiece to *Sylva Sylvarum* (1627).

This picture shows several extraordinary and cryptic matters. Firstly, Sir Francis Bacon is depicted framed in an oval frame — not itself unusual, but its original significance is to represent the auric field symbolised by the 'pearl of great price' and the *'vesica piscis'*. When seen in conjunction with the other cryptic signs, the use of this frame is quite deliberate and symbolic. Secondly, instead of casting a shadow, Francis' head (and hat) throws off a light or radiance of its own, illumining the shadow behind his head with a halo — a crown of light. Thus the artist is carefully depicting the 'countenance of the Lord' symbology, in a way that is sufficiently cryptic both to avoid the charge of heresy and yet to draw the attention of earnest questors. Cryptically, the artist is portraying an enlightened or sainted man, who is anointed ('Christed') with spiritual light, having direct knowledge of the Lord (*i.e.* wisdom). Taken in conjunction with the oval frame, the picture portrays the holy grail — the pure vessel that contains the shining wine of true knowledge.

Completing the cryptic picture are the two 'Quest' mottoes by which Francis was guided thoughout his life — *"Moniti Meliora"*, which is derived from *"Moniti Meliora Sequamur"* ("Let us, being admonished, follow better things") of Virgil's *Aeneid*, and *"Mediocria Firma"* ("Mediocrity/the Middle Path is safe/firm/strong") — and a ribband about his neck on which is suspended a badge or jewel that is deliberately concealed from our view by a piece of paper held in Francis' left hand. What this mystery badge is was not revealed openly until 1640.

(39) SIR FRANCIS BACON, Baron Verulam, Viscount St. Alban — frontispiece to his *Sylva Sylvarum*, 1st edition (1627).

(40) SIR FRANCIS BACON — frontispiece to *Of the Advancement and Proficience of Learning* (1640).

In this picture the concealed badge or pendant of the *Sylva Sylvarum* frontispiece is revealed clearly. It is the **Lesser George.** Several things are striking about this. Firstly, that this badge was deliberately veiled in the earlier frontispiece, but now revealed blatantly, drawing specific attention to it and what it means. Symbolically it is the badge of the Grail Knight, the Knight of the Garter whose archetype is St. George, the Rose Cross Knight, Shaker-of-the-Spear and Dragon-Slayer. On the one hand, the artist is announcing that Francis Bacon is/was such a knight — esoterically, a Rosicrucian (in the true sense, hence the halo around his head in the *Sylva Sylvarum* portrait). On the other hand the artist is stating that Francis Bacon was a Knight of the Order of the Garter, rather than just an ordinary knight. But exoterically history tells us only that Francis Bacon was given an ordinary knighthood by James I on July 23rd, 1603. An engraver of the stature of William Marshall would not be likely to risk hostile criticism and possible prosecution for false representation of the premier English Order of Chivalry, neither would the publisher — yet the Lesser George is deliberately shown. Why?

There are two persons in the realm of England who are automatically deemed to be Knights of the Order of the Garter by right of position: the reigning Sovereign and the Prince of Wales, heir to the throne.

The other features in this picture are all noteable, and caballistically designed. As usual Francis Bacon is shown wearing his black hat — a well-known symbol to many, representing the secrecy and 'invisibility' of the Rosicrucian 'light' to those incapable of seeing or understanding. The hat is the Baconian representation of what the Greeks called Pallas Athena's helmet of invisibility given her by Pluto. It is the helmet of knowledge, of light. Above the hat is a plaque of light surrounded with a wreath of olives — the symbol of illumination. The words read *"Tertius A Platone, Philosophiae Princeps"* ("The third after Plato, the Chief/Leader of Philosophy").

Turned towards the left of the picture (with the curtain or veil behind him on the right of the picture, as seen 'heraldically') Francis is writing the 7th volume of the Great Instauration. Each volume is a clear representation of a 'Part' of the Great Instauration. The first two 'Parts' lie flat on the table (*i.e.* the 'Advancement of Learning' and the 'New Method'). The 3rd, 4th, 5th and 6th "Parts" stand upright on the shelf (*i.e.* the Natural History, the works of Poesie and Drama, the 'Anticipations', and the 'Second Philosophy' or true Axioms). The 7th 'Part' is that divine Sabbatical Mystery that only the truly illumined have access to. On the open pages of this mystery book Francis has written *"Mundus"* and *"Mens"* ("the World/Universe" and "the Mind") and *"Conubio jungam stabili . . ."* ("the Connection made firm in Marriage") — a reference to the necessary marriage between thought and action (*i.e.* Mercury and Mars) in order to achieve anything of real value, which in turn bestows the true knowledge or illumination that constitutes the completion of the Great Work. This was the principal teaching that Francis came to emphasise for humanity. Thought is not enough: we must put our ideas into action and prove their value by results. If our thoughts are good and true, then our actions will produce goodness and usefulness. If not, then think again: but do not divorce thought from action. A man is judged by what he does, rather than by what he thinks; but, as action follows thought, we need to purify and perfect our thoughts, whilst at the same time not stopping in beautiful thoughts but putting them into practice. This is **science** (thought) and **art** (action) — science being accurate ideas concerning truth, and art being the subsequent actions performed with grace and expertise. Francis Bacon was the Instaurator or Renovator of all science and art, that man might become proficient in them and thereby achieve the illumined state of **PEACE.** He sits enthroned as **Hemetes,** the *"bucinator novi temporis"* — "the herald/trumpeter of the new age", whose trumpet (*bucina*) is the shepherd's horn; for he is the hermetic Pastor, the Shepherd of humanity and Messenger of Light.

(40) SIR FRANCIS BACON, Baron Verulam, Viscount St. Alban — from frontispiece to his *Of the Advancement and Proficience of Learning* (1640), engraved by William Marshall.

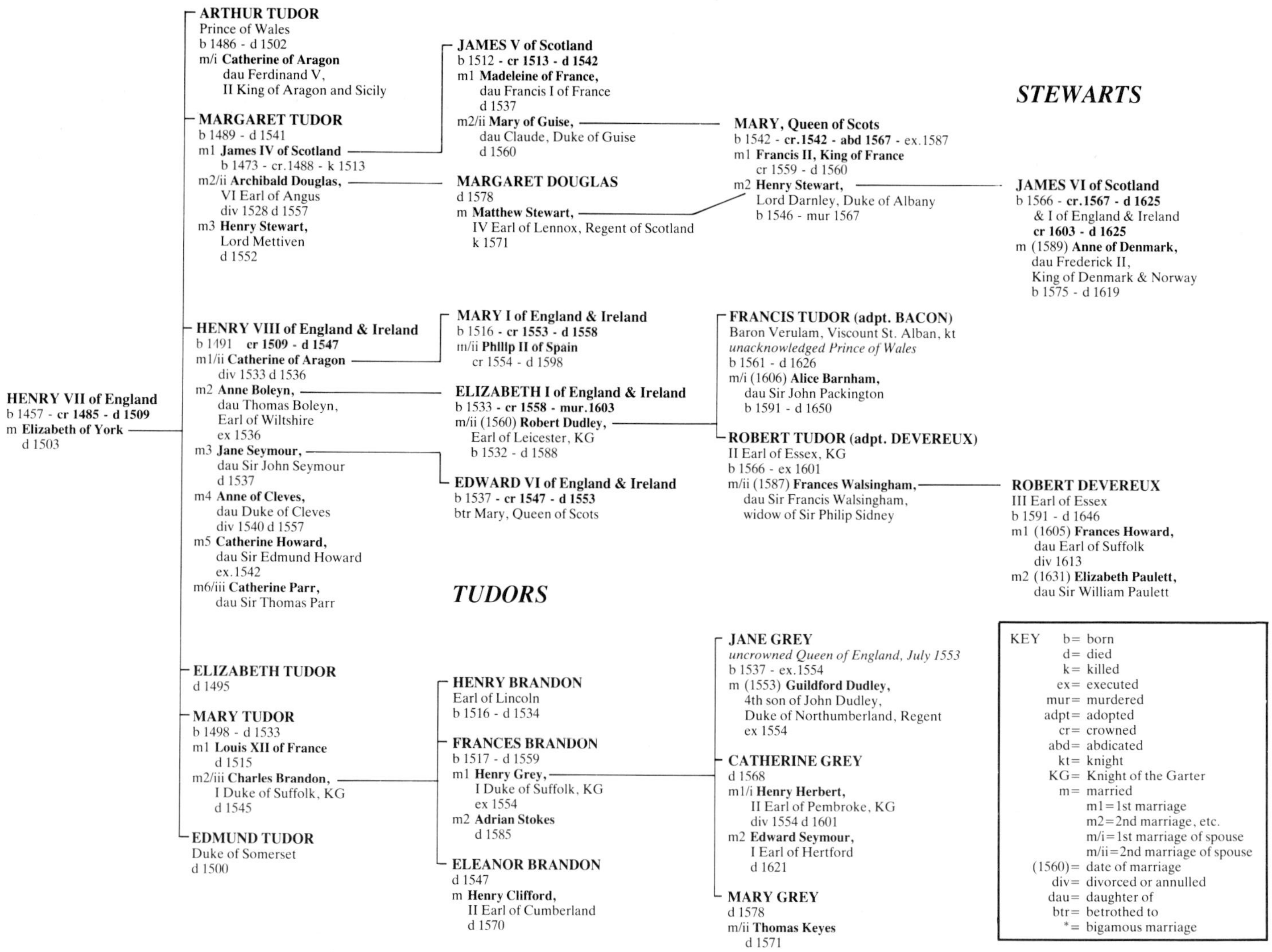

Diagram I: TUDOR-STEWART FAMILY TREE

TUDORS

HENRY VII of England
b 1457 - cr 1485 - d 1509
m **Elizabeth of York**
d 1503

- **ARTHUR TUDOR**
  Prince of Wales
  b 1486 - d 1502
  m/i **Catherine of Aragon**
  dau Ferdinand V,
  II King of Aragon and Sicily

- **MARGARET TUDOR**
  b 1489 - d 1541
  m1 **James IV of Scotland**
  b 1473 - cr 1488 - k 1513
  m2/ii **Archibald Douglas**,
  VI Earl of Angus
  div 1528 d 1557
  m3 **Henry Stewart**,
  Lord Methven
  d 1552

- **HENRY VIII of England & Ireland**
  b 1491 cr 1509 - d 1547
  m1/ii **Catherine of Aragon**
  cr 1533 d 1536
  m2 **Anne Boleyn**,
  dau Thomas Boleyn,
  Earl of Wiltshire
  ex 1536
  m3 **Jane Seymour**,
  dau Sir John Seymour
  d 1537
  m4 **Anne of Cleves**,
  dau Duke of Cleves
  div 1540 d 1557
  m5 **Catherine Howard**,
  dau Sir Edmund Howard
  ex.1542
  m6/iii **Catherine Parr**,
  dau Sir Thomas Parr

- **ELIZABETH TUDOR**
  d 1495

- **MARY TUDOR**
  b 1498 - d 1533
  m1 **Louis XII of France**
  d 1515
  m2/iii **Charles Brandon**,
  I Duke of Suffolk, KG
  d 1545

- **EDMUND TUDOR**
  Duke of Somerset
  d 1500

From Margaret Tudor / James IV:

- **JAMES V of Scotland**
  b 1512 - cr 1513 - d 1542
  m1 **Madeleine of France**,
  dau Francis I of France
  d 1537
  m2/ii **Mary of Guise**,
  dau Claude, Duke of Guise
  d 1560

  - **MARY, Queen of Scots**
    b 1542 - cr 1542 - abd 1567 - ex.1587
    m1 **Francis II, King of France**
    cr 1559 - d 1560
    m2 **Henry Stewart**,
    Lord Darnley, Duke of Albany
    b 1546 - mur 1567

    - **JAMES VI of Scotland**
      b 1566 - cr.1567 - d 1625
      & I of England & Ireland
      cr 1603 - d 1625
      m (1589) **Anne of Denmark**,
      dau Frederick II,
      King of Denmark & Norway
      b 1575 - d 1619

- **MARGARET DOUGLAS**
  d 1578
  m **Matthew Stewart**,
  IV Earl of Lennox, Regent of Scotland
  k 1571

From Henry VIII:

- **MARY I of England & Ireland**
  b 1516 - cr 1553 - d 1558
  m/ii **Philip II of Spain**
  cr 1554 - d 1598

- **ELIZABETH I of England & Ireland**
  b 1533 - cr 1558 - mur.1603
  m/ii (1560) **Robert Dudley**,
  Earl of Leicester, KG
  b 1532 - d 1588

  - **FRANCIS TUDOR (adpt. BACON)**
    Baron Verulam, Viscount St. Alban, kt
    unacknowledged Prince of Wales
    b 1561 - d 1626
    m/i (1606) **Alice Barnham**,
    dau Sir John Packington

  - **ROBERT TUDOR (adpt. DEVEREUX)**
    II Earl of Essex, KG
    b 1566 - ex 1601
    m/ii (1587) **Frances Walsingham**,
    dau Sir Francis Walsingham,
    widow of Sir Philip Sidney

- **EDWARD VI of England & Ireland**
  b 1537 - cr 1547 - d 1553
  btr **Mary, Queen of Scots**

From Mary Tudor / Charles Brandon:

- **HENRY BRANDON**
  Earl of Lincoln
  b 1516 - d 1534

- **FRANCES BRANDON**
  b 1517 - d 1559
  m1 **Henry Grey**,
  I Duke of Suffolk, KG
  ex 1554
  m2 **Adrian Stokes**
  d 1585

  - **JANE GREY**
    uncrowned Queen of England, July 1553
    b 1537 - ex.1554
    m (1553) **Guildford Dudley**,
    4th son of John Dudley,
    Duke of Northumberland, Regent
    ex 1554

  - **CATHERINE GREY**
    d 1568
    m1/i **Henry Herbert**,
    II Earl of Pembroke, KG
    div 1554 d 1601
    m2 **Edward Seymour**,
    I Earl of Hertford
    d 1621

  - **MARY GREY**
    d 1578
    m/ii **Thomas Keyes**
    d 1571

- **ELEANOR BRANDON**
  d 1547
  m **Henry Clifford**,
  II Earl of Cumberland
  d 1570

STEWARTS

ROBERT DEVEREUX
III Earl of Essex
b 1591 - d 1646
m1 (1605) **Frances Howard**,
dau Earl of Suffolk
div 1613
m2 (1631) **Elizabeth Paulett**,
dau Sir William Paulett

KEY
- b = born
- d = died
- k = killed
- ex = executed
- mur = murdered
- adpt = adopted
- cr = crowned
- abd = abdicated
- kt = knight
- KG = Knight of the Garter
- m = married
  - m1 = 1st marriage
  - m2 = 2nd marriage, etc.
  - m/i = 1st marriage of spouse
  - m/ii = 2nd marriage of spouse
- (1560) = date of marriage
- div = divorced or annulled
- dau = daughter of
- btr = betrothed to
- * = bigamous marriage

ROBERT KNOLLYS, kt
m1 Lettice Peniston

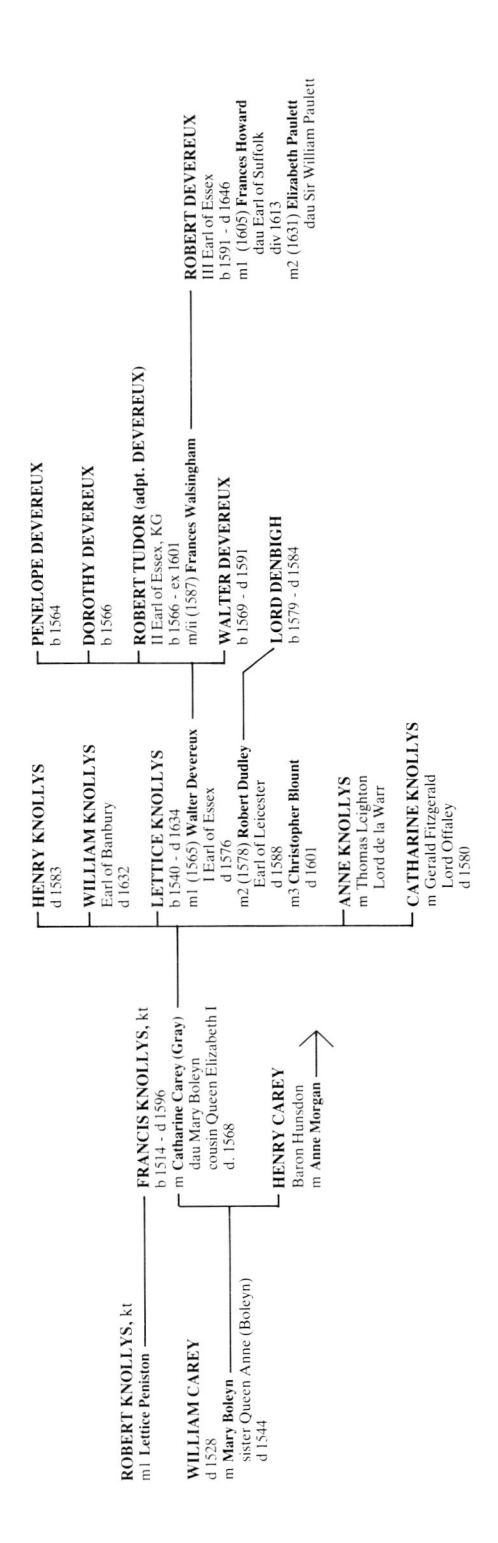

FRANCIS KNOLLYS, kt
b 1514 - d 1596
m **Catharine Carey (Gray)**
dau Mary Boleyn
cousin Queen Elizabeth I
d. 1568

WILLIAM CAREY
d 1528
m **Mary Boleyn**
sister Queen Anne (Boleyn)
d 1544

HENRY CAREY
Baron Hunsdon
m **Anne Morgan**

- **HENRY KNOLLYS**
  d 1583
- **WILLIAM KNOLLYS**
  Earl of Banbury
  d 1632
- **LETTICE KNOLLYS**
  b 1540 - d 1634
  m1 (1565) **Walter Devereux**
  I Earl of Essex
  d 1576
  m2 (1578) **Robert Dudley**
  Earl of Leicester
  d 1588
  m3 **Christopher Blount**
  d 1601
- **ANNE KNOLLYS**
  m Thomas Leighton
  Lord de la Warr
- **CATHARINE KNOLLYS**
  m Gerald Fitzgerald
  Lord Offaley
  d 1580

From Lettice Knollys & Walter Devereux:
- **PENELOPE DEVEREUX**
  b 1564
- **DOROTHY DEVEREUX**
  b 1566
- **ROBERT TUDOR (adpt. DEVEREUX)**
  II Earl of Essex, KG
  b 1566 - ex 1601
  m/ii (1587) Frances Walsingham
- **WALTER DEVEREUX**
  b 1569 - d 1591
- **LORD DENBIGH**
  b 1579 - d 1584

From Robert Tudor (Devereux):
- **ROBERT DEVEREUX**
  III Earl of Essex
  b 1591 - d 1646
  m1 (1605) **Frances Howard**
  dau Earl of Suffolk
  div 1613
  m2 (1631) **Elizabeth Paulett**
  dau Sir William Paulett

Diagram J: KNOLLYS-CAREY FAMILY TREE

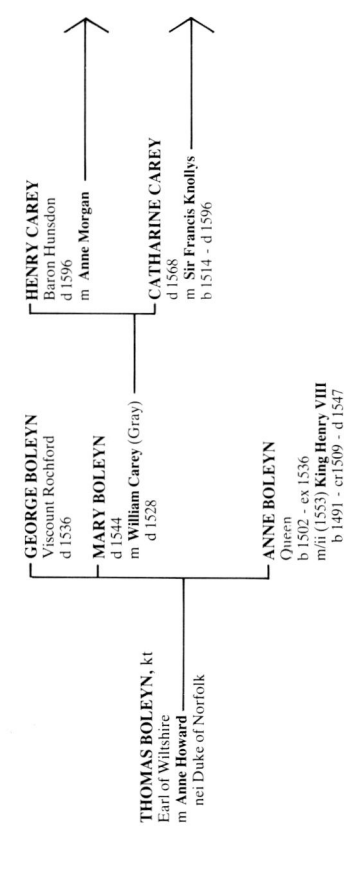

Diagram K: BOLEYN-CAREY FAMILY TREE

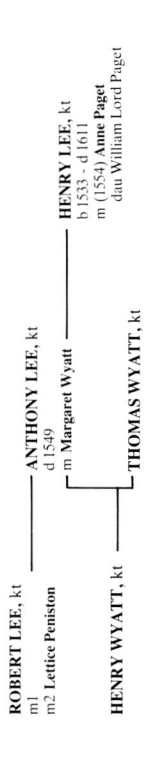

THOMAS BOLEYN, kt
Earl of Wiltshire
m **Anne Howard**
nei Duke of Norfolk

- **GEORGE BOLEYN**
  Viscount Rochford
  d 1536

- **MARY BOLEYN**
  d 1544
  m **William Carey** (Gray)
  d 1528
  - **HENRY CAREY**
    Baron Hunsdon
    d 1596
    m **Anne Morgan** →
  - **CATHARINE CAREY**
    d 1568
    m **Sir Francis Knollys**
    b 1514 - d 1596 →

- **ANNE BOLEYN**
  Queen
  b 1502 - ex 1536
  m/ii (1553) **King Henry VIII**
  b 1491 - cr1509 - d 1547

Diagram L: WYATT-LEE FAMILY TREE

ROBERT LEE, kt
m1
m2 Lettice Peniston

- **ANTHONY LEE**, kt
  d 1549
  m **Margaret Wyatt**
  - **HENRY LEE**, kt
    b 1533 - d 1611
    m (1554) **Anne Paget**
    dau William Lord Paget

HENRY WYATT, kt
- **THOMAS WYATT**, kt

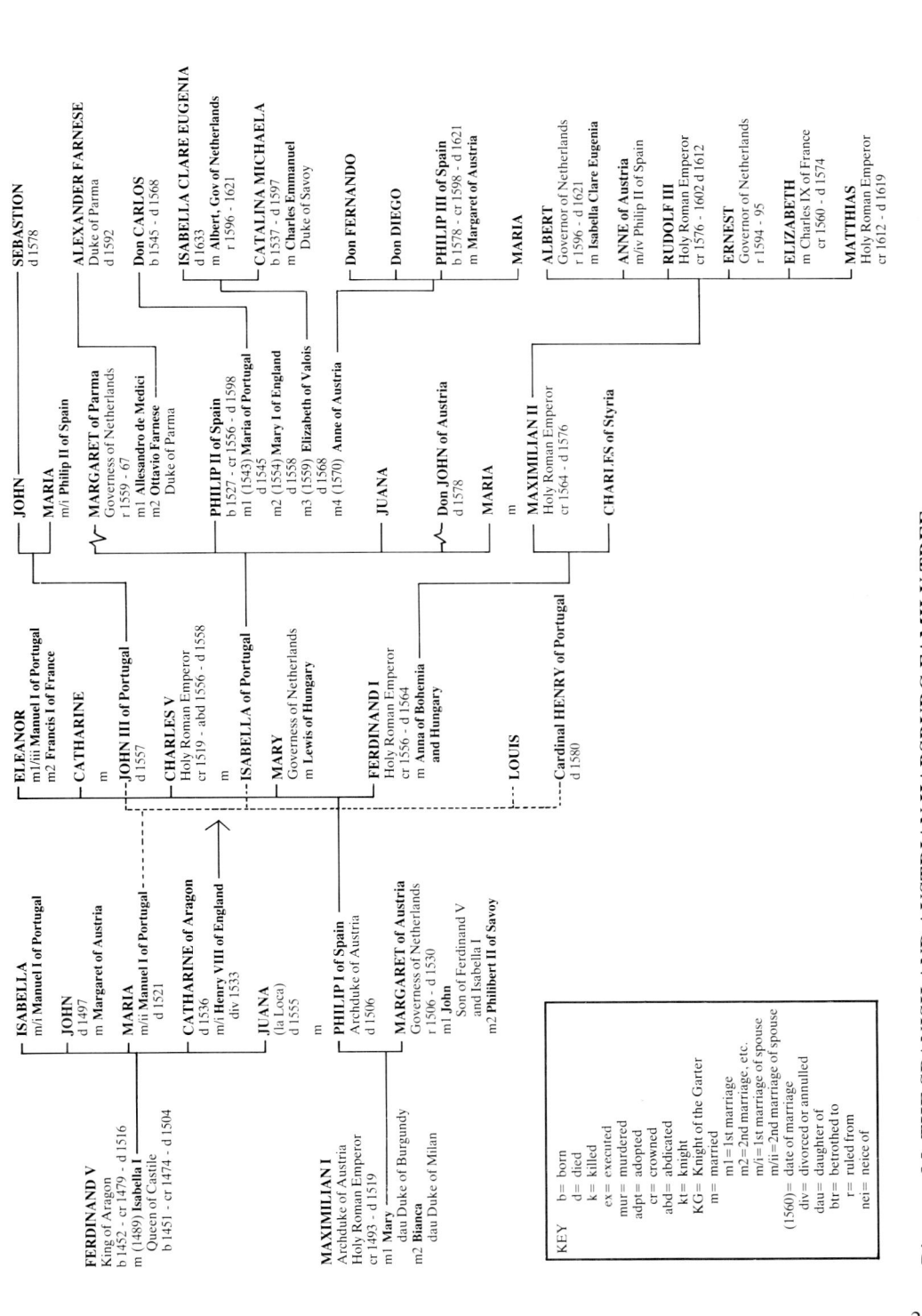

Diagram M: THE SPANISH AND AUSTRIAN HABSBURG FAMILY TREE

(Family tree diagram)

FERDINAND V
King of Aragon
b 1452 - cr 1479 - d 1516
m (1489) **Isabella I**
Queen of Castile
b 1451 - cr 1474 - d 1504

Children of Ferdinand V and Isabella I:

- **ISABELLA** m/i Manuel I of Portugal
  - **JOHN** d 1497
- **JOHN** d 1497 m Margaret of Austria
- **MARIA** m/ii Manuel I of Portugal d 1521
  - **ELEANOR** m1/iii Manuel I of Portugal m2 Francis I of France
  - **CATHARINE**
  - **JOHN III of Portugal** d 1557
  - **ISABELLA of Portugal** m CHARLES V
  - **MARY** Governess of Netherlands m Lewis of Hungary
  - **CATHARINE of Aragon** d 1536 m/i Henry VIII of England div 1533
- **JUANA** (la Loca) d 1555 m **PHILIP I of Spain** Archduke of Austria d 1506

MAXIMILIAN I
Archduke of Austria
Holy Roman Emperor
cr 1493 - d 1519
m1 **Mary** dau Duke of Burgundy
m2 **Bianca** dau Duke of Milan

Children of Philip I and Juana:
- **CHARLES V** Holy Roman Emperor cr 1519 - abd 1556 - d 1558
- **FERDINAND I** Holy Roman Emperor cr 1556 - d 1564 m Anna of Bohemia and Hungary
- **MARGARET of Austria** Governess of Netherlands r 1506 - d 1530 m1 **John** Son of Ferdinand V and Isabella I m2 **Philibert II of Savoy**

Children of Charles V and Isabella of Portugal:
- **PHILIP II of Spain** b 1527 - cr 1556 - d 1598
  - m1 (1543) Maria of Portugal d 1545
  - m2 (1554) Mary I of England d 1558
  - m3 (1559) Elizabeth of Valois d 1568
  - m4 (1570) Anne of Austria
- **MARIA** m/i Philip II of Spain
- **JUANA**
- **Don JOHN of Austria** d 1578

Children of Maria (sister of Philip II) line:
- **SEBASTION** d 1578
- **JOHN**
- **MARIA** m/i Philip II of Spain
- **MARGARET of Parma** Governess of Netherlands r 1559 - 67 m1 Allesandro de Medici m2 Ottavio Farnese Duke of Parma
  - **ALEXANDER FARNESE** Duke of Parma d 1592

Children of Philip II:
- **Don CARLOS** b 1545 - d 1568
- **ISABELLA CLARE EUGENIA** b 1633 m Albert, Gov of Netherlands r 1596 - 1621
- **CATALINA MICHAELA** b 1537 - d 1597 m Charles Emmanuel Duke of Savoy
- **Don FERNANDO**
- **Don DIEGO**
- **PHILIP III of Spain** b 1578 - cr 1598 - d 1621 m Margaret of Austria
- **MARIA**

Children of Ferdinand I:
- **MAXIMILIAN II** Holy Roman Emperor cr 1564 - d 1576
- **CHARLES of Styria**

Children of Maximilian II:
- **ALBERT** Governor of Netherlands r 1596 - d 1621 m Isabella Clare Eugenia
- **ANNE of Austria** m/iv Philip II of Spain
- **RUDOLF III** Holy Roman Emperor cr 1576 - 1602 d 1612
- **ERNEST** Governor of Netherlands r 1594 - 95
- **ELIZABETH** m Charles IX of France b 1560 - d 1574
- **MATTHIAS** Holy Roman Emperor cr 1612 - d 1619

LOUIS — — — **Cardinal HENRY of Portugal** d 1580

KEY
b = born
d = died
k = killed
ex = executed
mur = murdered
adpt = adopted
cr = crowned
abd = abdicated
kt = knight
KG = Knight of the Garter
m = married
  m1 = 1st marriage
  m2 = 2nd marriage, etc.
  m/i = 1st marriage of spouse
  m/ii = 2nd marriage of spouse
(1560) = date of marriage
div = divorced or annulled
dau = daughter of
btr = betrothed to
r = ruled from
nei = neice of

249

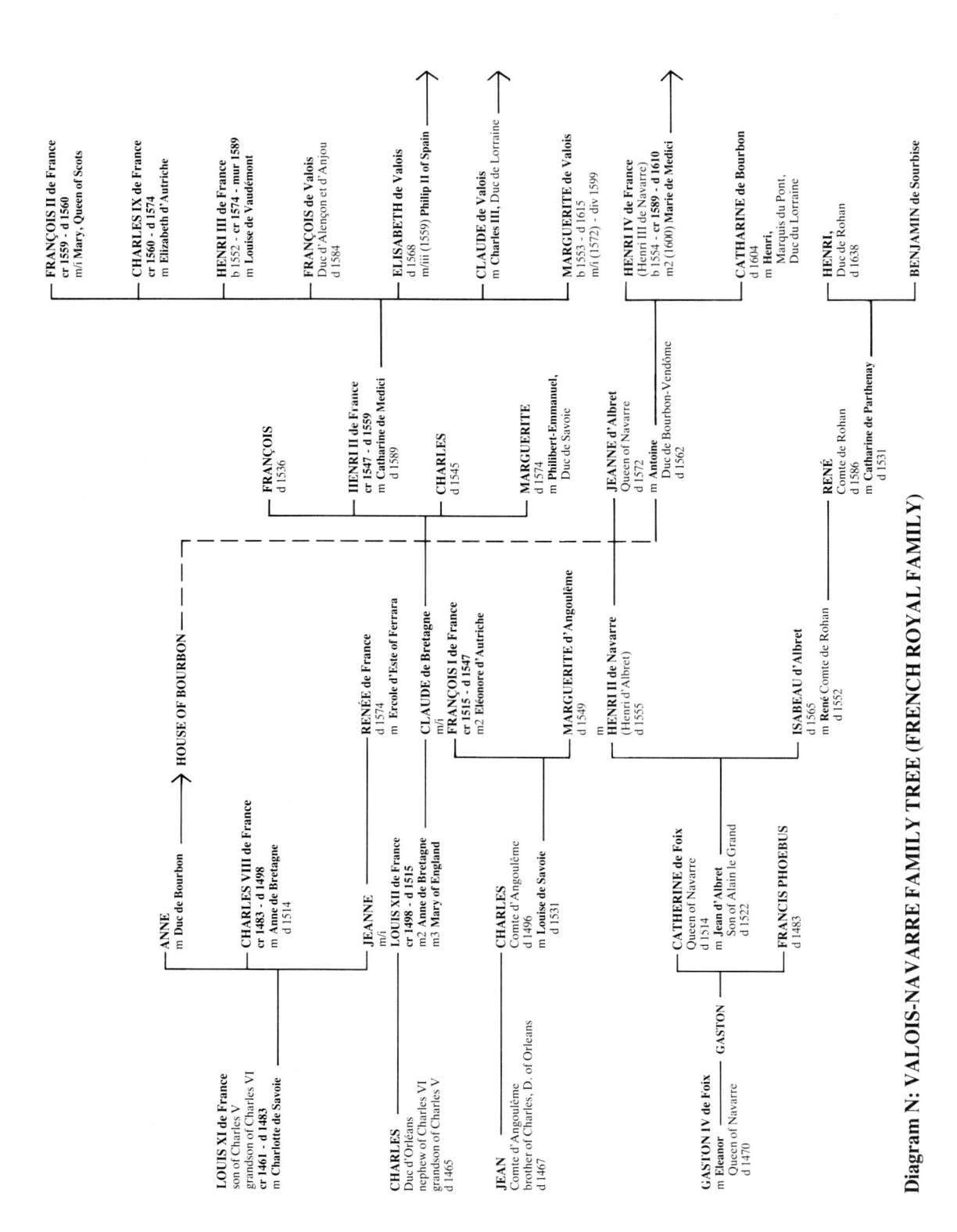

Diagram N: VALOIS-NAVARRE FAMILY TREE (FRENCH ROYAL FAMILY)

LOUIS XI de France
son of Charles V
grandson of Charles VI
cr 1461 - d 1483
m **Charlotte de Savoie**
d 1465

├── **ANNE**
│   m **Duc de Bourbon** → HOUSE OF BOURBON
│
├── **CHARLES VIII de France**
│   cr 1483 - d 1498
│   m **Anne de Bretagne**
│   d 1514
│
└── **JEANNE**
    m/i

CHARLES
Duc d'Orléans
nephew of Charles VI
grandson of Charles V

- **LOUIS XII de France**
  cr 1498 - d 1515
  m² **Anne de Bretagne**
  m³ **Mary of England**
  - **RENÉE de France**
    d 1574
    m **Ercole d'Este of Ferrara**
  - **CLAUDE de Bretagne**
    m/i
    - **FRANÇOIS I de France**
      cr 1515 - d 1547
      m² **Eléonore d'Autriche**
      - **FRANÇOIS**
        d 1536
      - **HENRI II de France**
        cr 1547 - d 1559
        m **Catharine de Medici**
        d 1589
        - **FRANÇOIS II de France**
          cr 1559 - d 1560
          m/i **Mary, Queen of Scots**
        - **CHARLES IX de France**
          cr 1560 - d 1574
          m **Elizabeth d'Autriche**
        - **HENRI III de France**
          b 1552 - cr 1574 - mur 1589
          m **Louise de Vaudémont**
        - **FRANÇOIS de Valois**
          Duc d'Alençon et d'Anjou
          d 1584
        - **ELISABETH de Valois**
          d 1568
          m/iii (1559) **Philip II of Spain** ↑
        - **CLAUDE de Valois**
          m **Charles III, Duc de Lorraine** ↑
        - **MARGUERITE de Valois**
          b 1553 - d 1615
          m/i (1572) **HENRI IV de France**
          (Henri III de Navarre)
          b 1554 - cr 1589 - d 1610
          m² (1600) **Marie de Medici**
          div 1599
      - **CHARLES**
        d 1545

JEAN
Comte d'Angoulême
brother of Charles, D. of Orleans
d 1467

CHARLES
Comte d'Angoulême
d 1496
m **Louise de Savoie**
d 1531

- **MARGUERITE d'Angoulême**
  d 1549
  m
  - **HENRI II de Navarre**
    (Henri d'Albret)
    d 1555
    - **JEANNE d'Albret**
      Queen of Navarre
      d 1572
      m **Antoine**
      Duc de Bourbon-Vendôme
      d 1562
      → **HENRI IV de France** (see above)
      → **CATHARINE de Bourbon**
         d 1604
         m **Henri**, Marquis du Pont, Duc du Lorraine

GASTON IV de Foix ── **GASTON**
m **Eleanor**
Queen of Navarre
d 1470

- **CATHERINE de Foix**
  Queen of Navarre
  d 1514
  m **Jean d'Albret**
  Son of Alain le Grand
  d 1522
  - **FRANCIS PHOEBUS**
    d 1483
  - **ISABEAU d'Albret**
    d 1565
    m **René Comte de Rohan**
    d 1552
    - **RENÉ**
      Comte de Rohan
      d 1586
      m **Catharine de Parthenay**
      d 1531
      - **HENRI**, Duc de Rohan
        d 1638
      - **BENJAMIN de Sourbise**

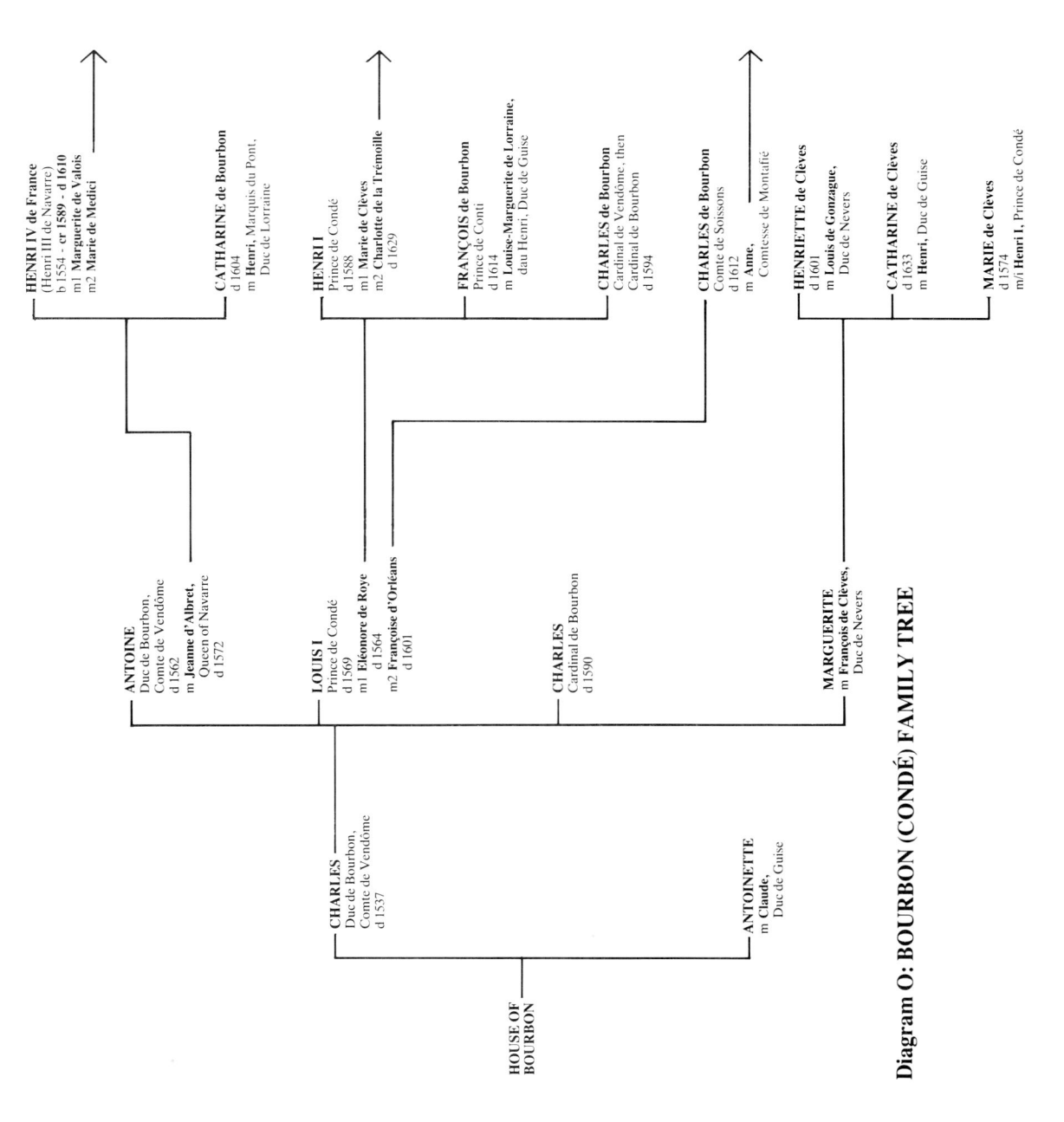

HENRI IV de France
(Henri III de Navarre)
b 1554 - cr 1589 - d 1610
m1 **Marguerite de Valois**
m2 **Marie de Medici**
↑

CATHARINE de Bourbon
d 1604
m **Henri**, Marquis du Pont, Duc de Lorraine

HENRI I
Prince de Condé
d 1588
m1 **Marie de Clèves**
m2 **Charlotte de la Trémoille**
d 1629
↑

FRANÇOIS de Bourbon
Prince de Conti
d 1614
m **Louise-Marguerite de Lorraine**, dau Henri, Duc de Guise

CHARLES de Bourbon
Cardinal de Vendôme, then Cardinal de Bourbon
d 1594

CHARLES de Bourbon
Comte de Soissons
d 1612
m **Anne**, Comtesse de Montafié
↑

HENRIETTE de Clèves
d 1601
m **Louis de Gonzague**, Duc de Nevers

CATHARINE de Clèves
d 1633
m **Henri**, Duc de Guise

MARIE de Clèves
d 1574
m/i **Henri I**, Prince de Condé

ANTOINE
Duc de Bourbon,
Comte de Vendôme
d 1562
m **Jeanne d'Albret**, Queen of Navarre
d 1572

LOUIS I
Prince de Condé
d 1569
m1 **Eléonore de Roye**
d 1564
m2 **Françoise d'Orléans**
d 1601

CHARLES
Cardinal de Bourbon
d 1590

MARGUERITE
m **François de Clèves**, Duc de Nevers

CHARLES
Duc de Bourbon,
Comte de Vendôme
d 1537

ANTOINETTE
m **Claude**, Duc de Guise

HOUSE OF BOURBON

Diagram O: BOURBON (CONDÉ) FAMILY TREE

Diagram P: GUISE-LORRAINE FAMILY TREE

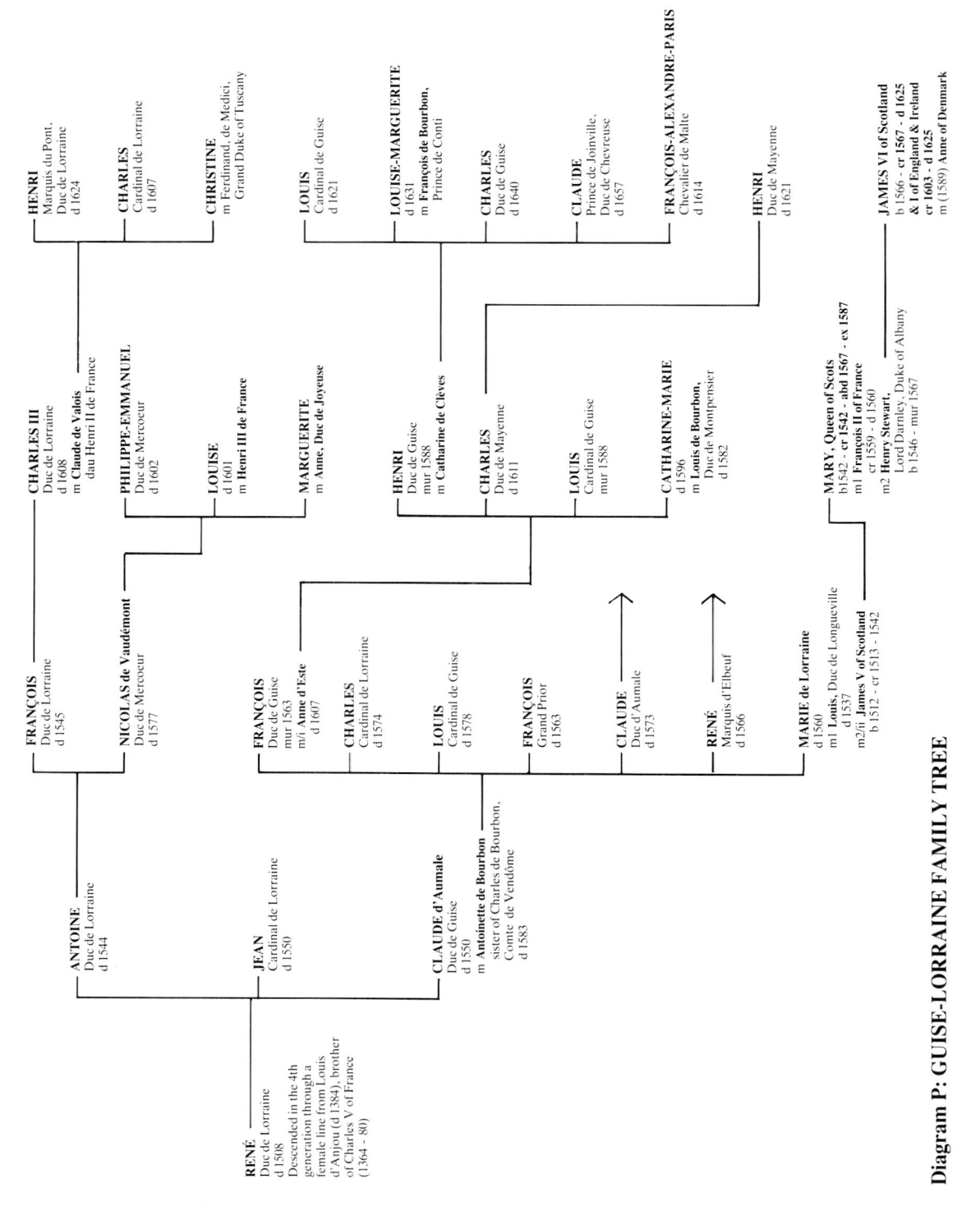

RENÉ
Duc de Lorraine
d 1508
Descended in the 4th generation through a female line from Louis d'Anjou (d 1384), brother of Charles V of France (1364 - 81)

- **ANTOINE** Duc de Lorraine d 1544
  - **FRANÇOIS** Duc de Lorraine d 1545
    - **CHARLES III** Duc de Lorraine d 1608 m **Claude de Valois** dau Henri II de France
      - **HENRI** Marquis du Pont, Duc de Lorraine d 1624
      - **CHARLES** Cardinal de Lorraine d 1607
      - **CHRISTINE** m Ferdinand, de Medici, Grand Duke of Tuscany
  - **NICOLAS de Vaudémont** Duc de Mercoeur d 1577
    - **PHILIPPE-EMMANUEL** Duc de Mercoeur d 1602
      - **LOUISE** d 1601 m **Henri III de France**
      - **MARGUERITE** m **Anne, Duc de Joyeuse**

- **JEAN** Cardinal de Lorraine d 1550

- **CLAUDE d'AUMALE** Duc de Guise d 1550 m **Antoinette de Bourbon** sister of Charles of Bourbon, Comte de Vendôme d 1583
  - **FRANÇOIS** Duc de Guise mur 1563 m/i **Anne d'Este** d 1607
    - **HENRI** Duc de Guise mur 1588 m **Catharine de Clèves**
      - **LOUIS** Cardinal de Guise d 1621
      - **LOUISE-MARGUERITE** d 1631 m **François de Bourbon, Prince de Conti**
      - **CHARLES** Duc de Guise d 1640
      - **CLAUDE** Prince de Joinville, Duc de Chevreuse d 1657
      - **FRANÇOIS-ALEXANDRE-PARIS** Chevalier de Malte d 1614
    - **CHARLES** Duc de Mayenne d 1611
      - **HENRI** Duc de Mayenne d 1621
    - **LOUIS** Cardinal de Guise mur 1588
    - **CATHARINE-MARIE** d 1596 m **Louis de Bourbon, Duc de Montpensier** d 1582
  - **CHARLES** Cardinal de Lorraine d 1574
  - **LOUIS** Cardinal de Guise d 1578
  - **FRANÇOIS** Grand Prior d 1563
  - **CLAUDE** Duc d'Aumale d 1573
  - **RENÉ** Marquis d'Elbeuf d 1566
  - **MARIE de Lorraine** d 1560 m1 **Louis, Duc de Longueville** d 1537 m2/ii **James V of Scotland** b 1512 - cr 1513 - 1542
    - **MARY, Queen of Scots** b1542 - cr 1542 - abd 1567 - ex 1587 m1 **François II of France** cr 1559 - d 1560 m2 **Henry Stewart, Lord Darnley, Duke of Albany** b 1546 - mur 1567
      - **JAMES VI of Scotland** b 1566 - cr 1567 - d 1625 & I of England & Ireland cr 1603 - d 1625 m (1589) **Anne of Denmark**

APPENDIX

Appendix

KENILWORTH FESTIVITIES (JULY 1575): LANEHAM'S LETTER

The following essay concerning Laneham's Letter describing the Kenilworth Festivities is by Thomas D. Bokenham.

In July 1575, took place the great festivities at Kenilworth which the Earl of Leicester organised for the Queen. These were designed to woo Her Majesty into acknowledging him as her Consort, and were devised on a lavish scale. This magnificent pageant was reported in two accounts, *The Princes Pleasures of Kenilworth* attributed to George Gascoigne, and a companion piece known as *Laneham's Letter* which many Baconians attribute to young Francis Bacon, then aged fourteen. Nothing is known of any writer of this name which is an interesting one, because an old word "laine" meant "concealment": a concealed *ham* or Bacon. The letter itself purports to have been written by a City merchant, staying at the castle, to another merchant in London. It was studied in detail by E. G. Harman, a Baconian scholar, in his book *Edmund Spenser and the Impersonations of Francis Bacon* (1914). He argued correctly that this letter "was not written by a man, but reflects the impressions of a sensitive and delighted spirit; in short, of a young poet on his entry, freed for the first time from academic pupilage, into the great world". That poet was Francis Bacon, who left Cambridge that very year. Much of this letter is quoted by Harman and it shows the youthful exuberance and delight which its author took in everything he saw and did on this colourful occasion. It also shows the great powers of observation and attention to detail which had developed in this remarkable young man. Harman never accepted the fact of Francis' royal birth, nor did he explain how it was that this young student from the University was able to effect an invitation to these festivities; but the Queen and Leicester, who both delighted in these shows, knew where their son's tastes lay and it is almost certain that Francis himself played a part in the activities. The letter is amusing and quite informative and I must include some excerpts for your entertainment.

> "For about nine o'clock, at the hither part of the Chase, where torchlight attended, out of the woods on her Majesty's return, roughly came there forth Hombre Salvagio with an Oken plant plucked up by the roots in his hands, himself forgrone [that is, overgrown] all in moss and ivy: who for personage, gesture and utterance besides, counter mandst the matter to very good liking: and had speech to effect: 'That continuing so in these wild wastes, wherein oft he had fared both far and near, yet hapt he never to see so glorious an assembly afore: and now cast into great grief of mind, for that neither by himself could he guess, nor knew where else to be taught, what they should be or who bare estate. Reports some had he heard, of many strange things but broyled thereby

so much the more in desire of knowledge. Thus in great pangs bethought he and called he upon all his familiars and companions, the Fawns, the Satyres, the Nymphs, the Dryads and the Hamadryads; but none making answer, whereby his care the more increasing, in utter grief and extreme refuge, called he aloud at last after his old friend Echo that he wist would hide nothing from him, but tell him all if she were here."

This letter, I should add, is written with atrocious spelling designed, no doubt, to convey the untutored inelegance of the erstwhile merchant author. Now follows a description of an Elizabethan acrobat who must have produced roars of applause from the crowd of spectators:

"Now within also in the meantime was there shewed before her highness, by an Italian, such feats of agility, in goings, turnings, tumblings, castings, hops, jumps, leaps, skips, springs, gambols, somersaults, capretties and flights; forward, backward, sidewise, adownward, upward, and with sundry windings, gyring and circumflexions; all so lightly and with such easiness as by me in few words it is not expressible by pen or speech, I tell you plain. It blessed me, by my faith, to behold him and I began to doubt whether he was a man or a spirit; and I ween had doubted me till this day of men that can reason and talk with two tongues, and with two persons at once, sing like birds, courteous of behaviour, of body strong and in joints so nimble withal, that their bones seem as lythe and pliant as sinews. They dwell in a happy Island, as the book terrms it, four month's sailing southward beyond Ethiop. Nay, Master Martin, I tell you no jest, for both Diadorus Siculus, an ancient Greek historiographer, in his third book of the acts of the old Egyptians, and also from him, Conrad Gesnerus, a great learned man and a very diligent writer in all good arguements of our time, but deceased, in the first chapter of his *Mithridates* reporteth the same. As for this fellow, I cannot tell what to make of him, save that I may guess his back be metalled like a lamprey, that has no bone but a line like a lute to a string."

Harman then quotes the colourful description of the ancient minstrel he thought might well have been an imaginary self-portrait of Francis himself; indeed, to me, it is probable that he actually impersonated this character. The impersonator, we are told, was a lord of 14 years. It is too long to quote here, but it is of great interest because it includes a reference to the escutcheon worn on his breast "of the ancient arms of Islington". "As one that was well schooled and could his lesson parfit without book to answer at full, upon question, the minstrel answered, 'how the worshipful village of Islington in Middlesex, well known to be one of the most ancient and best towns in England next London at this day'." He goes on to talk of its produce, its milk, cream, butter, *etc.*, which were the best in the world. It is known that this place was loved not only by Francis but by the Queen and Leicester, where, it is said, they spent some of their childhood together; and I suggest that its inclusion here was a ploy to remind the Queen of her former childhood with the man who sought her recognition.

Finally, one further quotation must be given which I am convinced concerns Francis' activities while at Kenilworth:

"In afternoons and nights, sometime am I with the right worshipful Sir George Howard, as good a gentleman as any lives, and sometime at my good Lady Sidnei's chamber, a noblewoman that I am as much bound to as any poor man may be to so gracious a lady [she was Mary, the Earl of Leicester's sister and mother of

Philip] and sometime in some other place [perhaps with his parents]. But always among the gentlewomen by my own good will (O, ye know that come always of a gentle spirit) and when I see company according, then can I be as lively too. Sometime I foot it with dancing, now with my Gittern and else with my cittern, then at virginals: Ye know nothing comes amiss to me; then caroll I up a song withal that by and by they come flocking about me like bees to honey, and ever cry 'another, good Laneham, another!' Shall I tell you? when I see Mistress — (Ah see a mad knave, I had almost told all) that she gives once but an eye or an ear, why then man, am I blessed! My grace, my courage, my cunning is doubled; she says sometime she likes it and then I like it much the better; it doth me good to hear how well I can do. And to say truth, what with mine eyes, as I can amorously gloit it, with my Spanish sospires, my French heighs, mine Italian dulcets, my Dutch houes, my double release, my high reaches, my fine feigning, my deep diapason, my wanton warbles, my running, my timing, my tuning and my twinkling, I can gracify the matters as well as the proudest of them, and was yet never stayed, I thank God. By my troth, countryman, it is sometimes by midnight ere I can get from them. And thus have I told ye most of my trade, all the live long day; what will ye more? God save the Queen and my Lord! I am well, I thank you".

This letter, dated "at the City of Worcester, the xx of August, 1575," ends in this way:

"Your countryman, companion and friend assuredly; mercer, merchant adventurer, the Clerk of the Council-chamber door and also keeper of the same, *El Principe negro.* par me, R.L. Gent. Mercer."

If the words *"El Principe negro"* (Spanish) were intended as a veiled reference to the Black Prince, it should be remembered that that gentleman was the Prince of Wales and the word "black" could also relate to one tarnished with a stain on his escutcheon — that is, the proverbial *'bend sinister.'*

PIERRE AMBOISE'S "LIFE" OF SIR FRANCIS BACON

In 1631 there was published in Paris, by the firm of Antoine de Sommaville and André de Soubron, a book entitled *Histoire Naturelle de Mre. Francoise Bacon, Baron de Verulan, Vicomte de Sainct Alban et Chancelier d' Angleterre.* This was an unique publication of Sir Francis Bacon's *Natural History*, in French. The translator gives his name as "Pierre Amboise" and tells us, in the 'Address to the Reader', that he had been aided for the most part in his translation by the author's manuscripts; but how he obtained those manuscripts is not explained. It is not a direct translation of *Sylva Sylvarum*, published by William Rawley (Bacon's chaplain) in 1626, as one might have expected, but a translation compiled from the original manuscripts and placed in an order different to that used by Rawley, in a manner explained by Amboise in his 'Address'. The license to print was issued to "Pierre Amboise, Escuyer, sieur de la Magdelaine", with a statement that he had translated into French a book entitled the "Natural History of Mr. Francis Bacon, Chancellor of England , with some letters of the same author; together with the life of the said Mr. Bacon, prepared by the same applicant", which he desires to bring into light. There are, in fact, no letters of Bacon's included in the publication, but there is the 'Life' and, at the end of the book, after the 'Natural History,' a translation of *New Atlantis* is included. The book is dedicated by

"D.M." to the Monseigneur de Chasteauneuf, who was Ambassador Extraordinary to England from France in 1629 and 1630, and published "Avec Privilege du Roy". It is not known who "D.M." was.

Amboise's 'Life' was the first biography of Francis Bacon to be published openly, and gives many more details about Francis' extraordinary life than Rawley does in his very unsatisfactory 'Life' published with the *Resuscitatio* of 1657. In the 1640 edition of *The Advancement of Learning*, Gilbert Wats, the translator of the *Advancement of Learning* (from Latin into English), writes of Amboise's translation of Bacon's *Natural History* and quotes from the 'Life' of Bacon by Amboise, referring to Amboise's "just and elegant discourse upon the life of our Author". But it was not until the ninth and final edition of *Sylva Sylvarum* was published in 1670 that Amboise's 'Life' of Bacon was actually allowed to appear in print in the English edition of Bacon's *Natural History*.

ADDRESS TO THE READER

This work of Mr. Bacon's though posthumous, does not the less deserve to be recognized as legitimate, since it has the same advantages as those that have been brought to light whilst he was living. If the Author had had the desire to see it there, we should have seen this work in the press at the same time as his other books, but having designed that it should grow more, he had intended to defer the printing until the completion of all his works. This is a Natural History where the qualities of metals and minerals, the nature of the elements, the causes of generation and decay, the different actions of bodies upon each other, are treated with so much brilliancy, that he seems to have learned the science at the school of the first man. Of a truth, if in this he has rivalled Aristotle, Pliny, and Cardan, he has nevertheless borrowed nothing from them, as though he had intended to make it plain that these great men have not treated the subject so fully, but that there still remain many things to be said, For my part, though I have no intention of establishing the reputation of this Author at the expense of Antiquity, I think I may always truly say that in this subject he has had a certain advantage over them; since the greater number of the Ancients, who have written upon nature, have been content to retail to us that which they have learnt from others: and, without reflecting that very often that which has been given to them as a true description is very far from the truth, they have preferred to bolster up with their arguments the tales of others rather than themselves to make original research. But Mr. Bacon, instead of stopping at the same boundaries as those who have preceded him, will have Experience joined with Reason. And to effect this he had a country house somewhat close to London, which he retained only in order to carry on his Experiments. In this place he had an infinite number of vases and phials; some of which were filled with distilled waters, others with plants and metals in their native state; some with mixtures and compounds; and leaving them exposed to the air throughout all seasons of the year, he observed carefully the different effects of cold or of heat, of dryness and of moisture, the simple productions and corruptions, and other effects of nature. It is in this way that he has found out so many rare secrets, the discovery of which he has left to us; and that he has exposed as false so many axioms that until now have been held as inviolable among the Philosophers. If, in order to make the meaning clear, I have used in this translation many words more Latin than French, the Reader should lay the blame chiefly upon the sterility of our language, which is so defective that many things often remain unexpressed unless we have recourse to foreign languages.

I shall be pleased also if the Reader will take notice that in this translation I have not exactly followed the order observed in the original English, for I have found so much confusion in the disposition of the matter that it seemed to have been broken up and dispersed rather by caprice than by reason. Besides, having been aided for the most part by the Manuscripts of the Author, I have deemed it necessary to add to, or to take from, many of the things that have been omitted or augmented by the Chaplain of Mr. Bacon, who after the death of his Master, printed in a confused manner all the papers that he found in his cabinet.

I say this, so that those who understand English will not accuse me of inaccuracy, when they encounter in my translation many things that they do not find in the original.

DISCOURSE ON THE LIFE OF M. FRANCIS BACON CHANCELLOR OF ENGLAND

Those who have known the quality of M. Bacon's mind from reading his works, will — in my opinion — be desirous to learn who he was, and to know that Fortune did not forget to recompense merit so rare and extraordinary as was his. It is true, however, that she was less gracious to his latter age than to his youth; for his life had such happy beginning and an end so rough and strange, that one is astonished to see England's principal Minister of State, a man great both in birth and in possessions, reduced actually to the verge of lacking the necessaries of life.

I have difficulty in coinciding with the opinion of the common people, who think that great men are unable to beget children similar to themselves, as though nature was in that particular inferior to the art which can easily produce portraits that are likenesses: especially as history teaches us that the greatest personages have often found in their own families heirs of their virtues as well as of their possessions. And indeed, without the need of going to search for far away examples, we see that M. Bacon was the son of a father who possessed no less virtue than he: his worth secured to him the honour of being so well-beloved by Queen Elizabeth that she gave him the position of Keeper of the Seals, and placed in his hands the most important affairs of her Kingdom. And in truth it pains me to say that soon after his promotion to the first-named dignity, he was the principal instrument that she made use of in order to establish the Protestant Religion in England.

Although that work ws so odious in its nature, yet if one considers it according to political maxims, we can easily see that it was one of the greatest and boldest undertakings that had been carried out for many centuries: and one ought not the less to admire the Author of it, in that he had known how to conduct a bad business so dexterously, as to change both the form of Religion, and the belief, of an entire Country, without having disturbed its tranquility. M. Bacon was not only obliged to imitate the virtues of such an one, but also those of many others of his ancestors, who have left so many marks of their greatness in history that honour and dignity seem to have been at all times the spoil of his family. Certain it is that no one can reproach him with having added less than they to the splendour of his race. Being thus born in the purple [*ne parmy les pourpres*] and brought up with the expectation of a great career [*l'esperance d'une grande fortune*], his father had him instructed in '*bonnes lettres*' with such great and such especial care, that I know not to whom we are the more indebted for all the splendid works [*les beaux ouvrages*] that he has left to us: whether to the mind of the son, or to the care the father had taken in making him cultivate it. But, however that may be, the obligation we are under to the father is not small. Capacity [*jugement*] and memory were never in any man to such a degree as in this one: so that in a very short time he made himself conversant with all the knowledge he could acquire at College. And though he was then considered capable of undertaking the most important affairs [*capable des charges les plus importants*] yet, so that he should not fall into the usual fault of young men of his kind (who by a too hasty ambition often bring to the management of great affairs, a mind still full of the crudities of the school), M. Bacon himself wished to acquire that knowledge which in former times made Ulysses so commendable, and earned for him the name of Wise; by the study of the manners of many different nations. I wish to state that he employed some years of his youth in travel, in order to polish his mind and to mould his opinion by intercourse with all kinds of foreigners. France, Italy and Spain, as the most civilised nations of the whole world, were those whither his desire for knowledge [*curiosité*] carried him. And as he saw himself destined one day to hold in his hands the helm of the Kingdom [*le timon du Royaume*] instead of looking only at the people and the different fashions in dress, as do the most of those who travel, he observed judiciously the laws and the customs of the countries through which he passed, noted the different forms of Government in a State, with their advantages or defects, together with all the other matters which might help to make a man able for the government of men.

Having by these means reached the summit of learning and virtue, it was fitting that he should also reach that of dignity. For this reason, some time after his return, the King, who well knew his worth, gave him several small matters to carry out, that might serve for him as stepping-stones to high positions: in these he acquitted himself so well that he was in due course considered worthy of the same position that his father vacated with his life. And in carrying out the work of Chancellor he gave so many proofs of the largeness of his mind, that one can say without flattery that England owes to his wise counsels, and his good rule, a part of the repose she has so long enjoyed. And King James, who then reigned, should not take to himself alone all the glory of this, for it is certain that M. Bacon should share it with him. We may truly say that this Monarch was one of the greatest Princes of his time, who understood thoroughly well the worth and value of men, and he made use to the fullest extent of M. Bacon's services, and relied upon his vigilance to support the greater part of the burden of the Crown. The Chancellor never proposed anything for the good of the State, or the maintenance of justice, but was carried out by the Royal power; and the authority of the Master seconded the good intentions of the servant; so that one must avouch that this Prince was worthy to have such a Minister, and he worthy of so great a King.

Among so many virtues that made this great man commendable Prudence, as the first of all the Moral virtues, and that most necessary to those of his profession, was that which shone in him the most brightly. His profound wisdom can be most readily seen in his books, and his matchless fidelity in the signal services that he continuously rendered to his Prince. Never was there man who so loved equity, or so enthusiastically worked for the public good as he: so that I may aver that he would have been much better suited to a Republic than to a

Monarchy, where frequently the convenience of the Prince is more thought of, than that of his people. And I do not doubt that, had he lived in a Republic, he would have acquired as much glory from the citizens, as formerly did Aristides and Cato, the one in Athens, the other in Rome. Innocence oppressed found always in his protection a sure refuge, and the position of the great gave them no vantage ground before the Chancellor, when suing for justice.

Vanity, avarice, and ambition, vices that too often attach themselves to great honours, were to him quite unknown, and if he did a good action, it was not from the desire of fame, but simply because he could not do otherwise. His good qualities were entirely pure, without being clouded by the admixture of any imperfections; and the passions that form usually the defects in great men, in him only served to bring out his virtues; if he felt hatred and rage it was only against evil doers, to show his detestation of their crimes; and success or failure in the affairs of his country, brought to him the greater part of his joys or his sorrows. He was as truly a good man, as he was an upright judge, and by the example of his life, corrected vice and bad living, as much as by pains and penalties. And in a word, it seemed that Nature had exempted from the ordinary frailties of men him whom she had marked out to deal with their crimes. All these good qualities made him the darling of the people, and prized by the great ones of the State. But when it seemed that nothing could destroy his position, Fortune made clear that she did not yet wish to abandon her character for instability, and that Bacon had too much worth to remain so long prosperous. It thus came about that amongst the great number of officials such as a man of his position must have in his house, there was one who was accused before Parliament of exaction, and having sold the influence that he might have with his master. And though the probity of M. Bacon was entirely exempt from censure, nevertheless he was declared guilty of the crime of his servant, and was deprived of the power that he had so long exercised with so much honour and glory. In this I see the working of monstrous ingratitude and unparalleled cruelty; to say that a man who could mark the years of his life, rather by the signal services that he had rendered to the State, than by times or seasons, should have received such hard usage, for the punishment of a crime which he never committed; England, indeed, teaches us by this that the sea, that surrounds her shores, imparts to her inhabitants somewhat of its restless inconstancy. This storm did not at all surprise him, and he received the news of his disgrace with a countenance so undisturbed that it was easy to see that he thought but little of the sweets of life, since the loss of them caused him discomfort so slight. He had, fairly close to London, a country house replete with everything requisite to soothe a mind embittered by public life, as was his, and weary of living in the turmoil of the great world. He returned thither to give himself up more completely to the study of his books, and to pass in repose, the remainder of his life. But as he seemed to have been born rather for the rest of mankind than for himself, and as by the want of public employment he could not give his work to the people, he wished at least to render himself of use by his writings and by his books; worthy as these are to be in all the libraries of the world, and to rank with the most splendid works of antiquity.

The history of Henry VII is one of those works which we owe to his fall, a work so well received by the whole world, that one has wished for nothing so much as the continuation of the History of the other Kings. And even yet he would not have given opportunity for these regrets, had not death cut short his plans, and thus robbed us of a work that bid fair to put all the others to shame.

The Natural History is also one of the fruits of his idleness. The praiseworthy wish that he had, to pass by nothing but to connote the nature and qualities of all things, induced his mind to make researches which some learned men may perhaps have indicated to him, but which none but himself could properly carry out. In which he has without doubt achieved so great a success, that but little has escaped his knowledge, so that he has laid bare to us the errors of the ancient Philosophy and made us see the abuses that have crept into that teaching, under the authority of the first authors of the science. But whilst he was occupied in this great work, want of means forced him to concentrate his mind on his domestic affairs. The honest manner in which he had lived was the sole cause of his poverty; and as he was ever more desirous of acquiring honour than of amassing a fortune, he had always preferred the interests of the State to those of his house; and had neglected, during the time of his great prosperity, the opportunities of enriching himself: So that after some years passed in solitude he found himself reduced to such dire necessity that he was constrained to have recourse to the King, to obtain, by his liberality, some alleviation of his misery. I know not if poverty be the mother of beauty, but I aver that the letter he wrote to the King on that occasion is one of the most beautiful examples of that style of writing ever seen. The request that he made for a pension is conceived in terms so lofty and in such good taste, that one could not deny him without great injustice. Having thus obtained the means to extricate himself from his difficulties, he again applied himself, as before, to unravel the great secrets of nature. And as he was engaged during a severe frost in observing some particular effects of cold, having stayed too long in the open, and forgetting that his age made him incapable of bearing such severities; the cold, acting the more easily on a body whose powers were already reduced by old age, drove out all that remained of natural heat, and reduced him to the last condition that is always reached by great men only too soon. Nature failed him while he was chanting her praise: this she did, perhaps, because, being miserly and hiding from us her best, she feared that at last he would discover all her treasures, and make all men learned at her expense. Thus ended this great man, whom England could place alone as the equal [*en paralelle avec*] of the best of all the previous centuries.

HUMANISM

Humanism, from Latin *humanus*, "human", is a devotion to those studies which promote human culture and learning, asserting the intrinsic value of man's life and the greatness of his potentialities, distinguishing the human from the bestial and other forms of nature on the one hand, and from the divine on the other. It was the most characteristic attitude of the Renaissance in western Europe, and was inherited from the Ancient Mystery Schools, particularly those of Rome, Greece and Ancient Egypt. The essential doctrine from which Humanism is derived — and which is common to all the major Wisdom Traditions the world over — are the teachings of Hermes Trismegistus (the Greek name for the Word of God, both purely spiritual and also embodied and revealed in perfect human beings). This line of 'Hermetic' teaching descended *via* Ancient Egypt from the great 'Immortals' of antiquity, then *via* Orphism, the Mosaic Tradition of the Hebrews (now embodied in Judaism) and Druidism, to become enshrined in the 'inner' teachings and revelation of Christianity. The Egyptians called the Teacher and teachings, the Thrice-Greatest Thoth; the Druids called Him/them Merlin; and the Hebrews referred back to the Atlantean Enoch as the great Teacher on earth, from whom our sacred Tradition is derived.

A fundamental maxim of the teachings is that there are Three Principal Ideas or "Heads" which derive from the One Single and Absolute Idea of Truth: namely, God, Man and Cosmos.

> "First, God; second, the Cosmos; third, Man. The Cosmos for Man's sake; and Man for God's."[354]

> "From One Source all things depend; but the Source is from the One and Only. Three then are they: God the Father and the Good, the Cosmos, and Man. God doth contain the Cosmos; the Cosmos containeth Man. The Cosmos is the offspring of God; and Man, as it were, is the offspring of the Cosmos."[355]

In addition, these three principal Ideas of the One Idea are three-fold, God creating and the other two (Cosmos and Man) having Nous (Spirit), Soul and Body. But even though each of the Three Heads have Spirit, Soul and Body, yet at the same time the Three Heads can themselves be seen as being the principles of Spirit, Soul and Body: *viz.* God the Father is the Spirit or creative principle; Man is the Soul or Mind that is capable of thought and hence of being illumined by the Spirit (*Man* means "Thinker"); and the Cosmos is the Body of God.

> "The Creator, not with His Hands, but with His Word, made the Cosmos. So thou shouldst think of Him as everywhere, and ever-being, the Author of all things: One, and Only, Who by His Will framed the things that are. For this [the Cosmos] is the Body of Him, not tangible, nor visible, nor measurable, nor extensible, nor like any other body; nor fire, nor water, nor air, nor wind; but all these things are from Him, for, being the Good, He willed to consecrate this Body to Himself alone, and adorn it.

> "As an ornament of the divine Body, He sent down Man — a life that dies and yet cannot die. And over all other lives, and over the Cosmos, did Man excel, because of his Reason and the Nous. For Man became the spectator of the works of God, and marvelled, and did strive to know their Author."[356]

> "God is not ignorant of Man; nay, right well doth He know him, and willeth to be known. This is the sole salvation of Man — **the knowledge of God**. This is the way up to the Mount Olympus. By this alone the Soul becometh good: not sometimes good and sometimes evil, but good of necessity."[357]

In this last statement is summed up the purpose of Man and the meaning of Humanism. As the purpose of Man is to know God, and, in knowing God, to become all-good, then similarly the purpose of Humanism is to know God, or Good, and to practice it.

Everything that is cultural, humane, tending towards a development of man's love, understanding and his willingness and ability to serve others with gentleness and kindness, and with wisdom, is part of Humanism. Thus Humanism is another word for what was called, in the past, the Initiatory Path of mankind, and which was studied and taught in the Mystery Schools.

Man is able to come to know God either (a) by inspiration and revelation direct from God, or (b) by the careful observation of Nature (*i.e.* the Cosmos) and the discovery of its underlying laws. Francis Bacon refers to the former as Divinity, and the latter as Philosophy; the one referring to God's Word, and the latter to God's Works. Divinity is direct enlightenment by God's Word or Wisdom, the Light or Consciousness of Love; whereas Philosophy is the intelligent love of Wisdom, which develops the human understanding or comprehension of Wisdom. Both are necessary, and both have to be put into practice by man in order to fulfil Truth and really KNOW God.

The Humanism of the Renaissance, modelled to a large extent upon the teachings of earlier Academies of neo-Platonists, and of Plato himself, tended to be somewhat over-intellectual, leaving to the side the more intuitive and emotional nature of man (which is capable of direct inspiration and revelation *via* the heart), and to a large extent ignoring the actual practice of the axioms discovered or supposed to exist. Fundamentally, Christianity, which was born into this Humanistic environment, set out not only to show the living example of the perfect human being (*Hu-man* = "the man of Light", or "the illumined man", also known as Jesus) but also the *life* of that perfect man — not just what he thought and knew, but also what he did. Christianity set out to teach the original Wisdom Teachings, that all good things concerning man begin with **Faith** (which is fundamentally unshakeable trust arising from a complete and fearless love for God, or Truth), develop into **Hope** (which is our vision and understanding of Truth), and are fulfilled in **Charity** (which is putting our love and understanding of Truth into practice); and that of these three Charity is supreme, because it is the final goal of our becoming "like unto God". Francis Bacon set out to re-affirm and re-establish this fundamental truth in the world.

These three — Faith (or Love), Hope (or Understanding), and Charity (or Service) — have an analogy with the Three Heads (God the Creator, Man the Thinker, and the Cosmos of God's Works), as also with the three aspects of Spirit, Soul and Body. Truth lies hidden, and yet revealed, in all this — and Man is capable of eventually knowing it and expressing it perfectly. This is the goal of Humanism, of Christianity, and of the Great Instauration.

THE PLÉIADE

The following essay on the French Pléiade of the sixteenth century is written by William T. Smedley and was first published in Baconiana *Vol IX, Third Series, No.34. It is reproduced here by kind permission.*

The French Renaissance of literature had its beginning in the early years of the sixteenth century. It had been preceded by that of Italy, which opened in the fourteenth century, and reached its limit with Ariosto and Tasso, Macchiavelli and Guicciardini during the sixteenth century. Towards the end of the fifteenth century modern French poetry may be said to have had its origin in Villon and French prose in Comines. The style of the former was artificial and his poems abounded in recurrent rhymes and refrains. The latter had peculiarities of diction which were only compensated by the weight of thought and simplicity of expression. Clement Marot, who followed, stands out as one of the first landmarks in the French Renaissance. His graceful style, free from stiffness and monotony, earned for him a popularity which even the brilliancy of the Pléiade did not extinguish, for he continued to be read with genuine admiration for nearly two centuries. He was the founder of a school which Mellia de St Gelais, the introducer of the sonnet into France, was the most important member. In fiction Rabelais and his followers concurrently effected a complete revolution. Marguerite of Navarre (sister of Francis I, b.1492 — d.1549), who is principally known as the author of *The Heptameron*, maintained a literary court in which the most celebrated men of the time held high place. It was not until the middle of the sixteenth century that the great movement took place in French literature which, if that which occurred in the same country three hundred years subsequently be excepted, is without parallel in literary history.

The Pléiade consisted of a group of seven men who, animated by a sincere and intelligent love of their native language, banded themselves together to remodel it and its literary forms on the methods of the two great classical tongues and to reinforce it with new words from them. They were not actuated by any desire for gain. In 1549 **Jean Daurat**, then 49 years of age, was professor of Greek at le Collège de Coqueret in Paris. Amongst those who attended his classes were five enthusiastic youths whose ages varied from seventeen to twenty-four. They were **Pierre de Ronsard**, **Joachim du Bellay**, **Remy Belleau**, **Antoine de Baif** and **Etienne Jodelle**. They and their Professor associated themselves together and received as a colleague **Pontus de Tyard**, who was twenty-eight. They formed a band of seven renovators, to whom their countrymen applied the cognomen of the **Pléiade**, by which they will ever be known. Realising the defects and the possibilities of their language, they recognised that by appropriations from the Greek and Latin languages, and from the melodious forms of the Italian poetry, they might reform its defects and develop its possibilities so completely that they could place at the service of great writers a vehicle for expression which would be the peer if not the superior of any language, classical or modern. It was a bold project for young men, some of whom were not out of their teens, to venture on. That they met with great success is beyond question; the extent of that success it is not necessary to discuss here. The main point to be emphasised is that it was a deliberate scheme, originated, directed, and matured by a group of little more than boys. The French Renaissance was not the result of a spontaneous bursting out on all sides of genius. It was wrought out with sheer hard work, entailing the mastering of foreign languages, and accompanied by devotion and without hope of pecuniary gain. The manifesto of the young band was written by Joachim de Bellay in 1549 and was entitled *La Defense et Illustration de la Langue Francaise*. In the following year appeared Ronsard's Ode — the first example of the new method. Pierre de Ronsard entered Court life when ten years old. In attendance on French Ambassadors he visited Scotland and England, where he remained for some time. A severe illness resulted in permanent deafness and compelled him to abandon his profession, when he turned to literature. Although Du Bellay was the originator of the scheme, Ronsard became the director and the acknowledged leader of the band. His accomplishments place him in the first rank of the poets of the world. Reference would be out of place here to the movement which was after his death directed by Malherbe against Ronsard's reputation and fame as a poet and his eventual restoration by the disciples of Sainte Beuve and the followers of Hugo. It is desirable, however, to allude to other great Frenchmen whose labours contributed in other directions to promote the growth of French literature. **Jean Calvin**, a native of Noyon, in Picardy, had published in Latin in 1536, when only twenty-seven years of age, his greatest work, both from a literary and theological point of view, *The Institution of the Christian Religion*, which would be accepted as the product of full maturity of intellect rather than the first fruits of the career of a youth. What the Pléiade had done to create a French language adequate for the highest expression of poetry Calvin did to enable facility in arguement and discussion. A Latin scholar of the highest order, avoiding in his compositions a tendency to declamation, he developed a stateliness of phrase which was marked by clearness and simplicity. **Théodore Beza**, historian, translator and dramatist, was another contributor to the literature of this period. **Jacques Amyot** had commenced his translations from the *Theagenes* and *Chariclea* three years before Du Bellay's manifesto appeared. Montaigne, referring to his translation of Plutarch, accorded to him the palm over all French writers, not only for the simplicity and purity of his vocabulary, in which he surpassed all others, but for his industry and depth of learning. In another field **Michel Eyquen Sieur de Montaigne** has arisen. His moral essays found a counterpart in the biographical essays of the **Abbé de Brantome**. **Agrippa D'Aubigné**, prose writer, historian and poet; **Guillaume de Saluste du Bartas**, the Protestant Ronsard whose works were more largely translated into English than those of any other French writer; **Philippes Desportes** and others might be mentioned as forming part of that brilliant circle of writers who had during a comparatively short period helped to achieve such a high position for the language and literature of France.

In 1576, when Francis Bacon arrived in France, the fame of the Pléiade was at its zenith. Du Bellay and Jodelle were dead, but the fruits of their labours and of those of their colleagues was evoking the admiration of their countrymen. The popularity of Ronsard, the prince of poets and the poet of princes, was without precedent. It is said that the King had placed beside his throne a state chair for Ronsard to occupy. Poets and men of letters were held in high esteem by their countrymen. In England for a gentleman to be amorous of any learned art was held to be discreditable and any proclivities in this direction had to be hidden under assumed names or the names of others. In France it was held to be discreditable for a gentleman **not** to be amorous of the learned arts. The young men of the Pléiade were all of good family and all came from cultured homes. Marguerite of Navarre had set the example of attracting poets and writers to her Court and according honours to them on account of their achievements. The kings of France had adopted a similar attitude. During the same period in England Henry VIII, Mary and Elizabeth had been following other courses. They had given no encouragement to the pursuit of literature. Notwithstanding the repetition by historians of the assertion that the good Queen Bess was a munificent patron of men of letters, literature flourished in her reign in spite of her action and not by its aid.

What must have been the effect on the mind of this brilliant young Englishman, Francis Bacon, when he entered into this literary atmosphere so different from that of the Court which he had left behind him?

COPY OF THE WILL OF SIR NICHOLAS BACON, LORD KEEPER OF THE GREAT SEAL

In the name of God, Amen. The thre and twentithe daie of December in the yere of our Lord God a thousand fyve hundred the seaventye and eighte, and in the one and twentithe yere of the reigne of our Sovereigne Laydie Elizabeth by the Grace of God Queene of England, Fraunce and Ireland, Defender of the Faithe, etc. (I) Sir Nicholas Bacon Knighte Lord Keper of the Greate Seale of England beynge of whole mynde and memorie doe make this my present testament in manner and forme follywynge revokynge all former wills and testaments made by me, before the date hereof.

First. I comyte my sowle to the hands of Almightie God whoe of his omnipotencie did create yt and of his infinite mercie redemed yt and nowe as my undoubted hope ys by the same mercifull redemcon will glorifie yt and save it.

My desier ys to be buried at Pawles where my tombe is. And because I geve noe blackes to the riche that have noe neede therefore I geve to the poore that have neede fyve hundred the marks to be distributed accordynge as by a sedule subscribed wythe my hand dothe appeare. I will notwithstandyng blackes be geven to my householde folkes both at London and Gorhamburie and to all my children their husbands and wiffes.

Item I geve to my deare and welbeloved wiefe one thowsande fyve hundredthe ounces of my plate whereof thone haulfe guylte and thother haulfe parcel guylte and white, to be chosen by hir oute of all my plate excepte soutche parcels as I geve away by speciall name.

I give hir also all my lynen, naperie, hangynges, coches, lytters, and all other my howshold stufe and howsholde stoore remayninge at London excepte my readie money, plate and armor and excepte suche evidence as apperteyne to eny lands or hereditaments as be assigned to eny of my children by my former wief, and excepte suche things as remayne in my studie and suche things as I geve awaye by speciall woordes requyringe my wief in consideracon of the same provision and stoore to kepe so many of my howsholde together at her charges during a monethe after my deathe as will tarrie so longe for the better doinge whereof I give hir in readie money c^{li} I give hir also suche jewells and golsmythes worke (excepte plate) as remaynethe with hir. I will also to my said wief all my horses and geldynges. And also all my intereste in all my stockes of sheepe goynge at Ingham or Tymwoorthe or within eny of my sheepe courses there. To possesse and use Furynge hir life uppon condicon that within one yeare nexte after my decease and before her marriage agayne she become bounde to my executors in the some of twoo hundredthe pounds that at the tyme of her deathe she shall leave to suche person or persons as oughte then to possesse the same mannor and stocke of sheepe goynge uppon the same mannor and within the same sheepe courses of like goodnes and of as greate a nomber as she shall receave.

And this is donne because I ame bounde uppon covenaunts of marriage of my eldest sonne to leave suche a stocke after the deathe of my said wief. An I will that the the stockes letten with Stifkey goe as the lands is there appoynted to goe and remayne. And I will that the one haulfe of all the howsholde stufe that shall remayn at Gorhamburie at the tyme of my deathe (excepte my plate, tent and pavylion) to Anthonie at thage of 24 years. And if he die before then to Fraunces at the same age. And thother haulfe I will to Anthonie after the deathe of my wief And in the meanetyme my wief to have the use of it. To whome also I geve all my greene store of howseholde remayninge either at Redburn or Windridge and all my other goodes and cattalls remayninge there (except my plate and money and other things before geven or excepted).

Item I will that all my lease of Aldenham and all copiehold lands or tenements lyinge in the parishes of Sainte Michall or Sainte Stephens nighe Staint Albones or joyninge to any lands of Westwicke, Gorhamburie or Praye shall remayne and goe accordyne as my howse of Gorhamburie is appoynted to goe and remayne.

Item I geve to my said wief all my intereste in Yorke Howse in consideracon of which legacies and in consideracon of suche assurances of mannors lands and tenements as I have assuered unto my said wief and for all loves that have benne betwene us I desier her to see to the well bringing upp of my twoo sonnes Anthonie and Frauncis that are nowe left poore orphans without a father.

And further I will bequethe to the said Anthonie my sonne all that my lease and tearme of yeres and all my intereste and demaunde which I have of or in all those woodes comonly knowne or called by the name or names of Brittelfirth alias Brighteighfirthe alias Brighteighe woode and Burnet Heathe lyinge nd beynge in the parish of Sainte Stephens in the countie of Hertforde. And also all that yerely rente of £26 13 4 due and payable for the said woodes. And also all my righte tittle and possession which I have of and in eny lands tenements and heriditaments assuered to my said (son) Sir Nicholas for the true payment of the said rente of £26 13 4. And also all that my lease and tearme of yeres and all my tittle and intereste and demaunde which I have of or in the fearme of Pynner Parke lying in the parrishe of Harrowe in the County of Middlesex. And also of and in all my other landes tenements and heriditaments lying in the said parrishe of Harrowe. To have and to houlde to the said Anthonie the said woodes lying within the said parrishe of Sainte Stephens. And all the said fearme called Pynner Park and all the said landes and heriditaments in Harrowe for and duryinge so maney yeres as yt shall happen the said Anthonie to live. And if yt shall fortune the said Anthonie to die before the full ende and expiracon or determinacon of the said leases and tearmes therein contained then my will and intent is that the eldest sonne of the bodie of the said Anthonie for the tyme beynge and the heyres mayles of his bodie for the tyme beynge shall have houlde occupie and enjoye successively during their severall lyves all the said woodes and fearme and other the premysses before bequeathed to the said Anthonie for so maney yeres as the said eldeste sonne of the said Anthonie for the time beinge or the heyres males of the bodie of the said eldeste sonne shall severallye and successivelie fortune to live and yf it fortune the said Anthonie and his said eldeste sonne and the heyres males of the said eldeste sonne and everie of them to die without issue male of their bodies and of the body of every of them before the full ende and determination of the saide leases and termes of yeares therein contained, then my will and full meanynge is further that Francis my sonne shall have houlde occupy and enjoe the said woodes fearme and other the premysses before bequeathed to the said Anthonie. To hym the said Frauncis his executors and assignes for ever.

Item I geve also to my eldeste sonne and his heyres all my fearmes in Mildenhall and of Langerfearme and of the lands and tenements in Ilketeshall and of my howse in Silver Streete that I have of the House of Westminster and of my fearme of Dullynghams.

And further I will to my said heyre my tent and pavilyon remayninge at Gorhamburie and all my apparrell armor and weapon remaininge eyther at Redgrave or at any howse in London and all my howshoulde stufe stocke stoore and other goodes remayning at Redgrave, and all things remayninge in my studie at London excepte suche as be geven awaye by speciall wordes.

Item I geve to Robert Blackeman my nephewe all my intereste in the lease of the meadowes and grounde at Hame.

And to Nathaniell my sonne towardes the buildynge of his howse at Stifkey twoo hundredthe poundes and besides all my lease of the lands in Stifkey and my stocke of sheepe goeing uppon them.

Item I give to the Master and Fellowes of Bennet Colledge in Cambridge to the buildinge of a chappell there cc[li].

And I geve to every of my freendes and to my servantes and suche other person as be named in a pagyne hereafter follwynge subscribed with my hand all suche thyngs and somes of money as beene in the same appoynted.

Provided alwayes that iff Ann my said wief doe not make or cause to be made within one yere next after my decease and before she be married agayne to everie of my sonnes Nicholas, Nathaniell, Edwarde, Anthonie and Frauncis, a sufficient release in lawe of all her right tittle intereste and demaundes of dower of and in all the mannors landes tenements and hereditaments whereof by reason of my seysin she is or then shalbe dowable and deliver or cause to be delivered to everie of my said sonnes one suche release within the said yere and before she be married, then I will all my legacies guifts, and bequests to her made shalbe voied and then I will the same together with the reste of my goodes debtes and catttalles after my debtes paied funeralls discharged and legacies performed to my eldeste sonne Nicholas.

Item I will that the hundredthe poundes stocke remayninge witht he Mayor of Sainte Albones and his brethern's handes for the setting of the poore of woorke be continued in their handes so long as they performe the covernauntes agreed uppon beteene them and me otherwise that my wief or heyres to Gorhamburie receave and kepe the same.

And of this my will I make my executors Sir Nicholas Bacon Knyghte and Nathanyell Bacon, and overseer my Lorde Treasorer my brother in lawe to whon I geve a standynge cuppe with a cover garnyshed with christall weighting 53 ounces 3 quarters, and to my Ladie Burghleye my sister in lawe a deepe bowle with cover haveyng my cognizaunce weighing 21 ounces and a half.

To Anthonie my jewell that I weare and to my daughter Bacon my eldeste sonne's wief my cheaste in my study made by Albert and my little boxe with ringes to Mistress Butts my ringe with the beste turquois.

In wittnes whereof I have subscribed everie pagyne of this my will with myne owne hande and set to my seale the daie and yeree firste above written.

NOTES and BIBLIOGRAPHY

NOTES, Part I

1. See F.B.R.T. Journal I/3, *Dedication to the Light*.
2. *I Peter ii. 5*.
3. The word 'beginning' is translated from the Greek word *'Arche'*, which has the fuller meaning of 'the beginning without end', 'the circle'. It conveys the idea of the ever-present NOW of Eternity. Divine Creation is beyond time: it is past, present, future, and more, all as one. It is a transcendant state or condition that our time-based perceptions cannot grasp rationally. The word *'Bereshith'*, meaning 'In the beginning (without end)', was used in the original Hebrew, and this word was also used as the title of the whole book of *Genesis* because it is the book concerning all beginnings. 'Genesis' is from a Greek word, meaning 'Creation', and is a translation of the Hebrew *'Beriah'*, which refers to the creative aspect of Divinity in conceiving or imagining the form of manifest Divinity — the divine thinking process — and relates specifically to the first chapter of *Genesis*. The original Mosaic title of *Genesis* was *'Sepher'*, which means the book of "First Causes manifesting Fundamental Principles". Moses is said to have written the first Hebrew version of *Sepher*, his original Hebrew script containing much of the Ancient Wisdom teaching that he had learnt in Egypt — it forming part of the secret doctrine of the Ancient Egyptians, in which Moses, as High Priest of On (Heliopolis), was fully learned.
4. 'God' is a word used by English translators of the Bible to stand for various different names of the Divine Being written in the Hebrew text. The different names originally used referred to precise and definite aspects, characteristics and powers of Deity, thus helping the readers of the Scriptures to comprehend what was being written about. To translate all these various names under one general name, 'God', is to lose the richness and scientific precision of the original text, and a loss of understanding. Our word, 'God', is related to the word 'Good', and is a definition of the Absolute Deity as the All-Goodness, this indeed being the essential characteristic of Divine Being. In this particular verse of *Genesis* i, the original Hebraic word was *AELOHIM*, which is both a singular and plural word. The *AELOHIM* are the Seven Creative Spirits, which constitute the One Holy Spirit, the Word of God. The name refers to the creative aspect of God, which is God's Divine Thought. Hence, when the name *AELOHIM* is used, we know that the text is referring to the creative Thought of God. The whole of *Genesis* i and the first three verses of *Genesis* ii uses the name *AELOHIM*, as these verses refer to God's creative Thought, the Logos-Christos-Sophia. From *Genesis* ii. 4 to the end of *Genesis* iii the name of God is *JIHOVAH AELOHIM*, as these verses refer to the generative aspect of God — that aspect of the spiritual Thought that incarnates into Matter and generates the living soul, and thereby becomes the manifest Deity. *JIHOVAH* is linked with *AELOHIM* as this takes place before the complete incarnation of *JIHOVAH* into Matter (*i.e.* before the 'Fall'). But in *Genesis* iv the 'Fall' takes place, and so the name of God becomes simply *JIHOVAH* in this respect, and refers to the realm of formation, the generation of living forms out of matter that will eventually be able to manifest Divinity.
5. *Genesis* i. 1.
6. Jesus said:

    > "Pray every day to your **Heavenly Father** and **Earthly Mother**, that your soul become as perfect as your Heavenly Father's holy spirit is perfect, and that your body become as perfect as the body of your Earthly Mother is perfect. For if you **understand**, **feel**, and **do** the commandments, then all for which you pray to your Heavenly Father and your Earthly Mother will be given to you. For the **wisdom**, the **love**, and the **power** of God are above all.
    >
    > "After this manner, therefore pray to your Heavenly Father:
    >
    >> Our Father, which art in heaven, hallowed be thy name.
    >> Thy kingdom come.
    >> Thy will be done on earth as it is in heaven.
    >> Give us this day our daily bread.
    >> And forgive us our debts, as we forgive our debtors.

> And lead us not into temptation, but deliver us from evil.
> For thine is the kingdom, the power, and the glory, for ever.
> Amen.

"And after this manner pray to your Earthly Mother:

> Our Mother, which art upon earth, hallowed be thy name.
> Thy kingdom come,
> Thy will be done in us, as it is in thee.
> As thou sendest every day thy angels, send them to us also.
> Forgive us our sins, as we atone all our sins against thee.
> And lead us not into sickness, but deliver us from all evil,
> For thine is the earth, the body, and the health.
> Amen.

(From *The Gospel of Peace of Jesus Christ*, translated by Edmond Szekely and Purcell Weaver from the original Aramaic texts that constitute accurate notes made by John the Beloved of his Master's personal teachings.)

7. *Brahm*, as the first existing, Absolute Being, is more properly written *Brahman* or *Brahma* (neuter), the unchanging, pure, free, undecaying Supreme Root and Essence of all existence. *Brahman* becomes *Brahmâ* (masculine) when the Absolute Being becomes active as Creative Being or Thought. *Brahmâ* is the living Word or Divine Thought.

8. *Genesis* i. 2.

9. *i.e.* As the Word or Archetype of all Creation.

10. *i.e.* "Born of the Virgin Mary".

11. *Genesis* i. 2.

12. *i.e.* Creation.

13. *i.e.* Emanation.

14. *Genesis* i. 3-4.

15. "The essential form of knowledge . . . is nothing but a representation of truth: for the truth of being and the truth of knowing are one, differing no more than the direct beam and the beam reflected."
    (Francis Bacon, *Advancement of Learning*.)

16. *i.e. Hu*-Man. *Hu* = spiritual radiance, the light of love.

17. *e.g.* "Hallowed be thy Name."

18. Freemasonic teaching.

19. *c.f.* Jesus was crucified and resurrected at 33 years of age;
    King David ruled all Israel for 33 years;
    BACON = 33 in Simple Cipher (B = 2, A = 1, C = 3, O = 14, N = 13).
    There are 33 vertebrae in the human spine.

20. The Ancient Egyptians also wrote down only the consonant sounds.

21. *Aleph* is counted as a consonant, the sound 'Ah' being considered both a vowel and consonant. Thus the 'A' or 'Ah' stands for the unmanifest emanation of love from the divine heart, and also its manifest expression in a living form of matter. The so-called 'Double A' of the Baconian-Rosicrucians symbolises this. 'A', pronounced 'Ah', is pre-eminently the 'heart' sound.

22. *Genesis* ii. 7.

23. *J* or *Jod* is also written as *Y* or *Yod* which gives its proper pronunciation (as in 'yacht' or 'yogi').

24. Alternatively, Y.H.V.H.

25. For instance, the three Patriarchs, Abraham, Issaac and Jacob, represented the three stages of Emanation (the initial desire, love, the initiating or motivating force), Creation (the thought) and Formation (the generating, activating power). Jacob, when he had begot twelve sons, was called 'Israel', and 'Israel' is the name applied to all their generations

(past, present and future) that are "chosen" or perfected. He that is chosen of God is he that is anointed of God — the Messiah or Christ.

Is-Ra-El is an Ancient Egyptian word, meaning 'the visible/manifest face/soul of the Light of God'. Horus, or Jesus, has a similar meaning. *ISRAEL* is the manifest *JIHOVAH*, Who performs twelve generative acts or labours which beget the twelve sons which beget the twelve tribes, which make God manifest and known.

26. "Beginning" = *Arche* = the Circle of Eternity. (See note 3).

27. The Word = *Logos* = the Divine Thought that creates all things.

28. *i.e.* The Word created/creates all things. The Word was *AELOHIM* (God), which is the Divine Thought.

29. "Life" = the *Logos* or Will of God, which is Truth and which is essentially Love in its active or creative mode. (*i.e. Brahman* become *Brahmâ*.)
"Light" = the *Christos* or Bliss Consciousness, the Intuitive Wisdom of God.
"Men" = the *Sophia* or Intelligence, the Vision of Truth.

30. *i.e.* The Divine Intelligence (*Sophia*, Man) did not yet understand the Light or Wisdom of God.

31. Man = the incarnating Intelligence whose Name or Light is *IOA* (John), sent into matter from *AELOHIM*, the Divine Thought.

32. Incarnate man is the witness, revealing the Light (his Name) in his perfected living soul, so that the Light of Truth may be seen and known in all that he feels, thinks and does.

33. Man is not the Light (*Christos*), but is the Intelligence (*Sophia*) that can embody and reveal the Light in his soular form.

34. Through the intelligent soul of man, the Divine Idea (*IOA*) is embodied and revealed as grace and truth — the glory or light that is the same as that of the transcendent Light, the Only Son or Christ Consciousness of the Divine Mind.

35. John, the incarnate Man, bears witness (*i.e.* reveals) the Christ, the living Truth.

36. He of whom John speaks is the Christ, the Bliss Consciousness, which radiates from the Heart of God as sound and light within the heavenly Mind, before it can be reflected in the Mind as Intelligence. Thus, in terms of Divine Thought, the Christ comes before Man, the Intelligence. But in terms of manifestation in matter, the intelligent soul of Man precedes the Christ manifestation, for the human soul must first evolve to perfection before it becomes fully illumined or Christed with light.

37. *John* i. 1-15.

38. *John* x. 30.

39. *John* i. 18.

40. *John* xiv. 9.

41. *Genesis* ii. The Garden of Eden is part of the Cosmos or realm of form.

42. See *John* x. 1-18.

43. *Hyperborea* = 'Land beyond the North Wind'. The North Wind is the symbol of the creative Breath of God — the Divine Thought of God. The land that lies **beyond** the North Wind, from whence the North Wind is derived, is the very Heart of God, the Centre of Truth, Centre of Love, which breathes the Breath of Love and Life. It is from the heart that the divine thought radiates into the mind as light or intuition, to be reflected in the mind as intelligence. Thus *Hyperborea* is a name referring to the mystical heartland of Truth.

44. *THOR, TOR, TAU, THOTH, TOT* are all words designating Truth, the emanation of Divine Love from the Heart, which is also Divine Will (*i.e.* the will or desire-to-be). In its masculine or creatively active form it becomes the Word or Thought of God (*c.f. Brahman — Brahmâ*).

A, AH or *AR* is the sound that is peculiarly related to the heart centre and to the inbreath and outbreath of life itself. Hence the letter '**A**' or triangle, △, represents (i) the Source of

Life and Absolute Being of God, which is Trinitarian, and (ii) the full manifestation of that Divine Being as the perfected celestial soul, the Christ soul.

AR-THOR is thus the sacred name designating God, the All-Good, (i) as the Absolute Being, (ii) as Creator, and (iii) as the Manifest God, the perfect human soul.

45. *e.g.* The Glastonbury (Somerset) landscape zodiac. The dove holding the wafer of light (or fiery jewel) in its beak, and descending into the waters, is the Catholic symbol of Baptism. The wafer or jewel of light signifies the *Christos* or Bliss Consciousness that is the radiance of Truth, the Voice that speaks the Word of Truth; with the actual form of the wafer or jewel representing the archetypal Vision or Thought-form of Truth.

46. The cycle of Ages, each Age being approximately 2,160 years in duration.

47. E. W. Bullinger, *The Witness of the Stars*.

48. The Vedic term for the *Christos* or Christ Spirit is *Vishnu*, the All-Pervader who is *Ananda*, Pure Bliss.

49. The Supreme God of the Ancients — the God of all gods — was often designated as an Archer. The Supreme God means the Absolute Being, which is pure Life or Love, the Desire-to-Be. All other gods represent the various phases and aspects of the One God, the One and Only Life. All the ancient traditions knew and taught this, but not everyone was able to grasp what it means (a situation which still applies today, despite men's uncomprehending assertal that there is only one God).

> "They (the ancient poets) say then that Love was the most ancient of all the gods; the most ancient therefore of all things whatever, except Chaos, which is said to have been coeval with him; and Chaos is never distinguished by the ancients with divine honour or the name of a god. This Love is introduced without any parent at all; only, that some say he was an egg of Night. And himself out of Chaos begot all things, the gods included. The attributes which are assigned to him are in number four: he is always an infant; he is blind; he is naked; he is an archer".
> (Francis Bacon, *Cupid*, from *The Wisdom of the Ancients*.)

50. This was the true meaning of *ATON* (*ATEN* or *ATAN*, pronounced *áhtein*) the Ancient Egyptian name revealed openly for the first time by Akhenaten, the so-called "heretical" Pharoah. *ATON*, to which our words Atonement (*i.e.* At-one-ment), Attainment and Attunement are kin, signifies and describes the embodiment of the Divine Thought. *Á* (or *Ah*) describes the manifest and living Heart of God as the Christ soul of light (the 'spiritual body'); *T* reinforces this idea, representing the Truth (*Tot, Tau, etc.*) which is the original impulse of Desire, Love or Will — the emanation of Life from the sublime Heart, that motivates all things; *ON* or *AN* (pronounced *ain* or *ein*) signifies the Heavenly Mind that is filled with the Thought of Truth.

51. In the so-called 'Dogon' Tradition, associated with Ancient Egypt, Sirius is known as *NOMMO*; a name which is also applied to the Great Initiates who came from Sirius (or *via* Sirius) to this Earth. The *Nommos* are the Babylonean *Oannes* (*i.e. JOHANNES*), represented as amphibious creatures or fishes — hence one of the more important significances of the fish in the Christian tradition. They are recorded as being the founders of all civilisation and culture. Sirius is known as 'the Land of the Fish'. (See Robert K. G. Temple, *The Sirius Mystery*.)

52. Aetheric = fully spiritualised matter, the shining substance of the celestial soul, the body of light.

Ether = primal matter, fundamental and unspiritualised; but now also meaning matter in its denser forms of vibration or livingness.

53. *Genesis* v.

54. *Genesis* vi. 2. The disciples of the Sirian Initiates are masters and adepts in terms of this world.

55. *Genesis* vi. 2.

56. *Genesis* iv. 4.

57. The Secret Doctrine says 869,000 years ago.

58. With three main cataclysmic periods.

59. 1,050,000 years ago.

60. The disciples each represented individual parts or aspects of the JESUS soul; thus, as Jesus went through the initiations, so each disciple showed that particular aspect of the soul undergoing the initiation, demonstrating for mankind the type of reactions, difficulties and achievements that are likely to be experienced by any soul on the initiatory path.

61. *John* x. 1-18.

62. 8 × 8 = 64 = number of the Chess Board, and one of the principal numbers of the Rosicrucian Brotherhood. 8 is the symbolic number denoting the Holy Spirit, and is called the Cosmic Lemniscate. The Holy Spirit is the Source or '8th Point' of the Seven Spirits; all Seven being derived from and constituting the so-called Eighth, which is the Whole. It is also the number of Mercury, as the Incarnate Word of God — the Messenger and Teacher of Truth.

63. Horus = *Heru* = Jesu or Jesus.

64. *e.g.* St. George = Horus/Jesus.

65. *c.f.* Sire, Sir.

66. The Sign of the Heart. Leo rules the heart, the lion signifying the royalty and strength of the heart, which is divine emotion of love, the holy Will of God and its radiant light of Truth.

67. *Genesis* ix. 16.

68. *Matt.* v. 8.

69. *John* x. 24-25.

70. *John* xiv. 6-11.

71. See F.B.R.T. Journal I/2: *The Virgin Ideal*.

72. Joseph = 'He shall add (experience to learning)'.

73. Hermon = *Sirion* ('to glitter').
    = *Sion* ('elevated').
    = *Sirius* (the 'Crown' or 'Head')

74. *Matt.* xvii. 1-9. *Mark* ix. 2-8.

75. *John* xx. 11-17.

76. *John* xx. 26-29.

77. The olive symbolises illumination, which bestows (or is) peace, knowledge, immortality, fruitfulness, fertility and plenty. The olive branch carried in the beak of the dove specifically denotes the carrying of Divine Light into incarnation. The Light, and the whole Thought of God, is Divine Peace. To be given the olive branch denoted that one had reached a recognised state of illumination, and had become a wise man or 'man of peace'. The olive leaf denotes renewal of life. The oil or juice of the olive, used in the manufacture of the ointment used to anoint priests and kings, signifies the divine unction or blessing (also known as 'chrism') that baptizes the full initiate with light — with the bliss consciousness and knowledge of God.

78. *Luke* xxiv. 50-51. *Acts* i. 9-11.

79. *John* x. 30.

80. *John* xiv. 10

81. *I Cor.* xv. 45.

82. *John* xxi. 15-19. *Acts* ii. 5-43.

83. *John* xiv. 16-17.

84. *John* xiv. 26.

85. *John* xv. 26-27.

86. *Acts* ii. 1-4.

87. Hence David is born in Bethlehem, the secret heart chakra, and rules in Jerusalem, the major heart chakra. 'David', the *Christos*, designs the Temple of Light; Solomon, who is David's 'son' or soul, builds the Temple and **is** the Temple.
88. *Matt.* iii. 3.
89. *Matt.* iii. 16.
90. *Matt.* iii. 11-12.
91. *Luke* x. 18.
92. Silver and gold keys = knowledge of good and evil;
    = mastery of intuition and reason (radiance and reflection);
    = illuminated with wisdom and understanding;
    = mastery of heaven and earth (mind and matter);
    = mastery of spirit and body (heavenly thought and earthly nature);
    = the dual-aspected Key of Life.
93. Hermetic Teachings, from *The Secret Doctrine*, bk. II. p. 233. 'Peter', meaning 'the Jewel of Light', 'Revealer', 'Teacher', 'Interpreter', is thus the Doorkeeper, and is also referred to as "Satan" by Jesus. He is the one who sees, recognises and **knows** the Christ.
94. *Matt.* x. 16.
95. *Matt.* xxvii. 53-56. *Mark* xv. 40-41. *Luke* xxiii. 49.
96. See F.B.R.T. Journal I/2: *The Virgin Ideal*.
97. The Law or Wisdom of God is called *'Torah'* in Hebrew. It is pre-eminently the **revealed** Law, hence the word *Torah* is composed of *TOR*, the Truth, and *AH*, the manifest heart of God. *ARTHOR* and *TORAH* are, in this sense, one and the same.
98. *Matt.* xxii. 37-40. *Mark* xii. 29-31.
99. *Genesis* ix. 1-16.
100. *Genesis* xvii. 1-22.
101. *Gal.* v. 6.
102. *Exodus* xix-xxxi, xxxiii-xxxiv. *Deut.* v-xi.
103. I *Chron.* xvii. 3-14.
104. *Matt.* xxvi. 28.
105. *John* xiii. 34.
106. *John* xv. 14.
107. *Hebrews* viii. 8-11

NOTES, Part II

1. Frances Yates, — 'The Entry of Charles IX and his Queen into Paris, 1571' — *Astraea, The Imperial Theme in the Sixteenth Century.* (Routledge Kegan & Paul, 1975.)

2. *'Magnitude'* signifies the brightness of a star. *'Apparent magnitude'* indicates the brightness of a star as we see it; whilst *'absolute magnitude'* is a measure of its intrinsic luminosity, independant of the star's distance from us. In terms of apparent magnitude, some of the brightest stars in the sky have brightness of the first magnitude. This is defined as the brightness of a candle flame at a distance of 1,300 feet. Stars of the second magnitude are fainter than the first by a factor of two and a half; *etc.* A difference of five magnitudes means that one star is precisely a hundred times as brilliant as another. A few stars are brighter than the first magnitude, and the scale is extended to zero and beyond into negative magnitudes. Our sun's magnitude is –27. *(The New Space Encyclopaedia.)*

3. A.E. Waite, *The Brotherhood of the Rosy Cross.* (University Books, Secaucus, New Jersey — reprinted 1973.)

4. Paracelsus 'saw' the star and what it meant, prophetically. Paracelsus died in 1541, before the supernova made its physical appearance.

5. *i.e.* the major planetary conjunctions of 1602-4.

6. Extract from Michael Srigley's essay, *Francis Bacon, A Forerunner*, printed in *The Beacon*.

7. 'Circumpolar' is a description given to stars or constellations that appear to surround the north pole (star) and which describe the whole of their diurnal circles above the horizon; *i.e.* their apparent circular movement around the pole never takes them below the horizon and thus out of our sight, as is the case with the majority of the stars and constellations.

8. See F.B.R.T. Journal I/2, *The Virgin Ideal*, in which the mystical significance of *Cassiopeia* is discussed.

9. Because of the 'wobble' of the Earth's axis as it moves about the sun, although the general direction that the axis points in is constantly orientated towards the same portion of the celestial sphere, nevertheless the real poles of the Earth's celestial sphere appear to rotate about the imaginary fixed north and south poles. If there was no 'wobble', the position of the Earth's north pole in the celestial sphere would lie in the heart of *Draco*, the Dragon; but because of the 'wobble' the Earth's north pole tours around the Dragon 'heart' in a great circle which takes approximately 26,000 years to describe. At present the north pole is in a highly significant position, almost coinciding with the seventh star in the constellation of *Ursa Minor*, 'the Little Bear'. This is the brightest star in the constellation, called (in Arabic) *Al Ruccaba*, "the turned, or ridden upon" — the central star and hub of the zodiacal wheel (at present), and thus called (in Latin) *Polaris*.

   The Little Bear is not the original name for this constellation, although now its name (and that of its companion, the Great Bear) is built into the legends of Arthur the King, who is described as the mighty Bear. But the original name is more meaningful, being (in Hebrew) *Dōhhver*, 'a fold' (of animals, but particularly sheep). The name, besides meaning the 'sheepfold' which Jesus and others refer to, recorded in the Bible and elsewhere, also means 'rest' or 'security' — the place of peace into which the righteous are shepherded, where the Assembly of the Holy Ones takes place, and from whence the Saviours or 'Shepherds' of mankind come. But in Hebrew there is a similar sounding word, *Dōhv*, meaning 'a Bear', hence the confusion in the translation. Similarly the confusion with the English word, Dove, which this constellation also represents symbolically: for in the seven stars of *Dōhv* the Word is sounded and the Light is poured out upon the Earth to baptize, vitalise, guide, teach and illumine mankind. Arthur is derived from *Ar-Thor*, meaning 'the Word (of Love)', *i.e.* the Christ. This word is borne by and spoken in the holy Breath of God (*i.e.* divine Love), which Breath is symbolised by the Dove but can in fact be seen as a (deliberate?) borrowing phonetically of *Dōhv*, which essentially means the Place of Peace — which of course is the perfect abode of Love, the 'sheep-fold' of the Bible. The 'lambs of God' are they who come from this perfect place of Peace, of Love, bearing the divine love and its radiant wisdom to the world.

In the East the seven stars of *Dōhv* are called the Seven *Rishis* or Teachers, and represent the Seven Rays of the one White Light of Christ. It is significant that one of those seven stars or *Rishis* is the present Pole Star of our world. The 'North' is known as the Place or Seat of Government, and the 'Mount of Congregation' of the Lord is in the north, where the Assembly of the Holy Ones is to be found. The 'north wind' is the creative Breath of God; and *Hyperborea*, 'the land beyond the north wind', is the source from which the north wind comes. That source is the Heart of God; and this is the secret that is manifesting in the heavens, in *Dōhv*. On Earth, *Hyperborea* has a special relationship to the country of Britain, and thus Arthur is primarily the great king of Britain. Esoterically, Britain is the heart-abode of the incarnate *Dōhv* and the principal home-land of the great teachers of humanity from the most ancient times. The history and destiny of Britain is connected with this, and this important (but estoteric) fact lies behind the Baconian work.

Only at the junction of the Piscean and Aquarian Ages does the north pole touch *Al Ruccaba* so nearly. In 2,000 years the north pole will lie near to *Al Rai Deramin*, 'who bruises/breaks' the power of the Dragon, the third brightest star of *Cepheus*, 'the Branch' or Redeemer, who is the Crowned King. This future *Polaris* marks the King's left knee, whilst his foot lies over the present-day *Polaris*. He signifies the King of Kings who is coming to rule in the Second Coming, and marks the culmination of the Aquarian Age. The north pole will continue to traverse and lie within *Cepheus* for a further 3,500 years, during the Ages of *Capricornus* and *Sagittarius*, culminating with the north pole being marked by *Al Deramin*, 'Coming quickly', the brightest star of *Cepheus* which marks the King's right shoulder. 5,000 years ago (*c.* 3000 B.C.) the north pole was marked by *Thuban*, 'the subtle' (or 'subtile'), the brightest star in the tail of *Draco*, the Dragon. The importance of where we are now and where we are going can easily be seen in the story of these stars. (See F.B.R.T. Journal I/2, *The Virgin Ideal*, for further information)

10. David Lloyd, *The Statesmen and Favourites of England* (1665).

11. William Rawley, 'The Life of the Honourable Author' *Resuscitatio* (1657).

12. Anthony Bacon was "a Gentelman equal to him [Francis] in height of wit, though inferior to him in the endowments of learning and knowledge". (William Rawley).

13. See Appendix: *Humanism*.

14. 'Humour' is derived from the Latin word, *humor*, meaning 'moisture' and, in a special way, the fluids of the body and mind. There are four cardinal humours of the body or natural (lower) self of man which are said to constitute and determine a person's physical and psychological make-up. The four are defined psychologically as the **choleric**, the **sanguine**, the **phlegmatic** and the **melancholic** temperaments. They relate respectively to the choler (yellow bile), the blood, the phlegm and melancholy (black bile) of the physical body, and to the alchemical Elements of *fire, air, water* and *earth*. Each person is composed of a mixture of these four. A person with a greater proportion of choler is inclined to irritability, anger, argument and fighting. Someone who is sanguine rather than anything else tends to be mentally alert, active, hopeful, outward-looking and communicative. Phlegmatic people tend to be tranquil, somewhat lethargic, cool and even-tempered but emotionally sensitive, patient and self-possessed. A melancholy person has a usual tendency towards gloominess, dejection and general depression of spirits; but out of this comes first a tender or pensive sadness, leading to a deep contemplation and awareness of life with all its difficulties and sorrows, which in turn leads to an appreciation of beauty and joy. The four humours are also said to be ruled by certain 'planetary' influences: the choleric by Mars, the sanguine by Jupiter, the phlegmatic by the Moon, and the melancholic by Saturn.

15. See Frances A. Yates, 'The Occult Philosophy and Melancholy : Durer and Agrippa', *The Occult Philosophy in the Elizabethan Age* (1979).

16. *Ibid*.

17. *Ibid*.

18. William Rawley, 'The Life of the Honourable Author', *Resuscitatio* (1657).

19. Whitgift accounts, *Brit. Nag.* vol. xxx, iii, p. 444.

20. Commons = daily dinner in hall.

21. Philemon Holland was a minor Fellow in 1573, and became a Major Fellow in 1574.

22. *The Dictionary of National Biography.*
23. Goodby, *The England of Shakespeare.*
24. Mullinger, *History of Cambridge University,* vol. II.
25. *Ibid.*
26. *Ibid.*
27. *Ibid.*
28. Goadby, *The England of Shakespeare.*
29. *Ibid.*
30. Mullinger, *History of Cambridge University,* vol. II.
31. Goadby, *The England of Shakespeare.*
32. Strype's evidence, quoted in Mullinger's *History of Cambridge University,* vol. II.
33. *Ibid.*
34. Pierre Amboise, 'Discourse on the Life of M. Francis Bacon, Chancellor of England,' *Histoire Naturelle de Mre. Francis Bacon,* (1631).
35. William Rawley, 'The Life of the Honourable Author', *Resuscitatio* (1657).
36. Pierre Amboise, 'Discourse on the Life of M. Francis Bacon', *Histoire Naturelle* (1631).
37. James Spedding, *Lord Bacon's Letters and Life,* vol. I.4. (1862).
38. Francis Bacon, *Novum Organum,* I.xciii.
39. *Ibid.,* I. xc.
40. *Ibid.,* I. xcv.
41. *Ibid.,* I. xcvi.
42. *Ibid.,* I. xcii.
43. Francis Bacon, *De Interpretatione Naturae Proemium* (written *c.* 1603).
44. *Genesis* 1. 29.
45. Plut. in *Julius Caesar,* p. 735.
46. *Psalm* ix. 4.
47. Francis Bacon, *Historia Naturalis et Experimentalis ad Codendam Philosophiam, etc.*
48. Francis Bacon, *De Dignitate & Augmentis Scientarium* (1623).
49. "Books" = books, sculptures, paintings, architecture, emblems, *etc…*
50. Francis Bacon is using the word "men" as in man or mankind, male **and** female.
51. "*Principio sedes apibus statioque petenda, Quo neque sit ventis aditus,…*" (Virg. *Georg.* iv. 8).
52. Cicero, *Post. Red.* c.12.
53. I *Samuel* xxx. 24.
54. Virgil, *Georg.* iii. 128.
55. Francis Bacon, 'To the King', *De Dignitate & Augmentis Scientarium.*
56. Cicero, *Ep. ad Att.* ix. 8.
57. *James* i. 17.
58. Francis Bacon, 'To the King', *De Dignitate & Augmentis Scientarium.*
59. Francis Bacon, 'Proemium', *Instauratio Magna.*
60. *Maya* = 'illusion', 'veil'.
61. Francis Bacon, 'Preface' to *Novum Organum.*

62. Learning to LOVE Truth, UNDERSTAND Truth, and SERVE Truth corresponds to the three main paths of Yoga: (1) *Bhakti Yoga*, (2) *Ghani Yoga*, and (3) *Karma Yoga*, all of which are summed up in *Raja Yoga*.
63. Francis Bacon, *Natural and Experimental History*.
64. Letter from Francis Bacon to Lord Burghley, Jan. 1593.
65. Francis Bacon, *The Advancement of Learning* (1605).
66. "For each individual of us is not anger, nor fear, nor desire, just as he is neither pieces of flesh nor humours; but that wherewith we think is the soul". — Plutarch, *Morals:* 'On the Apparent Face in the Moon.' *Man* = 'the thinker.'
67. *Proverbs* 25 : 2.
68. Francis Bacon, *Advancement of Learning* (1605).
69. Francis Bacon, *'Of Truth' (Essays)*.
70. "*Opus quod operatur Deus à principo usque ad finem*" = "no man can discover the work that God maketh from the beginning to the end."
71. Francis Bacon, *Advancement of Learning* (1605).
72. Francis Bacon, *Novum Organum*, I: 120.
73. *Ibid.*, I : 124.
74. Orpheus.
75. *Ibid.*
76. *Ibid.*
77. Plato.
78. "The Son of God is the Christ or Divine Consciousness in man. No mortal can glorify God. The only honour that man can pay his Creator is to seek Him; man cannot glorify an Abstraction that he does not know. The 'glory' or nimbus around the head of the saints is a symbolic witness of their capacity to render divine homage." — Sri Yukteswar — *Autobiography of a Yogi* by Paramahansa Yogananda.
79. Francis Bacon, *Advancement of Learning* (1605).
80. Francis Bacon, *Controversies in the Church* (1590).
81. Francis Bacon, *Advancement of Learning* (1605).
82. *Ibid.*
83. *Ibid.*
84. *Ibid.*
85. *Ibid.*
86. *Proverbs* xx. 27.
87. I *Corinthians* viii. 1.
88. I *Corinthians* xiii. 1.
89. Francis Bacon, *Advancement of Learning* (1605).
90. "And above all these things, put on Charity, which is the bond of perfectness." *Coloss*. iii. 14.
91. *Isaiah* xiv. 14.
92. *Genesis* iii. 5.
93. *Matt.* v. 44-45; *Luke* vi. 27-28.
94. Francis Bacon, *Advancement of Learning* (1605).
95. *John* xiv. 6.
96. Francis Bacon, *Thoughts on the Nature of Things*.

97. Francis Bacon, *Advancement of Learning* (1605).
98. *Ecclesiastes* iii. 11.
99. Francis Bacon, *Cupid or the Atom* — transl. from *De Sapientia Veterum*.
100. *Eccles.* iii. 11.
101. Francis Bacon, *Principles and Origins according to the Fables of Cupid and Coelum* — transl. from *De Principiis Atque Originibus*.
102. Francis Bacon, *A Description of the Intellectual Globe*, ch. i. — transl. from *Descriptio Globi Intellectualis*.
103. Green, *Short History of the English People*, ch. i.
104. Linus, as mentioned by Paul in his second epistle to Timothy. (2 *Tim.* iv. 21.)
105. George F. Jowett, *The Drama of the Lost Disciples*.
106. *The Apostolic Constitutions*, Bk. I. xlvi.
107. *Gentiles*, meaning the peoples and nations of the world other than the people and nations of Israel, but later it came to mean 'foreigners' to Israel and, more specifically, those of other nations and peoples who were not brought up in or initiated into the religion and mysteries of Israel. St. Paul is often erroneously described as being the **only** apostle to the Gentiles, which is not true; many Christian initiates were sent to all countries, all nations. Paul was given the specific task of going to the Gentiles of Asia Minor and Aegea, and later to Rome and Britain; but even then he was not the only apostle to visit those nations.

    The inner teachings that Jesus' chief apostles had charge of also included certain powers and processes of initiation, and at first it was believed, when with Jesus, that these powers, initiations and knowledges could only be passed on to other circumcised Israelites. The decision to take the Christian teachings and revelations to all peoples of the world, and not just to the people of Israel, was initiated by Peter after receiving a clear vision and recognition of what needed to be done whilst in Joppa, which led him to the house of the Roman Centurian in Caesarea. He had to justify what he had done at Caesarea to the Christian Assembly in Jerusalem and, after some debate, the decision was ratified by the others and missions to the Gentiles as well as to the scattered Israelites were set in motion. (See *Acts* x-xi.) This was a momentous decision at that time in history, for up to then the rites and practices of the Hebraic/Mosaic tradition had been jealously guarded and reserved only for those who were born or adopted into that particular culture and tradition. For the first time the profound and powerful esoteric tradition (that later came to be called Judaism), with its initiations, was being offered to all nations regardless of colour, blood or creed. Only love, coupled with recognition of and faith in the Messiah — the Christ, the divine Love and Teacher of **all**, was required. It was an open and deliberate break with the strict sectarian rules of the Hebraic/Judaic Tradition that had operated up to that time, in which the actual physical blood tie was considered to be so important, and the Messiah was thought to be a strictly Jewish King, with the Israelites being the only chosen people of God. Other nations and traditions had applied similar discriminatory rules in the past, usually based on good enough reasons and for specific purposes (Ancient Egypt had been one), but every now and then a deliberate break had been made in that rule (*e.g.* Akhenaten in Egypt, Gautama Buddha in India, as well as Jesus in Palestine, had all 'rent' these exclusive veils). God's Love and Wisdom is all-inclusive, but for certain reasons it appears necessary for man's development that he (man) undergoes a period of exclusiveness or sectarianism before he is able to grasp and appreciate the universality of God. When the consciousness of this universality bursts upon a soul it literally rends the veil that divides his limited individuality from the cosmic universality. It is called 'cosmic illumination'.

    Jesus was a Jew and an Essene, brought up as an initiate of the 'elect of Israel' and in their prophetic tradition. He was a recognised master and prophet of Israel in his time — a 'holy man'. He called all his disciples from the ranks of the Tribes of Israel, many of them Essenes like himself, and they all observed the rites and practices of their tradition strictly. Jesus himself achieved and lived in the universal consciousness of Christ, as a Christ or 'Fully Illumined One', and his teachings embody the truths and symbology of other sacred traditions as well as Judaism (*e.g.* notably Orphism); but his disciples only came to this universality after the coming of the Holy Spirit of Love-Wisdom upon them, bestowing the higher consciousness upon them, which 'baptism' was only finally completed with Peter's vision and the subsequent realisation by the Christian Assembly at Jerusalem that "God hath also to the Gentiles granted repentance unto life." (*Acts* xi. 18).

108. The Roman Catholic hierarchy was founded during the 4th century A.D., after the reign of Constantine. Not until 610 A.D. was the title of Supreme Pontiff or Pope given to the Bishop of Rome. It was first given to the Bishop of Rome by the Emperor Phocas, in a purely political move, to spite Bishop Ciriacus of Constantinople who had justly excommunicated Phocas for causing the assassination of his predecessor, Emperor Mauritius. Gregory I, then Bishop of Rome, refused the title, but his successor, Boniface III, accepted.

The title, *Papas* or Pope, meaning 'Father', was originally the name given to any Bishop of the Christian Church, as being the chief of the Church 'Elders' in any one place. With the influence of the Roman Empire, the Christian Church modelled its administrative divisions established by Diocletian and continued by his successors. Bishops of each diocese, based in the chief cities, met in synod in the provincial capitals of the Empire. Before long the bishops from the great metropolitan centres were accorded special dignity. Rome came to be granted precedence in "honour" (but **not** in "authority"), and its Bishops shared rank and power with those of Antioch and Alexandria, to which were added Constantinople (in 381) and Jerusalem (in 451). Eventually the title of *Papas* or Pope was used only to refer to the Bishops of these five great metropolitan centres, and they were known as the Five Patriarchs. The domination of Church Council decisions by successive Roman Emperors, and the increasing imposition of 'Roman' religious orthodoxy, increasingly alienated large sections of the Christian Church. By the time of Gregory I, Bishop of Rome (590-604), orthodox 'Roman' Christianity was everywhere on the defensive because of the Germanic invasions, and later those of Islam. Islam overran three of the five patriarchates (Alexandria, Jerusalem and Antioch) in the 7th century A.D., leaving only the two quarrelling patriarchates of Rome and Constantinople, with Rome vying for pre-eminence in authority as well as in honour.

109. "Was never king more highly magnifyde nor dread of Romans was than Arviragus." — Ed. Spenser.

110. St. Edward the Confessor, King of England, had been England's previous patron saint.

111. The baptism of King Lucius took place on 28th May, A.D. 137. In 167 Lucius commemorated the event by building a chapel dedicated to St. Michael on the summit of Glastonbury Tor. In 170 he founded the church at *Caer Wynn* (Winchester) on the site of a previous Druidic sanctuary (now become the Cathedral of Winchester). In 179 he founded the church of *Lambedr* (St. Peter) on Tothill (Westminster) in *Caer Troia* (London), which he made the Metropolitan and Chief Church of the Kingdom. Three Archbishops were said to have been established by him — at *Caer Leon*, *Caer Troia* and *Caer Erroc* (York) — the three Archdruids becoming known as Archbishops.

112. *Caer Wynn* (Winchester) was made the principal royal capital of Britain in 500 B.C. by Dunwal Molmutius, the Great *Numa* (Law-maker) and one of the "Three Wise British Kings". The *Arviragus* or High King had previously had his royal capital at *Caer Troia*, but for a specific reason it was moved to *Caer Wynn*, where the supreme royal seat remained for 1500 years — except for the period when *Caer Meini* (Colchester — the capital of King Coel) was the principal seat of the High King, at the time of the Roman invasion of Britain under the Emperor Claudius (A.D. 42).

113. Sabellius, A.D. 250.

114. Francis Bacon, *Government of England*.

115. See note 111.

116. Not of course the present castle structure, but the original holy mound on which the later Norman castle was built. The earlier British name for this Druidic sanctuary and Gorsedd was *Win-de-Sieur*, meaning 'the White or holy mound of the Sieur or Lord.' It is also known as 'The Round Table Mound'. (See E.O. Gordon, *Prehistoric London*.)

117. *i.e.* Mary I.

118. Strype, *Annals*, vol. I.

119. *i.e.* "Supreme Head on earth of the Church of England."

120. Galahad, the 'bastard' son of Queen Guinevere and Sir Lancelot of the Lake, becomes the fully attained Grail Knight, achieving the Holy Grail in its entirety and thereby qualifying to be enthroned in the Siege Perilous (*i.e.* the Throne) of the Grail King, and to preside as Grail King over the assembly of knights.

121. The caste system was once used world-wide in all those cultures said to derive from the last *Manu*, the instigator and initiator of the present Root Race and Sub-Root-Race known as 'Aryan'. It may be found from the Vedic culture of India, to the Druidic culture of Britain and the Inca civilization of South America. Paramahansa Yogananda gives a good summary of the caste system in his remarkable book, *Autobiography of a Yogi*:

"The origin of the caste system, formulated by the great legislator Manu, was admirable. He saw clearly that men are distinguished by natural evolution into four great classes: those capable of offering service to society through their bodily labor (*Sudras*); those who serve through mentality, skill, agriculture, trade, commerce, business life in general (*Vaisyas*); those whose talents are administrative, executive, and protective — rulers and warriors (*Kshatriyas*); those of contemplative nature, spiritually inspired and inspiring (*Brahmins*). 'Neither birth nor sacraments nor study nor ancestry can decide whether a person is twice-born (*i.e.*, a Brahmin),' the *Mahabharata* declares, 'character and conduct only can decide.' Manu instructed society to show respect to its members insofar as they possessed wisdom, virtue, age, kinship or, lastly, wealth. Riches in Vedic India were always despised if they were hoarded or unavailable for charitable purposes. Ungenerous men of great wealth were assigned a low rank in society.

"Serious evils arose when the caste system became hardened through the centuries into a hereditary halter. India, self-governing since 1947, is making slow but sure progress in restoring the ancient values of caste, based solely on natural qualification and not on birth. Every nation on earth has its own distinctive misery-producing karma to deal with and honorably remove. India, with her versatile and invulnerable spirit, is proving herself equal to the task of caste reformation.

"Inclusion in one of these four castes originally depended not on a man's birth but on his natural capacities as demonstrated by the goal in life he elected to achieve,' an article in *East-West* for January, 1935, tells us. 'This goal could be (1) *kama*, desire, activity of the life of the senses (Sudra stage), (2) *artha*, gain, fulfilling but controlling the desires (*Vaisya* stage), (3) *dharma*, self-discipline, the life of responsibility and right action (*Kshatriya* stage), (4) *moksha*, liberation, the life of spirituality and religious teaching (*Brahmin* stage). These four castes render service to humanity by (1) body, (2) mind, (3) will power, (4) Spirit.

"These four stages have their correspondence in the eternal *gunas* or qualities of nature, *tamas*, *rajas*, and *sattwa*: obstruction, activity, and expansion; or, mass, energy, and intelligence. The four natural castes are marked by the *gunas* as (1) *tamas* (ignorance), (2) *tamas-rajas* (mixture of ignorance and activity), (3) *rajas-sattwa* (mixture of right activity and enlightenment), (4) *sattwa* (enlightenment). Thus has nature marked every man with his caste, by the predominance in himself of one, or the mixture of two, of the *gunas*. Of course every human being has all three *gunas* in varying proportions. The guru will be able rightly to determine a man's caste or evolutionary status.

" 'To a certain extent, all races and nations observe in practice, if not in theory, the features of caste. Where there is great license or so-called liberty, particularly in intermarriage between extremes in the natural castes, the race dwindles away and becomes extinct. The *Purana Samhita* compares the offspring of such unions to barren hybrids, like the mule, which is incapable of propogation of its own species. Artificial species are eventually exterminated. History offers abundant proof of numerous great races which no longer have any living representatives. The caste system of India is credited by her most profound thinkers with being the check or preventive against license that has preserved the purity of the race and brought it safely through millenniums of vicissitudes, though many other ancient races have completely vanished.' "

122. Masilio, Ockham and Dante — the great theorists of imperialism. See Dante's *Monarchia* (1559) and Marsilio's *Defensor*.

123. Petrarch — the most celebrated poet of the Italian Renaissance.

124. Pico della Mirandola — philosopher of Renaissance Neo-Platonism.

125. See the capital letter 'C' from John Foxe's *Acts and Monuments*, which was also used in John Dee's *General and Rare Memorials pertaining to the Perfecte Arts of Navigation* (1577).

126. In Cabbalistic terms, the 'Father' is *Chokmah* (Wisdom), the 'Mother' is *Binah* (Intelligence), and the Child is *Daath* (Knowledge). *Chokmah* is the *Christos* (Light) — the divine Idea or Wis-dom ('Knowledge of the Lord'); and *Binah* is the *Sophia* — the perfect Intelligence that contains and conceives the Light as the perfect Vision of Truth. *Daath* — the human knowledge and revelation of God — is the *Christos-Sophia* that is born or made manifest in matter in the form of the radiantly wise and loving soul. Jesus, or *Ies-Hu*, meaning 'the soul of Light' or 'embodied Light', is the initiatory name given to this manifest Christ.

127. "Justice is nothing else than the pious and religious worship of the one God" — Lactantius, *Div. Inst.* v. 5. '*De iustitia*'.

128. See title page of the Bishop's Bible (1569) — Justice and Mercy hold a crown over Elizabeth's head. Justice and Mercy correspond to the Cabbalistic *Geburah* (Righteousness, Discernment, Severity) and *Chesed* (Grace, Mercy, Generosity), which, when balanced and united, give rise to *Daath* (Knowledge) — the universal enlightenment that is the crown of man's endeavours and which does in fact bestow a corona or halo of light upon the head of the illumined person.

129. Roy Strong, *The Cult of Elizabeth*, iv — 'November's Sacred Seventeenth Day'.

130. *cf.* Francis Bacon's references to "the Light of God's Word" and "the Light of God's Works", both of which need to be studied so that we become knowledgeable in both. Bacon refers to these studies and knowledges as "Divinity" and "Philosophy" respectively. (See Appendix: *Humanism*.)

131. From 1581 onwards all Accession Day Tilts were held at Whitehall, with one exception in 1593 when, because of plague in London, the Tilt was staged beneath the walls of Windsor Castle.

132. The Queen arrived at Kenilworth Castle on 8th July 1575.

133. The Court arrived at Woodstock on 11th September 1975.

134. Carey, or Grey.

135. *i.e.* Henry VIII, Edward VI, Mary I, Elizabeth I, James I.

136. From Lee's Epitaph.

137. W. Segar, *Honor Military and Civill*, London, 1602, p. 197.

138. The Family of Love were a loosely knit 'family' or fellowship of many nationalities, in several countries, bound together by their common philosophy that it was each person's inner spiritual life that alone mattered, and that, by comparison, outer form was not so important but was (and should be) adaptable. The branches of a tree should be able to bend with the wind, whichever way the wind blows; but the tree itself must be healthy, not rotten inside. Thus a Familist was able to conform outwardly to whatever religion or government was officially established, whilst retaining inner virtue and peace. The Familist's philosophy was philanthropic and humanistic. Charity was its chief virtue. Their 'members' produced the **real** 'Renaissance' in terms of light and love, being in the forefront of the great illumined teachers, healers, writers, scholars, poets, artists and craftsmen; creating and sponsoring worthwhile translations and publications of all kinds (printing them as well), and organising charitable works and educational schemes.

139. 'Classical' — sometimes called 'Aristotlean'.

140. Frances Yates, 'Elizabethan Chivalry', *Astraea*.

141. Loricus and Lelius were two chivalric pseudonyms used by Lee in the tournaments.

142. 'Song in the Oak' — written by Edward Dyer.

143. *The Queen's Majesty's Entertainment at Woodstock, 1575,* p. xxviii. (See note 147).

144. Presented to the Queen on 1st January 1576 (modern reckoning).

145. Frances Yates, 'Elizabethan Chivalry' *Astraea.*

146. Dedication by George Gascoine to The Tale of Hemetes the Heremyte, presented to Queen Elizabeth, 1st Jan. 1576.

147. Only one copy of the print survives. It has been published in the book, *The Queen's Majesty's Entertainment at Woodstock, 1575,* edited by A. W. Pollard, Oxford, 1910. The first part of the Entertainment, *The Tale of Hemetes the Heremyte,* is printed from Gascoigne's copy in Nichols, *op. cit.* I. pp. 553-82.

148. Probably the Tournament of either 1579 or 1580.

149. *c.f.* Francis Bacon, *Advancement of Learning,* Bk. I:—

> "In the first event or occurrence after the fall of man, we see, (as the Scriptures have infinite mysteries, not violating at all the truth of the story or letter,) an image of the two estates, the contemplative state and the active state, figured in the two persons of Abel and Cain, and in the two simplest and most primitive trades of life; that of the shepherd, (who, by reason of his leisure, rest in a place, and living in view of heaven, is a lively image of a contemplative life,) and that of the husbandman: where we see again the favour and election of God went to the shepherd, and not to the tiller of the ground."

150. Among those present at the Kenilworth and Woodstock Entertainments were Mary Sydney and, almost certainly, her brother Philip, who had returned from the continent a few months previously, and their poet friend, Edward Dyer, who wrote and performed in at least one of the devices and who had been Steward of Woodstock Manor before the appointment of Sir Henry Lee. Together with Daniel Rogers, Gabriel Harvey and 'Immerito', plus a few others, these poets formed the literary nucleus called the 'English Areopagus'.

151. This is the first reference to 'The Shepheard's Calender', and shows that it was even then existing in some form or another.

152. According to various different ciphers of Francis Bacon's, which are categorical, and other evidence of different kinds, Robert Devereux was actually born the second natural son of Queen Elizabeth and the Earl of Leicester, and thus the younger brother to Francis Bacon. There is a manuscript horoscope drawn up for him, together with one for his own son, in the British Museum, which Jean Overton Fuller draws attention to in her book, *Sir Francis Bacon, A Biography.* The intimate details of his life recorded there suggest strongly that it was Robert Devereux himself who supplied the astrologer with the information. The date of his birth is given as Sunday November 10th, 1566, at 9.45 am. Jean Overton Fuller was able to verify the date from the data given in the horoscope, and in addition determine the place of birth: namely Nonsuch Palace, in Surrey. We have to thank Jean Overton Fuller for her excellent research in this, for previously it had been assumed that Robert had been born at Netherwood, Herefordshire; neither had his date of birth been known with accuracy. The horoscope data is precise, detailed and categoric, and heavily supports the thesis that Robert Devereux was indeed the son of Queen Elizabeth, whose palace Nonsuch was. There is no record of his birth in the church registers of the Netherwood area, as there should have been if he had indeed been born there as the son of Lord and Lady Hereford. Neither does there seem to be any record of the Queen's whereabouts at that time in question.

At the time of his birth, the Queen's cousin, Lettice Knollys, was the Chief Lady of the Bedchamber, and wife of Sir Walter Devereux, Viscount Hereford (afterwards created Earl of Essex). Like Lady Anne Bacon before her, Lettice was asked to adopt the Queen's son and keep the matter quiet. The baby boy was christened with his father's name, Robert, and seemed to become the special favourite of his royal mother and noble father. Lettice's own children — Penelope (b. 1563), Dorothy (b. 1565) and Walter (b. 1569) — were all born, as might be expected, at the Hereford's only country house, Chartley in Staffordshire. As for Lord Hereford, in 1569 he served as High Marshal of the field, under the Earl of Warwick and Lord Clinton, in the suppression of the northern insurrection. In 1572 the Queen made him a Knight of the Garter, created him Earl of Essex, and gave him an estate in the county of Essex. The Queen then sent him to Ireland to recover a barony in Ulster from the Irish rebels, which, when obtained, they were to divide between them. The Queen lent him £10,000 for the expedition, at 10% interest, on security of the Earl's estates which were to be forfeit to the Crown if the loan was not repaid. He set sail for Ireland with his troops in July 1573. At the end of 1575 he returned to England, during which occasion the Queen made him Earl Marshal of Ireland and sent him back again to Ireland. Soon after his arrival in Dublin he died of dysentry (some say of poison) on September 22nd, 1576. Robert became Earl of Essex on Sir Walter's death.

Whilst in Ireland, Essex wrote to Lord Burghley (on November 1st, 1573) offering him the "direction, education and marriage" of Robert — an unusual arrangement unless Robert was indeed the son of the Queen and Essex had been asked to give over the guardianship of the boy to the Queen's trusted friend and Lord Treasurer.

Robert remained in the country at Chartley during his boyhood, until January 11th, 1577, when he became a member of Burghley's family for a few months, prior to going up to Trinity College, Cambridge, in May, 1577, following in the footsteps of his elder brother, Francis. His Christmas vacation, 1577, was spent at Court, where the Queen showed some intimacy with him. On July 6th, 1581, at the age of fourteen, he received the degree of Master of the Arts. From that time he resided at Lanfey House, in Pembrokeshire, until 1584, when he went to live in London with his supposed step-father, the Earl of Leicester, who had secretly married his foster-mother, Lettice, in September 1578.

Robert may have resided with his father and foster-mother at Leicester House before this, during university vacations, from 1579-81. In December 1585 he accompanied his father, the Earl of Leicester, to the Low Countries, to join Sir Philip Sydney and the English army who were then helping the Dutch against the Duke of Parma's forces. He fought bravely at Zutphen, and was knighted on his return to England, residing mainly at Court from then on and shown great favour by his royal mother. In 1587 Leicester relinquished his privileged post as Master of the Horse in Robert's favour, and in 1588 was made a Knight of the Garter and General of the Horse in the army gathered at Tilbury to resist the Spanish Armada.

Leicester died that year, on September 4th, 1588, leaving by will his George and Garter to Robert. In 1589 Robert 'escaped' (against the Queen's orders) from Court to join the naval expedition sent against the Spaniards in Portugal. He returned and quickly made peace with his sovereign mother, but then secretly wed Sir Philip Sydney's widow, Frances (Walsingham), in April 1590. It was some months before the Queen discovered the marriage, and when she did she was furious, declaring that he had married not only without her consent but below his degree!

153. The third aspect of the Cross symbology was represented by the Aspen, signifying Sorrow and Fear, the counterpart to Joy. Joy and Sorrow are the two complementary aspects of Truth, thus the relationship of Oak, Holly and Aspen is that of Truth, Joy and Sorrow respectively. These three are embodied in the dramatic terms used for the Shakespeare plays — namely, History, Comedy and Tragedy — but the plays are not in reality categorised according to these types or aspects, but every single play has a mixture of all three within it. Cabbalistically Truth, Joy and Sorrow have a correspondence with the three 'pillars' of the Tree of Life — *viz.* the central pillar of Peace (or Equilibrium), the right-hand pillar of Mercy, and the left-hand pillar of Severity. In Vedic terms these three equate with and issue from the supreme principles of *Brahmâ* (the Creator, Truth — the 1st Principle) *Vishnu* (the Maintainer, the All-Pervading Light and Sustaining Life — the 2nd. Principle) and *Shiva* (the Generator, Destroyer, Regenerator and Transformer of all form — the 3rd. Principle). They have a further correspondence with man's three-fold nature of spirit (truth), soul (joy) and body (sorrow).

154. England is traditionally known as 'Mary's Dowry' — that is to say, her kingdom, her inheritance, her marriage offering. It is equated with the heart and symbolised by the white lily or rose, which becomes red and then golden as it gives birth to and is irradiated by the Christ child.

155. *Clas Myrddin* = 'the Enclosure of Merlin'.

156. Fate is an English word for *karma*, the natural law of cause and effect.

157. *Arcadia* was completed by Philip Sydney's sister, Mary, after his death in the Low Countries in 1586.

158. *i.e.* Francis Bacon, who used the pseudonym 'Immerito' for the poetic works later 'masked' by Edmund Spenser. 'Immerito' means 'the blameless one' — one who is innocent but who is made to bear the weight of blame or sin of others. For Francis it had two distinct allusions: one to the circumstances of his birth, with his royal mother ever blaming him and his arrival into this world as the cause of her woes, and the other to the more esoteric idea of the perfect and royal sacrifice — the 'lamb of God' — which he believed in as an ideal and acted out throughout his life to make it real and manifest. *The Shepheard's Calender* was first published anonymously in the winter of 1579-80. The first use of Spenser's name as the 'author' was in the publication of the first part of the *Faerie Queene* in 1590, for which the signature 'Ed. Spenser' was used. The *Complaints*, which followed in 1591, used the signature 'Ed. Sp.' — an unusual and pointed abbreviation which renders, in the Simple English Cabbala used by Francis Bacon, 57 (*i.e.* E.D.S.P. = 5.9.18.15 = 57), which is the cipher for one of his commonest signatures, FRA.BACON. The first allusion to 'Spenser' as a writer and poet in any published work seems to have been in 1589, shortly before the first part of the *Faerie Queene* was published, in the preface written by Nashe for Greene's *Menaphon*: "divine Master Spenser, the miracle of wit to bandie line for line, in the honor of England, gainst Spaine, France, Italie, and all the worlde." In Francis Bacon's ciphers, he categorically states that both Nashe and Greene, as well as Spenser, were 'masks' for his own authorship. Francis was well-known (and reported) for his ability to write in varying styles as required, suiting the style to the subject matter, and matching the 'style' of the chosen 'mask' or inventing one for that 'mask'. The seven principle 'masks' that he came to use were George Peele, Robert Greene, Edmund Spenser, Christopher Marlowe, Robert Burton (initially Timothy Bright), William Shakespeare and Francis Bacon ('Bacon' being only his foster-name, not his real one). These were not the only 'masks' that Francis concealed his authorship behind, but they were the principal ones that he used in a cabbalistic way, to form a sequence and pattern of writings that express and reveal divine law; the keys to which are deliberately scattered throughout his writings and 'completed' with the final key given in his *Abcedarium Naturae, or a Metaphysical Piece* (also known as *The Alphabet of Nature*), a 'fragment' of which was published for the first time in 1679 by Archbishop Thomas Tenison, in his book *The Lord Bacon's Physiological Remains*.

159. *Pallas Athena* literally means 'the Shaker of the Spear'. Many 'Gods' or archetypal Principles are known as "Spear-shakers". The Archangel Michael and Goddess Britannia are two associated with this country, being counterparts of each other. Apollo and Pallas Athena are their Greek equivalents. They represent the dual principles of Illumination — Wisdom and Intelligence — the one radiating and the other reflecting and making known the Light. The one is the pure knowledge or Idea of God (the holy Wisdom) that radiates from the omnipresent heart, whilst the other is that which receives and can know the Knowledge — the Mind or Intellect whose intelligence conceives (or holds within itself) the Light of Knowledge and gives birth to its reflection as Understanding (*i.e.* Revelation). The former is the Knowledge, the Thought, the Wisdom of God; the latter is the Knower, the Thinker, the wise or illumined Mind. It is the latter which is directly associated with 'Man', meaning 'the Mind' or 'Thinker', whilst the former is the 'Light'. Hence Man is to be made (*i.e.* to become in the course of evolution) the 'image' or reflection of the Light. The image or reflection is called the 'Hu-man', which means 'Man of Light' or 'the Illumined Man'. The word 'Jesus', from the Celtic *Ies-Hu*, has the same meaning, *Ies* being the Man, Mind or Soul and *Hu* being the Light.

Britannia actually means 'the Chosen of Anna', which is none other than Anna's 'daughter', Mary. *Anna* refers to the state of Being before the Light of God radiates within it (*i.e.* before God thinks). It is the state referred to in *Genesis:* "And Darkness lay upon the face of the Deep." Mary refers to the state of the mind or soul in which Light is shining and being reflected. The actual manifestation of Light in the soul is the Christ Child, born of the Virgin Mary, shining like a star or a jewel of light, a 'sun' or light. Both Anna and Mary are 'virgins', which refer to a condition of Being that is pure, unpolluted by wrong emotions and thoughts. Pallas Athena is the same as Britannia, which is Mary, the virgin Mother of the Christ manifestation. *Cassiopeia*, a title of Isis, meaning 'the Enthroned Lady' or 'Queen of the Heavens', is also the same as Mary or Athena. The 'throne' is the symbol of this Virgin Mother or Mind of Man, upon or in which the Christ Child 'sits' as King of the Heavens — as Solomon, the King of Peace (*Solomon* means 'Peaceful'), and

ERRATA

Page 286, Note 158:

The sixth sentence, beginning "The *Complaints*, which followed...", gives an incorrect statement of the cipher signature employed by Francis Bacon. E.D.S.P. does not of course equal 57 in Simple English Cabbala. The sentence should read as follows:

> The *Complaints*, which followed in 1591, used the signature 'Ed.Sp.' - an unusual and pointed abbreviation which renders, in Simple English Cabbala used by Francis Bacon, '9, 33' (*i.e.* E.D.S.P. = 5.4,18.15 = 9, 33), which is the cipher for "I. BACON' (*i.e.* I = 9; BACON = 33), whereas the fuller signature 'Ed.Spenser' renders, in the same Simple English Cabbala, '100' (*i.e.* E.D.S.P.E.N.S.E.R = 5.4.18.15.5.13.18.5.17 = 100), which is the cipher for 'FRANCIS BACON' (*i.e.* FRANCIS BACON = 67 + 33 = 100). By using this double cipher signature in such a definitive way, Francis Bacon was able to rule out chance when it came to the matter of others deciphering and recognising his signature and intention.

Page 293, Note 289:

This note should read:

> 289. Francis Bacon, Essay *Of Goodness and Goodness of Nature.*

Jedidiah, the 'Beloved of the Lord.' His temple is the illumined form of his own mind or soul, which temple is the same as the throne.

160. *Faerie Queene*, Bk. I: Introduction. The Nine Muses are all aspects of the one and only Supreme Muse, which is the perfect Intelligence called Pallas Athena, Britannia, Mary, the Faerie Queen, the Virgin Queen, *etc*. The sacred Muse of the Christian Rose Cross Knight is his own perfect and holy intelligence — his all-embracing 'genius'.

161. Faerie Queene, Bk. I, Canto I: 3.

162. *Advancement of Learning.*

163. *Ibid.*

164. *Ibid.*

165. *Ibid.*

166. *Ibid.*

167. *Ibid.*

168. *Ibid.*

169. *New Larousse Encyclopedia of Mythology.*

170. The falcon symbol was used in the same manner, with the same physical and metaphysical meanings, by other great civilizations all around the world. In the Andes of South America, for instance, the rising/setting sun is called *Inti*, meaning 'falcon'. We still have the remnants of this knowledge and symbol in the British tradition of the 'royal falcon'.

171. Earth = Universal Matter.

172. The 'air', 'sea' and 'earth' are the three major degrees or aspects of 'Earth', the Universal Matter or Ether. Universal Matter is organised and formed into these three basic grades or 'spheres' by the vibratory force of the Word of God. '**Air**' symbolises the sphere of thought (the mental planes), '**water**' represents the sphere of the emotions (the astral or emotional planes), and '**earth**' signifies the dense etheric and physical sphere of all outermost activity. They are called *manas, kama* and *sthula* in the Vedic teachings of the East. They comprise man's mortal, corruptible body, the very transformations of which gradually produce and build up the immortal soul. Man, the Universal Mind, has dominion over the whole realm of Nature, with its three spheres of existence. Mankind, individually and *en masse*, may once again achieve this domain over Nature, which once they had before the 'fall', by cleansing their emotions, thoughts and actions, and achieving in themselves the original purity of the Universal Mind.

173. Metaphysically, the herbs bearing seed and the trees bearing fruit and seed that are upon the "face of all the earth" refer to the Ideas or Thoughts of God which vibrate (as the Holy Spirit) "upon the face of the waters". "All the earth" refers to Universal Matter, as does "the waters". These Thoughts of God contain the archetypal ideas of seed, germination, growth, unfoldment or blossoming, achievement or fruiting, offering or sacrifice, death or dissolution and rebirth or reformation — the seeds of new life forms contained within the old forms of life which, through their metamorphoses or transformations, produce life forms transmuted from corruptible (*i.e.* ever changing) Nature into incorruptible souls of light. Other creatures "feed" off the "herbs", but man can "feed" off both "herbs" and "trees" — the 'Tree of Life' summing up his perfect food. This spiritual food, called the 'meat' or 'bread' of heaven (the divine Mind), is the same as the 'manna' which came (and still comes) from heaven to feed the Israelites in the Wilderness. 'Manna' are the divine Thoughts or Ideas that are inspired from the heart of God into the 'man' or 'mind', which, if the mind be pure and receptive enough, will feed and nourish the man, fulfilling all his needs and illumining his mind. The divine Ideas are the 'Words' of God.

174. *Genesis* i. 26-29.

175. Mind = Heaven.

176. "It is written that 'man shall not live by bread alone, but by every word of God,'" (*Luke* iv. 4).

177. *Genesis* i. 30.

178. "When the White Head decided to add to its beauty with an ornament, it sent out a spark from its own light.

"It breathed on the spark to cool it, and the spark grew firm.

"It expanded and hollowed itself out, like a blue, transparent skull enclosing thousands, myriads of worlds.

"This cavity is full of eternal dew, white on the side of the Father, red on the side of the Son. It is the dew of light and life, the dew that engenders universes and resurrects the dead.

"Some are resurrected in light, and the others in fire.

"Some in the eternal whiteness of peace, the others in the redness of fire and the torments of war.

"The wicked are the disgrace, so as to speak, of their father, and it is they who cover the face with its redness.

"In this skull of Universal Man, only begotten Son of God, resides Knowledge, with its thirty-two paths and its fifty gates."

('The Skull of the Microprosopopeia', *The Book of Splendours* — the *Zohar* of Jewish mysticism.)

179. Disciple = 'he who is disciplined' or 'who disciplines himself'.

180. *Psalm* xvii. 15.

181. *Isaiah* xiv. 12-14.

182. "They that see thee shall narrowly look upon thee, and consider thee, saying, 'Is this the man that made the earth to tremble, that did make kingdoms...' " (*Isaiah* xiv. 16.)

183. *Job* xli, *Psalm* lxxiv. 13-14, *Psalm* civ. 26, *Isaiah* xxvii. 1-2.

184. "Behold, I send you forth as sheep in the midst of the wolves: be ye therefore wise as serpents, and harmless as doves." (*Matthew* x. 16.)

185. "Piercing" and "crooked" refer to the strength and spiralling motion of the kundalini, the "dragon that is in the sea" of universal matter.

186. *Isaiah* xxvii. 1-2.

187. Symbolised for us in the Feast of Remembrance followed by the Crucifixion.

188. Oedipus was born the son of Laius, King of Thebes, and Jocasta. Told by the Delphic oracle that he would perish by the hands of his son, Laius ordered his wife to kill the babe. The Queen could not bring herself to do this, but gave the child to one of her attendants with orders to expose the child on the mountains. A shepherd discovered the child and carried Oedipus home, where he and his wife raised the child as their own. The accomplishments of the child soon became the admiration of the age. One of his companions, who envied Oedipus' talents, told him that he was an illegitimate child. Questioning this, he asked Periboea, his foster-mother, if this were true. Not wishing to hurt the boy, she replied that his suspicions were unfounded. Not satisfied with her answer, Oedipus went to Delphi to consult the oracle. The oracle told him not to return home, for if he did, he must be the murderer of his father and the husband of his mother.

This answer terrified Oedipus. He resolved not to return to the only home and parents he knew, but travelled away from Corinth and towards Phocis, where he was met and challenged by King Laius. Oedipus fought and slew the King, little knowing who he was. Then he made his way to Thebes, attracted by the fame of the Sphinx. This terrible monster, which Juno had sent to lay waste the country, resorted to the neighbourhood of Thebes and there devoured all those who attempted to explain without success the aenigma which he proposed. But the successful explanation of the aenigma would result in the death of the Sphinx. King Creon, who had ascended the throne of Thebes at the death of Laius, promised his crown and Jocasta to him who succeeded in the attempt.

The aenigma was: "What animal in the morning walks upon four feet, at noon upon two, and in the evening upon three?"

When Oedipus came to the Sphinx, he answered the Sphinx: "Man, in the morning of his life, walks upon his hands and his feet; when he attains the years of manhood, he walks upon his two legs; and in the evening of his life, he supports his old age with the assistance of a staff."

The monster, mortified at the correct answer, dashed his head against a rock and died. Oedipus ascended the throne of Thebes, and married Jocasta, by whom he had two sons and two daughters. Eventually Oedipus discovered that he had slain Laius, who was his real father, and was wed to his mother. In grief he put out his own eyes, as not fit to see the light, and banished himself from Thebes. He retired to Attica, led by his daughter Antigone, where he died and was swallowed up by the earth. As the oracle had also declared, his death became the prosperity of the country he died in. His tomb was near the Areopagus of Athens.

189. "And the disciples came, and said unto him, 'Why speakest thou in parables?'

"He answered and said unto them, 'Because it is given you to know the mysteries of the kingdom of heaven, but to them it is not given. For to whosoever hath, to him shall be given, and he shall have more abundance: but whosoever hath not, from him shall be taken away even that he hath. Therefore speak I to them in parables: because they seeing see not; and hearing they hear not, neither do they understand. And in them is fulfilled the prophecy of Esaias, which saith, *By hearing ye shall hear, and shall not understand; and seeing ye shall see, and shall not perceive: for this people's heart is waxed gross, and their ears are dull of hearing, and their eyes they have closed; lest at any time they should see with their eyes, and hear with their ears, and should be converted, and I should heal them.*

"'But blessed are your eyes, for they see: and your ears, for they hear. For verily I say unto you, that many prophets and righteous men have desired to see those things which ye see, and have not seen them; and to hear those things which ye hear, and have not heard them.'"

(*Matthew* xiii, 10-17)

190. *Advancement of Learning*, Bk. II.

191. *Ibid.*

192. *Faerie Queene*, Bk. I. 'Philosopher' = 'Lover of *Sophia*'. '*Sophia*' = the feminine aspect of Wisdom; that divine Intelligence and Consciousness which allows understanding to develop in man's own mind. *Sophia* is sometimes known as Truth. All true philosophers, or Rosicrucians, are wed to *Sophia*, the Truth. *Christos* and *Sophia*, the radiant and reflective aspects of divine Wisdom respectively, are 'wed' or united with the complementary aspects of man's nature when made ready or pure enough for the 'marriage' to take place.

193. The Biliteral and Word ciphers.

194. This rendering of the cipher is in a more modernised form of spelling to aid the reader.

195. *Biliteral Cipher*, pp. 90-91. (Cipher found in *Novum Organum*, 1620.)

196. See 'The Birth and Adoption of Francis Bacon', F.B.R.T. Journal I/3, *Dedication to the Light*.

197. *Biliteral Cipher*, pp. 137-139. (Cipher found in *Henry VII*, 1622.)

198. *Word Cipher*, pp. 89-98.

199. *Ibid.* pp. 100-101.

200. *Ibid.* p. 109.

201. *Ibid.* p. 138.

202. *Biliteral Cipher*, pp. 334-5 (Cipher found in *New Atlantis*, 1635.)

203. *Biliteral Cipher*, p. 88.

204. Here the Queen likens herself to King David.

205. Spirit.

206. *Word Cipher*, pp. 256-262.

207. *Biliteral Cipher*, p. 361.

208. *Ibid.* p. 88.

209. Robert Cecil's reports on Francis to the Queen.

210. Alice Barnham, who married Francis Bacon on May 10th, 1606.

211. *Biliteral Cipher*, pp. 335-6.
212. Letter to Essex (December or January 1595), published in *Resuscitatio*.
213. Letter to Robert Cecil (January 1595). Lambeth Library: Anthony Bacon's papers, vol. IV. folio 31.)
214. Flowers for Henri III's table were gathered from far and near, even being sent for from other countries. The flowers listed by Francis are all summer-flowering, except the spring-time daffodils which may have been sent from another country where they were still in season, thus reinforcing the date of Francis' first visit to Henri's Court as taking place in the **Summer** months. Francis' second visit — the publicly known one — took place in the **Autumn** (late September-October) of 1576.
215. *Word Cipher*, pp. 571-5.
216. Alençon's face was pitted with the results of smallpox scars.
217. *Word Cipher*, pp. 575-6.
218. *Ibid.* p. 587.
219. *Ibid.* p. 592.
220. *Ibid.* p. 593.
221. *Ibid.* p. 779.
222. *Ibid.* p. 785.
223. *Biliteral Cipher*, pp. 175-6. (1623 Shakespeare Folio.)
224. *Ibid.* p. 12. (*Midsummer Night's Dream*, 1600 Quarto.)
225. *Word Cipher*, p. 784.
226. *Ibid.* p. 268.
227. Henri de Navarre's mother was poisoned by Catherine de Medici.
228. Catherine de Bourbon.
229. *Word Cipher*, pp. 781-3.
230. Agrippa d'Aubigne, Navarre's faithful equerry, plus two or three other attendants.
231. Not all was entrusted to letter: "messages" were given by word of mouth.
232. *Word Cipher*, p. 874.
233. *Ibid.* pp. 879-880.
234. The Earl of Leicester, who was a warrior by nature.
235. *Word Cipher*, pp. 893-4.
236. *Ibid.* p. 917.
237. Pitched.
238. *Word Cipher*, p. 923.
239. *Ibid.* pp. 925-6.
240. *Ibid.* p. 931.
241. *Ibid.* p. 933.
242. *Ibid.* p. 934.
243. *Ibid.* p. 947.
244. A long section of love-talk between the two lovers follows here.
245. *Word Cipher*, pp. 989-994.
246. *Biliteral Cipher*, p. 337. (*New Atlantis*).
247. *Ibid.* pp. 361-2. (*Natural History*.)

248. The Family of Love — see note 138.
249. William T. Smedley, 'The French Renaissance and the Pleiades', *Baconiana* vol. ix. Third Series, no. 34.
250. Frances A. Yates, 'The Joyeuse Magnificences', *Astraea*.
251. D. P. Walker, *Spiritual and Demonic Magic from Ficino to Campenalla*, p. 100 (Warburg Inst., 1958.)
252. *Astraea*, p. 188.
253. Marguerite's diary.
254. Alençon was referred to as "Monsieur".
255. Marguerite's diary.
256. H. Noel Williams, *Queen Margot*, p. 192 (Harper & Bros., 1907).
257. Shakespeare plays.
258. *i.e.* 40 years.
259. *Biliteral Cipher*, pp. 79-80. (*Romeo and Juliet*, undated Quarto.)
260. A considerable amount of research on this subject has been done by Ewen MacDuff Strachen, to whom I am indebted for knowledge concerning this subject. His earlier books, discussing some of the principal examples which led him to his fuller discoveries, are *The Sixty Seventh Inquisition* and *The Dancing Horse Will Tell You*.
261. Letter to Burghley.
262. Letter to Burghley.
263. Letter to Essex.
264. See also: *Calendar of State Papers, Foreign, 1574-1577*.
265. *State Papers, Foreign*, 5th December 1577.
266. R. J. W. Gentry, *Shakespeare and the Italian Comedy* (*Baconiana* vol. XXXI no. 122.)
267. '*Librum Mundi*' (Book of the World'), also referred to as '*Librum Naturae*' ('Book of Nature') — the collected 'History' of Nature or God's Works. See the *Fama Fraternitatis or, A Discovery of the Fraternity of the Most laudable Order of the Rosy Cross*.
268. *Biliteral Cipher*, p. 88 (*Novum Organum*).
269. *Ibid.* p. 121.
270. *Ibid.* p. 122.
271. *Word Cipher*, pp. 32-34.
272. Marguerite's diary.
273. M. A. Mongez, *Histoire de la Reine Marguerite de Valois premiere femme du roi Henri IV*, (Paris, 1777), p. 204.
274. In the Tudor period each new year began on the 25th March, thus February 1577 by 16th century reckoning would be February 1578 by modern reckoning.
275. Joachim Gerstenberg, *Strange Signatures*, (The Francis Bacon Society, 1967), p. 13.
276. *i.e.* Back-conies, or Back-cony = Bacony, or Baconian.
277. William T. Smedley, *The Mystery of Francis Bacon*. (*Baconiana*, vol. IX, nos. 33-36.)
278. An account of the Gray's Inn Christmas revels of 1594 is given in the '*Gesta Graiorum*' set out in Nichol's *Progresses of Queen Elizabeth*, vol. III., p. 262. For some reason the annual Christmas revels, in which the students of Gray's Inn prided themselves, were intermitted for three or four years (1590-3), and in 1594 they resolved to redeem the time by producing "something out of the common way". Francis Bacon was called upon to assist in "recovering the lost honour of Gray's Inn". The result was an elaborate Entertainment or Revel designed to last for the Twelve Days of Christmas, beginning on the first day (or evening) of the Winter Solstice. It was called 'The Prince of Purpoole', in which chosen

members of Gray's Inn played the principal roles, and which included masques, mumming, dancing, performances of plays and acrobatics, and feasting. The 'Prince', with all his 'state', proceeded to the Great Hall of Gray's Inn on Friday 20th December, 1594, to open the proceedings. The initial entertainments were so successful that the members were encouraged to enlarge their plan and raise their style, and so they resolved to have certain "Grand Nights" in which something extra special should be performed for the entertainment of noble guests specially invited for those evenings. Accordingly, on Saturday 28th December (Innocents Day), having erected a stage for acting and dancing upon, and scaffolds for seating the spectators, the first Grand Night was held.

The play which was booked for that night was entitled *'A Comedy of Errors, (like to Plautus his Menechmus)'*, that was currently being acted by the Lord Chamberlain's Company. This play had originally been acted before Queen Elizabeth at Hampton Court, in January 1577, by "the children of Paul's" under the title of *Historie of Errors;* itself being an English adaptation of Plautus' *Menaechmi.* (*N.B.* There was no English translation of the *Menaechmi* published until 1595.) Subsequently it was taken up by the Lord Chamberlain's Company and, with a few additions and revisions, performed again before the Queen "on Twelf daie at night" at Windsor in 1581/2. The play remained with the Lord Chamberlain's Company until the time of James I, and was acted by them before the King in 1604. It was first published as *The Comedy of Errors* in the first Shakespeare Folio of 1623. It is the shortest Shakespeare play, and probably the earliest play of all to be written by Francis Bacon, probably during the Summer of 1576. It takes up the same basic theme as portrayed in *Hermetes the Heremyte*, that of the need to balance *Mercury* with *Mars*, thought with action, *etc.* — the very theme that underlay Francis' philosophy and work throughout his life.

The crowd of spectators and the excitement was so great on this first Grand Night that the student-actors were driven from the stage by the general disorder. When the tumult had subsided somewhat. the guests had to content themselves for a while with dancing and revelling until the play, *A Comedy of Errors,* could be put on by the professional actors as planned. The disorderliness and abandonment of "those very good inventions and conceits" which had been intended, was a great disappointment and brought about much criticism, and the night came to be known as "a Night of Errors". Accordingly, the next night a mock trial was arranged, in which the "conjurer" of the Revels was arraigned at the bar for supposedly being the cause of "that confused inconvenience". The "conjurer" or Master of Revels acquitted himself honourably, pointing out where the true cause of disorder lay.

After this 'trial' and 'acquittal', a grand consultation was held to determine how their lost honour might be redeemed, which ended in a resolution "that the Prince's Council should be reformed, and some conceits should have their places." To accomplish this, in a matter of a few days Francis Bacon was able to present an entertainment which is described as "one of the most elegant that was ever presented to an audience of statesmen and courtiers." It was performed on Friday 3rd January, 1595, the second Grand Night, and was called *'The Order of the Helmet'*. In this, the student-actors presented themselves as Knights of the Order of the Helmet (whose Muse was Pallas Athena, the 'Spear Shaker'), presided over by the Prince of Purpoole. After investiture, the knights vowed to keep nineteen articles, full of Baconian philosophy and precepts, including the vows to defend God and the State, to attack Ignorance, to defend Truth and Virtue ceaselessly and secretly. Then, after a speech by the Prince, six Counsellors delivered a speech each, charging the Knights to do certain things, namely:
1. the exercise of war (against vice and ignorance);
2. the study of philosophy (including the collecting of a perfect library, the making of a spacious botanical and zoological garden, a museum, and a laboratory or "palace fit for a philosopher's stone");
3. the winning of fame and eternizing of one's name "in the magnificence of goodly and royal buildings and foundations, and the new institution of orders, ordinances, and societies";
4. the absoluteness of state and treasure (see Bacon's Essays *Of Empire* and *Of the Greatness of Kingdoms*);
5. the exercise of virtue and gracious government;
6. the exercise of joyful pastimes and sports.

The night then concluded with dancing. Everyone was so delighted with the Entertainment that "thereby Gray's Inn did not only recover their lost credit, but got instead so much honour and applause as either the good reports of our friends that were present or we ourselves could desire."

279. "To teach Minerva" = to teach all that *Minerva*, the Goddess of Wise Intelligence, means. In Graeco-Christian terminology, *Minerva* is equated with *Sophia*, the perfect Vision and Intelligence of Truth which, by reflecting the Light of Truth (the *Christos*), can reveal Truth. *Minerva* is the Latin word for the Greek name, *Pallas Athena*, representing the more individualised aspect of *Sophia*, the Universal Intelligence of God. *Minerva* particularly corresponds to that understanding or illumination which is born, as it were, from the head of *'Zeus'* after much strife and head-aching in trying to comprehend what Truth is. She signifies, therefore, both the initial illumination or vision, and the final illumination or understanding.

Minerva, as Divine Intelligence, is the Patroness of the Liberal Sciences and Arts (in her aspect of Fame), the Goddess of Peace and Learning, Supreme Muse of Poetry and Drama, Protectress and Leader of all knight-heroes (*i.e.* the brave, valorous and righteous), and Upholder of Justice and Virtue. She inhabits Parnassus, the Mount of Poetic Inspiration and Illumination, being the consort of Apollo, God of Light (*i.e.* Divine Wisdom), who represents the individualised aspect of the *Christos*.

In the Mysteries, Minerva, like Isis, appears veiled to those who are still seeking her: Minerva's veil being the visor of her helmet. This signifies their temporary blindness or inability to recognise and understand Truth: the manifestation or 'face' of Truth appearing 'veiled' or concealed from them. To lift or rend this veil signifies illumination.

The idea of this veil or helmet was also linked with the realisation that Truth is not only hidden within nature, but works its powers of teaching under this 'mask', leading mankind covertly towards eventual illumination. The sages of old imitated this by teaching under the guise of drama and other pleasurable activities. Francis Bacon did the same.

280. *Academie Françoise* (1580), 'Dedication'.
281. *Ibid.* ch. i.
282. *The French Academie*, printed for Thomas Adams, published in London (1618).
283. *Ibid.*.
284. *Academie Françoise* (1580), ch. i.
285. *Ibid.*
286. See F.B.R.T. Journal I/2, *The Virgin Ideal,* and *The AA Headpieces* by Y. Ledsem (*Baconiana*, vol. VIII, no. 31).
287. *Oxford Dictionary*.
288. '*Wis-dom*' = 'Knowledge of the Lord God'.
289. Francis Bacon, Essay *Of Truth*.
290. William T. Smedley, *The Mystery of Francis Bacon*.
291. 'François Bacon' is used in French editions of Francis Bacon's literary works.
292. *The French Academie* (1618).
293. By modern reckoning the actual date of the book would be February 1577.
294. Brantôme, *Dames illustres*.
295. *Mémoires et lettres de Marguerite de Valois* (edit. Guessard). La Reole was one of the six surety-towns ceded to the Huguenots by the Peace of Bergerac.
296. Armand de Gontaut, Maréchal de Biron, one of the three principal leaders (with the Maréchaux de Montmorency and de Cossé) of the *Politiques,* who sympathised with many of the aspirations of the Huguenots and were hostile to the Guises. Biron was the French King's Lieutenant in Guyenne.
297. Guillaume Salluste du Bartas. He was born at Montfort, near Auch, in 1544, and became a soldier while still very young. He was entrusted by the King of Navarre with several diplomatic missions to England, Scotland and Denmark, and fell, fighting by his side, at Ivry.

298. Maximilien de Béthune, afterwards Duc de Sully.
299. The conference began on 4th February, 1579.
300. The Huguenot deputies adopted a very arrogant and bellicose tone, and Catharine felt obliged to address them "royally and very haughtily, even going so far as to declare that she would have them all hanged as rebels." Upon which the Queen of Navarre intervened and, with tears in her eyes, implored her mother to give them peace.
301. H. Noel Williams, *Queen Margot, Wife of Henry of Navarre.* (Harper & Bros., London & New York, 1907.)
302. Roy Strong, *Nicholas Hilliard.* (*Folio Miniatures:* Michael Joseph Ltd., London, 1975.)
303. In March, 1582, Pope Gregory XIII authorised the reform of the Julian calendar, and the revised 'Gregorian' calendar was adopted throughout most of Christendom. To bring the calendar up to date, ten days were added to the day following the feast of St. Francis, such that the 5th October 1582 became the 15th October 1582. This restored the Vernal Equinox, which had that year occurred on 11th March, to its correct position, 21st March. In 1561, the year of Francis' birth (New Style; the recorded year for this birth is 1560, Old Style, as one year ended and a new year began then on 25th March), the date of his birth would have been understood as being the 12th January. However, by modern reckoning, Francis would have been born on the 22nd January 1561.
304. Francis Bacon, *Sylva Sylvarum,* cent. X, exp. 986. See also F.B.R.T. Journal I/3, *Dedication to the Light.*
305. See Note 279.
306. 'A Pleasant and Conceited Comedie Called, Loues labors lost. As it was presented before her Highness this last Christmas. Newly corrected and augmented *By W. Shakespere.*' (Dated 1598; printed by William White for the publisher Cuthbert Burby.)
307. *Biliteral Cipher,* p. 203. (*Taming of the Shrew*, Shakespeare Folio, 1623.)
308. The stage-plays referred to are those in Word or Biliteral cipher, enumerated (in cipher) by Francis Bacon as follows:—

5 Cipher Histories:
(1) *The Life of Elizabeth** [Elizabeth I]
(2) *The Life of Essex*†
(3) *The White Rose of Britain* [Elizabeth of York]
(4) *The Life and Death of Edward Third*
(5) *The Life of Henry the Seventh*

5 Cipher Tragedies:
(1) *Mary, Queen of Scots**
(2) *Robert, the Earl of Essex**
(3) *Robert, the Earl of Leicester*
(4) *The Life and Death of Marlowe*
(5) *Anne Bullen** [Boleyn]

3 Cipher Comedies:
(1) *Solomon the Second*
(2) *Seven Wise Men of the West*
(3) *The Mouse Trap*

In addition he enciphered his own translations of:—

3 Cipher Epics:
(1) Homer's *Iliad*†
(2) Homer's *Odyssey*†
(3) Virgil's *Aenead*†

Besides these plays and epics, he also enciphered in narrative and poetic form the following:—

A Pastorall of the Christ†
Francis Tudor's Life†* [*i.e.* his own]
Green's Life†
Marlowe's Life†
The Earl of Essex's Life†
Queen Elizabeth's Life*†

> Two epistles on Cipher†
> Completion of *New Atlantis*†
> *Bachantes, A Fantasie*†
> Part of *'Thyrsis'*† [Virgil's *Aeclogues*]
> The Spanish Armada*
> The Massacre of St. Bartholomew's*
> The Public Trial of Queen Margaret* [of Navarre]
> The Assassination of the Duke of Guise and the Cardinal of Lorraine*
> The Declaration of Henry of Navarre as Heir to the Throne of France*
> The Assassination of Henri III*
> The Earl of Leicester†
> Mary, Queen of Scots†
> Marguerite of Navarre†
> Robert Cecil†

* = Word Cipher.   † = Biliteral Cipher.   Unmarked titles have not yet been deciphered, nor have his "French poems" or translations of Ovid.

309. *Biliteral Cipher,* p. 345. (*Natural History.*)
310. See F.B.R.T. Journal I/2, *The Virgin Ideal.*
311. Martin Hume, *Spanish Influence on English Literature.* (London, 1905), p. 61.
312. George Stronach, *Was Bacon Ever Abroad? When and Where?* (*Baconiana* vol. VI. no. 22.)
313. Martin Hume, *Spanish Influence on English Literature,* p. 61.
314. Alicia Amy Leith, *A Few Notes on Love's Labour's Lost.* (*Baconiana* vol. VIII, no. 32.)
315. *Ibid.*
316. Robert Greene, *The Royal Exchange* (1590).
317. Francis Bacon, *Cupid, or an Atom.*
318. Orphic Teaching.
319. *Ibid.*
320. *Ibid.*
321. *Ibid.* (-Plato).
322. *Ibid.*
323. Francis Bacon, *Thoughts on the Nature of Things.*
324. Francis Bacon, *Cupid, or an Atom.*
325. Francis Bacon, Essay *Of Truth.*
326. *Ibid.*
327. *Love's Labour's Lost,* I. i. 1.
328. *Ibid.* IV. i. 31-35.
329. *Ibid.* V. ii. 715.
330. *Matthew* x. 34.
331. *John* xv. 13.
332. *John* x. 10-11.
333. Francis Bacon.
334. Francis Bacon, *Advancement of Learning,* Book I.
335. *Venus and Adonis* was registered on 18th April 1593, and published soon after. Because of its blatantly esoteric meaning and the fact that it contained the first printed version of the symbolic name 'William Shakespeare', referring to both Pallas Athena (or Britannia) and

to St. George, the Rose-Cross knight-hero and 'shaker of the spear' of light against the dragon of ignorance and vice, it is almost certain that the poem was intended to be published on 23rd April, St. George's Day.

It should be well noted that 'Mr. William Shakespeare' (sometimes spelt 'Shake-speare') was made out to have both been born and died on St. George's Day. The word 'Shakespeare' is a fairly straightforward and well-used description of St. George (and all dragon-slayers in mythology). The name of 'William' is not always so readily understood; but the word is actually derived from *Hwyll*, the name of the Welsh Sun-god, plus *helm*, or helmet. In other words, 'William' means 'the Helmet of the Sun-god, *Hwyll* (or *Hu*).' This is the very same helmet of illumination that is worn by Pallas Athena, and which is given to her knight-heroes when they attain their goal. It is also known as the helmet of invisibility, as it cannot be perceived by those who have not the eyes to see, and the 'wearer' is, in his wisdom, secret or veiled in all his doings with those who do not understand truth. The 'helmet' of radiance, which is the Biblical "countenance of the Lord", is the veil itself, capable of being lifted and worn only by the true initiate. It is, in other words, a **'mask'** that veils the Mystery of Truth.

336. 'Printed for Henrie Tomes, and to be sold at his shop at Graies Inne Gate in Holborne. 1605.'

337. I *Timothy* vi. 20.

338. Hor. *Ep.* I. xviii, 69.

339. Tac. *Hist.* i. 51.

340. Cic. *Tusc. Disp.* v. 4, 10.

341. Divine and Natural philosophies.

342. Francis Bacon, *Advancement of Learning,* Book I (1605).

343. 68 = the Lord Keeper's age when he died.

344. *viz.* Possessions and revenues in Hertfordshire:— the manors of Abbotsbury, Minchinbury, Hores, Colney Chapel; the farm of the manor of the Priory of Redbourne, and the farm of Charings; in Middlesex:— woods in Brent Heath, Brightfaith Woods, Merydan Meads, and the farm of Pinner-Stoke.

345. Elizabeth Jenkins, *Elizabeth and Leicester,* ch. xix.

346. In 1583.

347. See F.B.R.T. Journal I/3, *Dedication to the Light,* pp. 33-34.

348. *viz.* the Sidney, Walsingham and Dudley families.

349. Sir Thomas Walsingham was knighted in 1597.

350. R. L. Eagle, *The Secret Service in Tudor Times.* (*Baconiana*, vol. XL, no. 154.)

351. 'British Museum', vol. I, p. 14.

352. 29th January 1579.

353. 5th January 1579.

354. *The Divine Pymander of Hermes Trismegistus,* Excerpt 1. 'Pymander' means 'Shepherd of Men'. "I am Pymander, the Shepherd of Men, the *Nous* of the Supreme; I know what thou desirest."
"Have me in thy mind, and whatever thou wouldst learn, I will teach thee."
"I am that Light, that *Nous,* thy God, who was before moist Nature appeared out of the darkness; and that Light-Word from the *Nous* is the Son of God."

355. *Ibid.,* X. 14.

356. *Ibid.,* IV. 1-2.

357. *Ibid.,* X. 15.

SELECT BIBLIOGRAPHY

The following publications are amongst those consulted and generally recommended, and are included in the following list as a possible selection of reading and reference matter for those who wish to study further. Bibliographies and reference material are included in many of these books listed, and the student will be able to follow his own trail once started.

This book list is additional to and forms a continuation of the Bibliography given in F.B.R.T. Journal I/3: Dedication to the Light.

> "Some books are to be tasted, others to be swallowed, and some few to be chewed and digested."

R. H. Allen, Star Names: *Their Lore and Meaning* (Dover Publ., N.Y., 1963).
A.M.O.R.C., *Sepher Yezirah* (C.H. Frank, New York, 1877).
H. Kendra Baker, *Glastonbury Traditions concerning Joseph of Arimathea* (Covenant Publ., London, 1928).
Shabaz Britten Best, *Genesis Revised: The Drama of Creation* (Sufi Publ., 1964).
H. P. Blavatsky, *The Secret Doctrine: The Synthesis of Science, Religion and Philosophy* (Theosophical Univ. Press, Pasadena, 1970).
E. W. Bullinger, *The Witness of the Stars* (Kregel Publ., Michigan, 1967).
Joseph Campbell, *The Mythic Image* (Bollingen Series C; Princeton Univ. Press, New Jersey, 1974).
E. Raymond Capt, *The Glory of the Stars: A Study of the Zodiac* (Artisan Sales, California).
E. Raymond Capt, *Jacob's Pillar: A Biblical Historical Study* (Artisan Sales, California, 1977).
Roger Cook, *Tree of Life: Image for the Cosmos* (Art and Cosmos series, Thames & Hudson, London, 1974).
J. C. Cooper, *An Illustrated Encyclopaedia of Traditional Symbols* (Thames & Hudson, London, 1978).
Timothy Daniel, *The Inns of Court in London* (Wildy, London, 1971).
A. G. Dickens, *The Age of Humanism and Reformation: Europe in the 14th, 15th and 16th Centuries* (Prentice-Hall Int., London, 1977).
Levi H. Dowling, *The Aquarian Gospel of Jesus Christ* (Fowler, London, 1964).
Robert Eisler, *Orpheus — The Fisher: Comparative Studies in Orphic and Early Christian Cult Symbolism* (Watkins, London, 1921).
Isabel Hill Elder, *George of Lydda: Soldier, Saint and Martyr* (Covenant Books, London, 1949).

J. H. Elliot, *Europe Divided 1559-1598* (Fontana, 1968).

R. L. Ellis and James Spedding, *The Works of Francis Bacon* (London, 1857).

F. A. Filby, Creation Revealed: *A Study of Genesis Chapter One in the Light of Modern Science* (Pickering & Inglis, Glasgow, 1963).

Joachim Gerstenberg, *Strange Signatures* (Francis Bacon Soc., London, 1967).

R. W. Gibson, Francis Bacon, *A Bibliography of his Works and of Baconiana to the year 1750* (Oxford, 1950).

Robert Graves and Raphael Patai, *Hebrew Myths: the Book of Genesis* (Cassell, London, 1963).

V. H. H. Green, *Renaissance and Reformation: A Survey of European History between 1450 and 1660* (Arnold, London, 1964).

James Hall, *Dictionary of Subjects and Symbols in Art* (Murray, London, 1974).

Manley P. Hall, Man: *Grand Symbol of the Mysteries* (Philosophical Research Society, Los Angeles, 1972).

Z'ev ben Shimon Halevi, *Tree of Life: An Introduction to the Cabala* (Rider, London, 1972).

Z'ev ben Shimon Halevi, *A Kabbalistic Universe* (Rider, London, 1977).

Z'ev ben Shimon Halevi, *Kabbalah: Tradition of Hidden Knowledge* (Art and Imagination series, Thames and Hudson, 1979).

Geoffrey Hodson, *The Concealed Wisdom in World Mythology* (Theosophical Publ. House, 1983).

George F. Jowlett, *The Drama of the Lost Disciples* (Covenant Publ., London, 1975).

A. J. Krailsheimer, *The Continental Renaissance 1500-1600* (Penguin, 1971).

Peter Lemesurier, *Gospel of the Stars: A Celebration of the Mystery of the Zodiac* (Compton Press, 1972).

Eliphas Levi, *The Book of Splendours* (Studies in Hermetic Tradition: Aquarian Press, 1973).

H. Spencer Lewis, *The Mystical Life of Jesus* (A.M.O.R.C., 1953).

H. Spencer Lewis, *The Secret Doctrines of Jesus* (A.M.O.R.C., 1965).

Lionel Smithett Lewis, *St. Joseph of Arimathea at Glastonbury. The Apostolic Church of Britain* (Clarke, Cambridge, 1922).

Ellen Conroy McCaffery, *An Astrological Key to Biblical Symbolism* (Thorsons, 1975).

K. E. Maltwood, *The Enchantments of Britain* (James Clarke, Cambridge, 1982).

K. E. Maltwood, *A Guide to Glastonbury's Temple of the Stars* (James Clarke, Cambridge, 1982).

W. M. H. Milner, *The Royal House of Britain: An Enduring Dynasty* (Covenant Publ., London, 1975).

Ajit Mookerjee, *Kundalini: The Arousal of the Inner Energy* (Thames & Hudson, 1982).

Alan and Veronica Palmer, *Who's Who in Shakespeare's England* (Harvester Press, 1981).

Charles Ponce, *Kabbalah: An Introduction and Illumination for the World Today* (Garnstone Press, London, 1974).

Mrs. Henry Pott, *Francis Bacon and his Secret Society* (Robert Banks, London, 1911).

John Pryce, *The Ancient British Church* (Longmans, London, 1878).

Oliver L. Reiser, *This Holyest Erthe: The Glastonbury Zodiac and King Arthur's Camelot* (Perennial Books, London, 1974).

Leo Schaya, *The Universal Meaning of the Kabbalah* (Allen & Unwin, London, 1971).

Carlo Suares, *The Song of Songs: the Canonical Song of Solomon deciphered according to the original code of the Qabala* (Shambala, London, 1972).

Edmond Szekely and Purcell Weaver, *The Gospel of Peace of Jesus Christ by the disciple John* (Daniel, London, 1937).

Edmond Szekely, *The Teachings of the Essenes from Enoch to the Dead Sea Scrolls* (Daniel, London, 1978).

Edmond Szekely, *The Gospel of the Essenes* (Daniel, London, 1976).
Florice Tanner, *The Mystery Teachings in World Religions* (Theosophical Publ. House, 1973).
David V. Tansley, *Subtle Body: Essence and Shadow* (Art and Imaginations series, Thames & Hudson, 1982).
Gladys Taylor, *Our Neglected Heritage* (Covenant Publ., London, 1969).
John W. Taylor, *The Coming of the Saints: Imagination and Studies in Early Church History and Tradition* (Covenant Publ., London, 1969).
Robert G. K. Temple, *The Sirius Mystery* (Sidgwick & Jackson, London, 1976).
Alan Watts, *Myth and Ritual in Christianity* (Thames & Hudson, 1954).
W. F. C. Wigston, *Bacon Shakespeare and the Rosicrucians* (George Redway, London, 1888).
H. Noel Williams, *Queen Margot, Wife of Henry of Navarre* (Harper, London, 1907).
Frank Woodward, *Francis Bacon's Cipher Signatures* (Grafton, London, 1923).
Baconiana, the Journal of the Francis Bacon Society.
The Companion Bible (Zondervan Bible Publishers).
The Divine Pymander of Hermes Trismegistus (The Shrine of Wisdom, 1955).
Muir's Atlas of Ancient, Medieval and Modern History (George Philip, London, 1982).
Mystical Theology and the Celestial Hierarchies by Dionysius the Areopagite (The Shrine of Wisdom, 1965).
The Virgin World of Hermes Trismegistus, trans. by Dr. Anna Kingsford and Edward Maitland (Secret Doctrine Reference Series, Wizards Book Shelf, Minneapolis, 1977).